PROCEDURES FOR LICENSING AUTHORITY OFFICERS

Licensing law is a wide ranging, detailed and complex body of law within the UK. This book comes at a time when local authorities are required to consider and approve, or reject, applications for an increasing number and very wide range of licences. The book provides easy to read and easy to follow procedures for a wide range of licences which local authorities and other public bodies are required by law to consider and issue.

Each chapter addresses a distinct topic and the book includes guidance on local authority and court procedures.

The main legal procedures used in the licensing field are presented as flow charts supported by explanatory text.

Licensing professionals and students will find this essential reading. It will also be a valuable reference for all those whose responsibilities demand they keep abreast of current licensing practices.

Tim Deveaux is an independent environmental health practitioner with 36 years of public sector experience in a wide range of areas in environmental health and sustainability. His experience includes developing and implementing strategies to change behaviour, and protecting the environment and human health, particularly in planning for, and implementing, climate change and sustainability solutions. He has edited the eighth edition of *Environmental Health Procedures* and written three books to support levels 1 to 4 health and safety courses.

PROCEDURES FOR LICENSING AUTHORITY OFFICERS

TIM DEVEAUX

Routledge
Taylor & Francis Group

LONDON AND NEW YORK

First published 2016 by Routledge

2 Park Square, Milton Park, Abingdon, Oxon OX14 4RN
605 Third Avenue, New York, NY 10017

Routledge is an imprint of the Taylor & Francis Group, an informa business

First issued in paperback 2021

British Library Cataloguing-in-Publication Data
A catalogue record for this book is available from the British Library

Library of Congress Cataloging in Publication Data
Deveaux, Tim, author.
Procedures for licensing authority officers / Tim Deveaux.
pages cm
Includes bibliographical references and index.
1. Trade regulation–Great Britain. 2. Licenses–Great Britain.
3. Administrative procedure–Great Britain. I. Title.
KD2200.D48 2016
343.4108–dc23
2015024379

ISBN: 978-1-138-90463-7 (hbk)
ISBN: 978-1-03-217959-9 (pbk)
DOI: 10.4324/9781315696249

Typeset in Times New Roman
by Wearset Ltd, Boldon, Tyne and Wear

Contents

Figures and flow charts

Subject index of procedures

Preface

The idea for this work came after I edited the eighth edition of *Bassett's Environmental Health Procedures* in 2013.

Rather than wade through Acts of Parliament or legal software in order to find the legal answer to a fairly simple practical problem which faces licensing officers, why not write a similar book for all the licences that local authorities and associated services (police, lawyers, fire and rescue services, etc.) come across every day.

Such a text could also be useful for those studying or beginning employment in licensing functions in all these services having procedures in a simple form in one place.

It has also been suggested that other people outside the licensing fraternity such as elected local authority members who become involved in licensing problems, would welcome a procedural analysis of this type.

This book attempts to meet all those needs and provide a means of identifying procedures relating to the many facets of licensing practice.

Each of the main procedures has been reduced to a diagrammatic form and included with it is text on the basic issues concerned with that procedure. By this means, the steps involved in the different procedures can be quickly followed and individual stages identified. Related procedures have been cross-referenced. At the beginning of each chapter, the reference, extent and scope of the procedure contained in that chapter are indicated. Definitions are usually left until the end of each chapter.

It should be stressed that this work is an interpretation of the law and is not the law itself. It is aimed at licensing which local authorities administer and enforce. Other professions may find it useful such as the police, fire and rescue service and licensing lawyers. The book may also be useful for commercial organisations which require a range of licences for their operations.

The writer would appreciate any comments on the work and suggestions as to alternative interpretations will be very helpful.

Acknowledgements

I would like to thank the following Environmental Health Professionals for their advice, assistance and invaluable input into this book: Graham Jukes and Tina Garrity for access to legal information; Elaine Rudman on Licensing; Julie Barrett on Law and legal procedure; Carol Deveaux and Jenny Morris on Food safety; Paul Christer, Steve Ramshaw and Howard Price on Environmental Permitting; Bob Mayho and Mandy Reed on Housing.

Chapter 1

LICENSING LEGAL STRUCTURE AND PROCESS

LICENSING LAW: AN INTRODUCTION

The procedures interpreted in this book are based on primary and secondary legislation as it applies in England. In some cases they will also apply in the rest of the UK but this varies because of the separate legislative powers of the National Assemblies in Northern Ireland, Scotland and Wales. In Northern Ireland and Scotland the primary legislation as well as the secondary legislation may be different but the Welsh Assembly has powers only in respect of the making of secondary legislation, e.g. regulations.

In each procedure the applicability of the procedure to the rest of the UK is indicated in the section headed 'Extent'.

THE FRAMEWORK FOR ENFORCEMENT POLICIES

In undertaking the enforcement of licensing legislation and forming their individual enforcement policies local authorities need to take account of government policy, government agencies and initiatives taken at national level which seek to guide their enforcement policies. The most significant are outlined below and it will be seen that in total they form a very substantial influence over local authority enforcement activity.

The Hampton Review

A review was commissioned by the Chancellor in 2004 to review regulatory inspection and enforcement with a view to reducing the administrative cost of regulation consistent with maintaining effective regulatory outcomes. The final report was made in March 2005 and is the basis of much of enforcement policy today. The principal findings included:

- entrenching the system of risk assessment throughout the regulatory system;
- reducing the need for inspections by up to one third;
- reducing the number of forms regulators send out by up to 25 per cent;
- making more use of advice by applying the principle of risk assessment;
- substantially reducing the need for form filling;

- applying tougher and more consistent penalties where these are deserved;
- reducing the number of regulators that businesses have to deal with;
- entrenching reform by requiring all new policies and regulations to consider enforcement and to use existing structures wherever possible; and
- creating a new business led body at the heart of government to drive implementation of the recommendations and challenge departments on their regulatory performance.

This resulted in the creation of the Better Regulation Commission, Better Regulation Executive and Local Better Regulation Office but the principles of better regulation have been embodied in a new body created to take better regulation forward (see below).

Better regulation

The Legislative and Regulatory Reform Act 2006 imposes a duty on any person exercising a specified regulatory function to have regard to the five principles of good regulation. The principles provide that regulatory activities should be carried out in a way which is

- transparent
- accountable
- proportionate
- consistent and
- targeted only at cases in which action is needed (section 21).

Section 22 of the Act enables a Minister of the Crown to issue a Code of Practice relating to the exercise of regulatory functions (the 'Regulators' Compliance Code'). This section imposes a duty on any person exercising a specified regulatory function to have regard to the Regulators' Compliance Code – April 2014, when determining general policies or principles and in setting standards or giving general guidance about the exercise of other regulatory functions.

The Legislative and Regulatory Reform (Regulatory Functions) Order 2007 specifies the regulatory functions to which these duties apply.

Regulators' Compliance Code

Regulators should:

- carry out their activities in a way that supports those they regulate to comply and grow;
- provide simple and straightforward ways to engage with those they regulate and hear their views;
- base their regulatory activities on risk;
- share information about compliance and risk;
- ensure clear information, guidance and advice is available to help those they regulate meet their responsibilities to comply; and
- ensure that their approach to their regulatory activities is transparent.

Better Regulation Delivery Office

The Better Regulation Delivery Office (BRDO) was created on 1 April 2012 as an independent unit within the Department for Business, Innovation and Skills. It is guided by the Representative Steering Group (RSG). It operates across the United Kingdom and support the Coalition Government's Strategy for Better Regulation.

BRDO promotes a regulatory environment in association with business in the delivery of regulation and developing practical tools for regulators.

This means:

- providing a centre of expertise for Primary Authority and extending the scheme to enable more businesses to participate;
- simplifying delivery of regulation for business, regulators and government;
- providing a forum for business engagement in shaping regulatory policy and delivery;
- supporting Local Enterprise Partnerships to tackle regulatory delivery issues at the local level; and
- providing policy advice to UK and Welsh Ministers on regulatory delivery.

The Enforcement Concordat

This agreement between the local authority associations and the government was reached in March 1998 and is supported by the British Retail Consortium. The Concordat is not mandatory but the vast majority of all organisations within the scope have adopted it. It deals with:

- standards of level of service and performance;
- openness in the provision of information and the use of plain language;
- helpfulness in working with those affected by environmental health laws;
- well publicised procedures for dealing with complaints about enforcement;
- proportionality to ensure that enforcement action is related to risk levels;
- consistency by enforcing in a fair, equitable and consistent manner; and
- procedures to guide the enforcement action of individual officers.

Most local authorities have embraced the Enforcement Concordat into their enforcement policies. The principles are embedded in operational procedures in local authorities and guidance has been produced to help local authorities – 'Applying the Regulators' Compliance Code and Enforcement Concordat – Local Better Regulation Office briefing for local authority regulatory services – March 2008'.

Best value

This concept involves continually improving how all functions, including those involving enforcement, are undertaken by an authority. The scheme is given statutory force in Part I of the Local Government Act 1999 and involves processes that ensure functional reviews include challenge, comparison, consultation and competition. Benchmarking, which is the continuous process of measuring services against leaders in the field allowing the identification of best practices, is central to all aspects of this review process. The principles of best value have to a large extent been embodied in a local government culture and framework of performance management.

Comprehensive Performance Assessment (CPA) and Comprehensive Area Assessment (CAA)

These schemes looked at how well services, including licensing, were delivered and how this impacted on the delivery of services. The schemes are based on audit and categorised local authorities into excellent, good, fair, weak and poor. The judgements were based on three separate elements:

- annual use of resources assessment;
- annual service assessment; and
- a periodic corporate assessment.

Performance management was supported by the use of performance information based around the use of Best Value Performance Indicators (BVPIs) which were rationalised to a suite of indicators many of which are retained by local authorities to measure their performance internally. CPA and CAA are no longer a priority for the government or local authorities.

Performance management and competence

The Local Better Regulation Office (LBRO) (now the Better Regulation Delivery Office, an independent unit within the Department for Business, Innovation and Skills) produced the Local Authority Regulatory Services Excellence Framework. Local authorities can measure service delivery quality based on the achievement of set targets, identified through consultation with their customers. Many have adopted quality management systems, some obtaining certificated systems through British Standards and ISO.

An Introductory Guide to Performance Management in Local Authority Trading Standards and Environmental Health Enforcement Services is a best practice guide published in 1999. It aims to assist local authorities with best value targeted performance management and includes matters relating to enforcement. The guide identifies the following themes as being central to good performance:

- protection of the wider agenda, i.e. councils leading and energising local communities;
- transparency and consistency;
- quality and value; and
- delivery and review.

The LBRO has also published the *Common Approach to Competence for Regulators* in 2011 which incorporates the Regulators Development Needs Analysis (RDNA) and Guide for Regulators Information Point (GRIP) tools are available to ensure professional competency in regulatory activities.

Local authority licensing and enforcement policies

Working within the directions and advice contained in the framework described above, each enforcing authority should adopt its own licensing and enforcement policies to guide departments and individual officers as to the way in which the licensing function and enforcement are to be carried out. This should include how decisions are to be reached in individual cases as to which enforcement option to adopt.

Such policies will also include provisions for dealing with disputes and complaints arising from licensing and enforcement activity. See Licensing Act 2003 (page 112).

ENFORCEMENT OPTIONS

The flow charts in this book indicate only the statutory provisions prescribed in the legislation that contains the particular procedure. There will also be a number of other provisions available depending upon the particular circumstances and the options are given below. In addition, there are a number of statutory remedies that are available generally and these are also included here.

No action

In some procedures once a local authority is satisfied that a particular set of circumstances exists it is required by law to take the next step of statutory action, usually the service of a statutory notice. These situations are usually identified in the legislation by the use of the word 'shall' or 'must' in respect of action required of the local authority. In other cases the action to be taken, or a decision to take no action, is at the discretion of the local authority.

Oral warning

It may be felt that, in the circumstances of a particular contravention, it can be adequately remedied by speaking to the person responsible and asking for action to be taken. Such actions should be recorded.

Informal letters/notices

These may be used to confirm the existence of contraventions and to ask for them to be remedied. Informal notices are in effect a letter given the form of a notice but nevertheless are not part of a statutory procedure and no offences are committed by not complying with them.

Statutory notices

These are notices served under the provisions of a particular procedure. They must be in accordance with the requirements of that procedure, and with the general provisions covering such notices, and are legally enforceable.

Non-compliance will lead to the possibility of prosecution and/or the taking of further steps specified in the legislation. They are the most usual form of statutory remedy used in environmental health law.

Formal cautions

These are applicable to the enforcement of all criminal law and may sometimes be appropriate in the enforcement of licensing legislation. Their purpose is to deal quickly and simply with less serious offences, diverting them from unnecessary court action and reducing the chances of re-offending. The detailed provisions are contained in the Ministry of Justice guidance Simple Cautions for Adult Offenders – November 2013 (and formally Home Office circulars 18/1994 and 30/2005). The essential elements are:

1. Cautioning decisions are at the discretion of the enforcing authority.
2. Cautions should be used only where they are likely to be effective and their use is appropriate for the offence.
3. They should not be used for indictable-only offences.
4. Consideration should be given to:
   • the nature and extent of the harm caused by the offence;
   • whether the offence was racially motivated;
   • any involvement of a breach of trust;
   • the existence of a systematic and organised background; and
   • the views of the persons offended against.
5. Cautions should be recorded.
6. Multiple cautioning of the same offender should not generally be used.
7. Before a caution is administered
   • there must be evidence of the offender's guilt;
   • the offender must admit the offence;
   • the offender must understand the significance of the caution and give informed consent to being cautioned; and
   • consideration must be given to the public interest.
8. Officers giving the caution, which is of course done in person, should hold a position of seniority.

Whilst continued offending after the receipt of a formal warning is not an offence in itself, the fact will both guide subsequent decisions regarding that individual person, organisation or company and will also be taken into account by a court in any later proceedings.

Prosecutions

Licensing law is a branch of criminal law and prosecution of offenders in the courts is the usual last stage for most of the procedures dealt with here. The decision to prosecute or not is a matter of discretion for the local authority, guided by its own enforcement policy which will have taken into account statutory and other guidance from the government and government agencies. The Code for Crown Prosecutors issued by the Director of Public Prosecutions, whilst mainly aimed at police enforcement, provides useful guidance on the process of deciding if prosecution should be taken.

Cases taken for offences against Housing Acts procedures are normally a matter initially for the Residential Property Tribunal whereas most other licensing procedures are dealt with at the initial stages by the magistrates' court. Higher courts will become involved in the event of appeals against decisions made by these courts.

Regulatory penalties and sanctions

Part 3 of the Regulatory Enforcement and Sanctions Act 2008 gives regulators sanctioning powers as follows:

- fixed monetary penalties – such fines will be imposed by a regulator in respect of low-level instances of non-compliance;
- discretionary requirements which include:
  - variable monetary penalties – requiring a person to pay a monetary penalty the value of which is determined by the regulator;
  - compliance notices – requiring a non-compliant business to undertake certain actions to bring themselves back into compliance; and
  - restoration notices – requiring a person to undertake certain actions to restore the position, as far as possible, to the way it would have been had regulatory non-compliance not occurred.
- stop notices – requiring a person to cease an activity that is causing serious harm or presents a significant risk of causing serious harm and has given rise, or is likely to give rise to regulatory non-compliance; and
- enforcement undertakings – an agreement offered by a person to a regulator to take specific actions related to what the regulator suspects to be an offence.

Fixed penalty offences

There has been a move towards this type of action particularly in the Environmental Protection Act 1990 and, more recently, in the Clean Neighbourhood and Environment Act 2005. The use of fixed penalty notices is a discretionary power given to authorised officers in some enforcement procedures as an alternative to prosecution and fine. The local authority may specify the amount of the fine to be collected within statutory bands as detailed in the Environmental Offences (Fixed Penalties) (Miscellaneous Provisions) Regulations 2007. Where the local authority does not take this option a default amount is specified in the regulations. The receipts from such notices may be used to defray the cost of designated environmental functions and is detailed in the Environmental Offences (Use of Fixed Penalty Receipts) Regulations 2007.

Closure/withdrawal of licence, etc.

It is sometimes the case that, in addition to prosecution to exact a penalty, the local authority may give consideration to the cancellation of the licence/approval it may have issued authorising the activity where the offence has taken place, e.g. pet shop licences.

Delegation of authority

The Local Government Act 1972 allows a local authority to delegate its powers and responsibilities to committees, sub-committees and officers. In respect of licensing law enforcement, a clear scheme of delegation to officers is vital if investigation and remedial action is to be swift and effective. Such a scheme will require the formal approval of the individual local authority. It will include matters like the exercise of powers of entry and inspection, the service of statutory notices and the institution of legal proceedings. There is also a need for the scheme to include the designation of officers to whom the Act or regulations concerned give authority to act, e.g. the designation of proper officers, authorised officers and officers who will be Inspectors under the Safety in Sports Grounds Act 1975.

Powers of entry

The powers of entry and investigation that relate to each procedure are given in the text relating to the procedure itself.

The commissioner for local administration (the Ombudsman)

Each procedural flow chart indicates the stages at which the courts may become involved in dealing with appeals, prosecutions, etc. In addition to this, any local authority action, or lack of it, is subject to investigation by the Local Government Ombudsman. Whilst the Ombudsman cannot challenge the decision itself, e.g. not to serve a statutory notice, he may do so in respect of the way in which the decision was reached. Whilst he does not have power to change the local authority decision, the Ombudsman can issue a public report criticising the authority and is also able to indicate a level of compensation which he feels ought to be made.

LICENSING AUTHORITIES

Responsibility for the enforcement of each procedure is identified either in the particular piece of primary legislation or in the regulations made under it. However, there are only minor differences in the definitions of local authorities in the many pieces of legislation. References to licensing authorities, food authorities, etc. usually mean a local authority.

Generally, the lowest tier of local authority for the area, excluding parish councils, enforces the licensing law dealt with in this book.

The authorities charged with the implementation of most of the procedures in the book are:

In England

(a) Unitary authorities

(b) District councils in areas where there is a county council
(c) County Borough councils
(d) The Common Council of the City of London
(e) The London Boroughs
(f) Transport for London

In Wales and Scotland, the unitary councils.

In Northern Ireland, the city councils, borough councils and district councils.

The council of the Isles of Scilly for:

* Caravan Sites and Control of Development Act 1960
* Local Government (Miscellaneous Provisions) Act 1976 and 1982
* Environmental Protection Act 1990
* Food Safety Act 1990
* Criminal Justice and Public Order Act 1994
* Licensing Act 2003
* Environmental Permitting Regulations 2010
* Anti-social Behaviour, Crime and Policing Act 2014.

In London, the Sub Treasurer of the Inner Temple or Under-Treasurer of the Middle Temple for:

* Control of Pollution Act 1974
* Clean Air Act 1993
* Licensing Act 2003
* Anti-social Behaviour, Crime and Policing Act 2014.

Where there is a substantial difference in definition, it is given in the relevant chapter.

A list of the definition sources in legislation can be found at Appendix 1.

PROFESSIONAL PRACTICE

Legal process

Legal process is to make sure that before something is done by 'authority' whether a local council or the police that the views, concerns and objections of individuals and groups who are affected by the decision are heard and the rights of those whose freedoms, whether personal or over property or goods may be curtailed by the action are protected. Licensing officers need to understand their responsibilities and duties in order to comply with the requirements of the legal process.

Enforcement policy

Local authorities must have an enforcement policy but it must be noted that an enforcement policy guides but does not direct. It is among other issues to be taken into account when considering how to decide on action. A local authority should have regard to its Enforcement Policy, as a guide to promoting consistency and equity of treatment. Decisions about prosecution are for the prosecutor to make. Enforcement policies are drafted and agreed taking into account the various government and agency policies and guidelines.

Local authorities should agree, through its Cabinet, a policy to demonstrate that their procedures comply with the right to a fair trial (Article 6 of the European Convention on Human Rights).

The policy should describe the manner in which defendants will be dealt with, and the considerations that will be taken into account when determining disposal of matters.

Examples of matters to be taken into account are:

- previous convictions for like offences;
- previous Home Office Cautions;
- the severity of the offence alleged;
- previous convictions by other departments of the Council which may lead to the view that the individual will not co-operate with the Council; and
- the degree of criminality involved or likelihood of re-offending.

A policy will ensure that the local authority itself is satisfied that all those against whom it is considering taking action are fairly treated.

When the policy has been approved and adopted there needs to be a good and compelling reason for deviation from its policies.

The local authority can clearly demonstrate that its reasoning is Article 6 compliant if challenged by a defending solicitor, or by an aggrieved 'victim' who considers that an alternative course to that adopted should have been taken.

This policy is for guidance only, and in every case officers must use discretion. In most cases, the discretion will determine that the course identified in the policy is appropriate, but where it is not, the reasoned justification for deviation should be recorded, and can then be cited where allegations of rigid adherence to policy without consideration of other factors are made.

Consider the risk to complainants in the event of them being identified by those they have complained of, and the steps they have taken to prevent this happening. The policy should clearly outline how you will protect identities.

The policy should consider the wording to be used in letters, indicating that the complainant's name and address should not be revealed and that all letters should be worded in such a way as to prevent identification of complainants – geographical references to the location of complainants should be as oblique as possible, or, for preference, make no mention of the complainant at all and reliance should be placed on the evidence of Licensing Officers.

Competence

Officers who are authorised by the local authority must hold a relevant qualification and be sufficiently competent in the particular field of operation. Local authorities must ensure that the officers operating on their behalf are competent to do whatever is required of them. It will be necessary to show knowledge and skills are held and practiced, refresher training done, how knowledge is maintained and that practical skills are up to date.

Inspections leading to potential prosecution

When carrying out an inspection which is likely to lead to a prosecution to establish matters of fact, speak to no-one, but note all of the potential offences. Questions as to identity of persons and possibly ownership of the premises can be asked, depending on the charges being contemplated, but nothing about the offences observed.

The potential defendant can then be asked to attend for an interview under caution where questions about each of the offences observed can be put to him. There is no guarantee of results from this process as defendants may not accept the invitation, or may give no comment interviews. The alternative is to lay informations citing all of the offences observed and run the risk of defences being laid out in court.

Police And Criminal Evidence Act (PACE)

PACE notebooks and contemporaneous notes

The PACE (Police And Criminal Evidence) page numbered notebook helps to demonstrate that contemporaneous notes were made when the writer says that they were.

The role of the PACE notebook is to tie the inspection sheet to the date of the inspection. The inspecting officer should record in his or her PACE notebook the date and time and address of the inspection and the fact that an inspection sheet was used for the details of the inspection. The inspection sheet can be given a unique code which should appear both on the sheet and in the PACE notebook and which will tie the two together. The PACE notebook should record time in and time out, the inspection sheet should show the detail of the inspection and its findings. The content of the PACE notebook can either be exhibited or alternatively listed as unused material in a court case. The defence should know of its existence and they can see it if they want to.

In court a well kept PACE notebook shows that the officer has credibility, knows what they are doing and why it is important. Well presented and completed PACE notebooks also protect officers from interrogation about when notes were made. They protect not just the evidence but also the officer giving it.

Good contemporaneous notes allow officers to refresh their memory before giving evidence about what they saw, did, heard and are a valuable resource when writing witness statements.

Witness statements are disclosed, contemporaneous notes are disclosable. Descriptions recorded in the note should appear in the same way in the statement.

Note: Dating and timing of input of information into a data collector is not usually a problem; data collectors usually have a date and time recording mechanism which, subject to the usual rules on the admissibility of computer generated information will be sufficient to establish when the information was input.

Caution

A person who is suspected of an offence must be cautioned before any questions about it (or further questions, if it is his answers to previous questions that provide grounds for suspicion) are put to him for the purpose of obtaining evidence which may be given to a court in a prosecution.

It is only necessary to caution before asking questions that may give rise to answers of evidential value, whether directly or indirectly. It is not necessary to caution before asking questions for other purposes, for example, to establish identity, or ownership or responsibility for a vehicle.

Wording of the caution

'You do not have to say anything, but it may harm your defence if you do not mention, when questioned, something you later rely on in court. Anything you do say may be given in evidence.'

The Code of Practice on PACE notes that minor deviations from its exact wording do not constitute breaches of the requirements of the Code of Practice, provided the general sense of the caution remains.

It remains at the absolute discretion of the person being questioned as to whether they say anything. What do you do if the defendant clearly does not understand? Do not, under any circumstances, try to explain the caution to the defendant yourself. If he has a solicitor with him, let the solicitor explain it. If the defendant does not have a solicitor, terminate the interview and let him get himself one.

The purpose of the caution is to protect the defendant. The duty of the Judge in court is to ensure that the prosecutor has obtained the evidence with which he intends to convict the defendant in a fair and lawful manner. A court which feels that a lack of understanding on the part of the defendant was exploited, whether intentionally or not, is likely to rule that the evidence has been unfairly obtained and is inadmissible.

Correspondence

E-mails

Unlike a telephone conversation which is not recorded, an e-mail has permanent form; once generated they are discoverable, whether as part of a case file or under the Freedom of Information Act.

The e-mail may contain information that is defamatory or damaging to the local authority case. It may be personal comments about the defendant or more importantly it may be information that may be damaging to the substance of the council case.

E-mails should be treated with the same care as a letter or memo. Language should be formal and polite, and the normal rules of professional courtesy should apply. Generally, they should relate to a single issue, so that they can be printed off, or retained in the relevant file. Multi-issue e-mails should be avoided, since they will have to be filed on a number of files, and will introduce unrelated material into those files. E-mails need to be treated as what they are, documents of potential evidential value and given the respect they merit in consequence.

Without prejudice

A letter written and headed 'Without prejudice' will contain something which, were to come into the public domain, could potentially be damaging for the writer, such as an admission against his interest, or a proposal for settlement of a matter. If the letter is accepted, this is the end of the matter, if the letter is not accepted the fact that it is written 'without prejudice' means that the recipient cannot later use the letter in evidence against the writer to demonstrate the offer made, or the weakness of the writer's position.

The use of 'without prejudice' correspondence or discussions means that lawyers can deal with each other on a pragmatic basis, and can discuss matters candidly without damaging their client's position whilst doing so.

Covert surveillance

In some situations it may be necessary for local authority enforcement officers to act in a covert manner in the investigation of possible breaches of licensing legislation, e.g. the operation of private hire vehicle without a licence.

Regulation of Investigatory Powers Act 2000 (RIPA)

The law governing the use of covert surveillance techniques by public authorities is contained in the Regulation of Investigatory Powers Act 2000 (RIPA). When public authorities, such as local authorities, need to use covert surveillance techniques to obtain private information about someone, they do it in a way that is necessary, proportionate and compatible with human rights.

RIPA applies to a wide range of investigations in which private information might be obtained including terrorism, crime, public safety and emergency services.

RIPA's guidelines and codes apply to actions such as:

- intercepting communications, such as the content of telephone calls, e-mails or letters;
- acquiring communications data – the 'who, when and where' of communications, such as a telephone billing or subscriber details;
- conducting covert surveillance, either in private premises or vehicles (intrusive surveillance) or in public places (directed surveillance);
- the use of covert human intelligence sources, such as informants or undercover officers; and
- access to electronic data protected by encryption or passwords.

The Regulation of Investigatory Powers (Directed Surveillance and Covert Human Intelligence Sources) Order 2010 states that such surveillance carried out by local authorities must be by officers of at least Director, Head of Service, Service Manager or equivalent status or by an officer designated as being responsible for the management of an investigation. Guidance to local authorities is given in the Home Office guidance to local authorities in England and Wales on the judicial approval process for RIPA and the crime threshold for directed surveillance.

From 1 November 2012 local authorities are required to obtain judicial approval prior to using covert surveillance techniques. Local authority authorisations and notices under RIPA will only be given effect once an order has been granted by a Justice of the Peace in England and Wales, a sheriff in Scotland and a district judge (magistrates' court) in Northern Ireland.

Additionally, from this date, local authority use of directed surveillance under RIPA will be limited to the investigation of crimes which attract a 6 month or more custodial sentence, with the exception of offences relating to the underage sale of alcohol and tobacco.

Generally, local authorities are only concerned with carrying on Directed Surveillance usually '*for the purpose of preventing and detecting crime or preventing disorder*', for example, making observations in respect of scrap metal dealing without a licence.

If an officer wants to carry out an investigation he/she must first ask him/herself if what he wants to do constitutes Covert Directed Surveillance. If it does, RIPA applies and he/she will need to have that activity authorised. He/she then needs to consider whether the purpose of the surveillance is to prevent or detect crime or to prevent disorder. If it is not, the proposed surveillance cannot be authorised under RIPA and therefore cannot lawfully be carried out. The simplest test is to ask whether you could commence proceedings based on the evidence you would obtain. If you can, RIPA authorisation is required. If you cannot, you cannot legitimise the proposed Covert Directed Surveillance and should not therefore carry it out.

Directed surveillance must be

    (i) covert;
   (ii) intrusive;
  (iii) for the purpose of a specific investigation; and
  (iv) carried out in such manner as to result in obtaining private information about a person, otherwise than by way of an immediate response to circumstances.

Covert surveillance – RIPA advises that surveillance is covert only if it is carried out in such manner that the subjects are unaware that it is, or may be taking place (section 26(8)(a)).

So, suppose you tell the subject that surveillance *will* be taking place and how it will be carried out. Let us suppose that a complaint has been received about underage drinking; the local authority would write a covering letter advising that they will be carrying out monitoring to ensure compliance with the licence, and this monitoring will take the form of periodic visits, to be carried out at any time of day or night, by inspectors who will engage in non-intrusive monitoring. The 'target' could not complain that the surveillance was covert, since he not only knows it will be taking place, he has been advised how it will be done and by whom. All he does not know is when. This knowledge is enough to take the surveillance out of the covert category and make it overt, and hence preclude the need for authorisation. It must be noted this procedure has not been tested in court and may be subject to challenge.

It is essential for authorising officers to ensure that they give proper consideration to the reasons given for seeking authorisation to conduct directed surveillance. The authorising officer must be able to say why they believed what they were told about the complaint, what convinced them and what further inquiries, if any, they required to be made. Did they consider whether there was an alternative to directed surveillance, and if so, did they require that it was explored? If not, why not? It is essential that there is a clear paper trail that demonstrates the thinking of the authorising officer.

There is a higher threshold for directed surveillance with the exception of two trading standards offences. Local authority applications will only be granted for offences that attract a maximum custodial sentence of 6 months. Magistrates must approve any RIPA regulated activity before it begins. This applies in England, Wales and Northern Ireland and to communications data in Scotland.

Home Office guidance says licensing officers can continue to use covert techniques provided they have considered other less intrusive avenues first. Officers need to present magistrates with relevant documentation and respond to questions by the officer with most knowledge of the case and demonstrate that covert surveillance is both proportionate and necessary. Consider the questions that may be put by the magistrates, in other words, have less intrusive measures been used and subsequently failed?

Stakeouts

The Court of Appeal in the case of *R* v. *Johnson* declared that there are two requirements for stakeouts.

First is that the officer in charge of the inquiry must be in a position to give evidence that he visited the premises to be used and ascertained the attitude of the occupiers of the fact that the premises was to be used, and their attitude to the fact that a surveillance operation had been in place would have to be disclosed, which may then lead to the identification of both the premises and the occupants. He/she may also inform the court of any local difficulties in obtaining assistance from members of the public that may have caused the use of the particular premises chosen.

The second requirement is that a more senior officer should be in a position to give evidence that immediately before the trial he had visited the premises and ascertained whether the occupiers were still of the same view and, whether they were or not, what their attitude was to the possible disclosure of the use made of the premises and of facts which may lead to the identification of both premises and occupiers.

The lessons of *R* v. *Johnson* for local authority officers are:

1. Make sure the local authority's enforcement policy takes account of the rules in *R* v. *Johnson*.
2. Be pragmatic, and use local authority premises or vehicles for surveillance where possible.

The Protection of Freedom Act 2012 includes provision for making codes of practice on surveillance, regulation of surveillance using CCTV and powers of the Secretary of State to amend powers of entry.

Obstruction

Officers often encounter being partially obstructed by being delayed just enough to mean that gaining access to a premises is pointless.

Delayed consent, however willingly given, does not constitute real consent, and taking up such an invitation to enter makes later pleading an allegation of obstruction in court difficult.

First approach

When immediate consent to enter is not given, officers could walk away on the grounds that delayed entry is pointless, and to prosecute for obstruction, based on the facts of the initial refusal. The fact that the defendant did not allow immediate access can be regarded as an offence.

Second approach

Stage 1 – If officers want to enter a premises and are being frustrated by being given only delayed consent, they could write to the owner and/or occupier telling them that they want to enter, they have power to do so and that when they ask to be allowed entry, access is required immediately. Owners, etc. are then forewarned of what officers expect and can take legal advice should they want to.

Importantly, officers also have a paper trail that can be produced in evidence showing the court that they have advised owners/occupiers of the position and their claims not to know or to understand are groundless. Stage 2 is to turn up unannounced and ask to be allowed in. If officers are delayed, even slightly, they should just go away. Do not wait. Then prosecute, alleging obstruction. Then repeat the process until the message gets through. Fines for obstruction can be significant, commonly in the order of £1000. Being fined once may cause the defendants to consider their position, being fined several times will usually make a real difference.

Warrants

Obstruction of an officer in the course of their duties under legislation is an offence. Furthermore, the same Acts make provision for officers who have been refused entry, or who apprehend that entry will be refused, to make application to the magistrates' court for a warrant authorising entry.

Obtaining a warrant is a straightforward procedure. An application is made to the magistrates' court, evidence as to why the warrant is required is given on oath, before the magistrates, usually in camera (in private), and subject to the magistrates being satisfied that the application has merit, the warrant will be issued. Where the premises in question is domestic, the property owner must be given 24 hours notice of the hearing application, in order that they may attend, should they wish. This requirement does not apply to commercial premises. Where a licensing officer can make out a sound case for the issue of a warrant, it will be granted. To execute a warrant, an officer is required to show it to the owner of the premises. If the property owner still refuses entry in the face of a valid warrant, they are in contempt of court and should be prosecuted.

Where 'have reasonable belief that an offence is being committed' applies, the requirement that must be satisfied when seeking a warrant is that the officer seeking to execute the warrant must have reasonable belief that an offence is being committed in the premises or the land to which the warrant relates. The reasonableness of this view will be tested by the magistrate granting the warrant, therefore must be sufficiently robust to satisfy that inquisition.

Reasonable belief is that attributed to a reasonable man with reasonable judgement. Where there is dispute as to what is and is not reasonable the Judge in court will be asked to arbitrate, and it is for the Judge to decide if, on the facts as known at a particular point in time, a prosecutor had reasonable grounds to believe that an offence had been committed.

Deciding to prosecute

Prosecutors must ask themselves two questions,

(1) On the evidence is there a *prima facie* case to answer, and subject to the answer being yes, (2) Is it in the public interest to prosecute?

The person making the decision to prosecute must be independent of the case, and must demonstrably be so.

Is there a prima facie case?

A lawyer will normally decide this – legal advice as to whether a certain series of facts constitutes an offence is definitive. A lawyer who is a member of an investigation team is not independent from the case, so a different, uninvolved lawyer, or licensing manager or whoever is considered to be the appropriate person, is going to have to review the file when it is completed and consider whether to prosecute. Your independent case review must demonstrably be completely independent.

The Code for Crown Prosecutors 2013, which contains the Full Code Test, is a more sophisticated test, with guidance, as to whether it is appropriate in a particular case to commence proceedings.

Is it in the public interest to prosecute?

This is for the officer in the case. Sometimes there will be good reasons for prosecuting a case, for instance the publicity generated by doing so may be disproportionate to the offence itself but may influence businesses. However, it may also have a detrimental effect on the way the Council is viewed by the public. Overall, it is the officer who must decide how best and most appropriately to deal with the offence.

Things to take into consideration

Courts will usually look at the overall offending when they impose sentence. This will include the number and seriousness of offences, however the court will not look favourably on a large number of Informations.

Where the defendant has indicated that he will be pleading guilty, the prosecutor can pick out a couple of specimen offences and then ask if the defendant is prepared to ask the court to take the remaining counts into consideration. Where he is, they are tabulated, and after his plea is accepted, they are passed to the court to be taken into consideration when sentence is considered. The time and resource saved by this practice attracts discount off sentence, but the offences to be taken into consideration (TiCs) remain recorded against the defendant, and can be cited as part of his previous offending record in any future cases.

Where the defendant has pleaded guilty he will usually accept the TiCs, which will have been put to him on his plea being indicated. It is an option for the prosecutor, even in the face of a not guilty plea to list a number of TiCs and offer them to the defence on the basis that the defendant will accept them in the event of his being convicted against his plea. If he is not convicted, the main offences and the TiCs will fall, but where such an offer is made, the prosecutor will generally be confident that he will succeed at trial, and therefore the risk of major loss is minimal. This provision should be in every local authority's enforcement policy.

'Shall have regard to'

You must have regard to something when the circumstances which cause it to be relevant actually exist. Therefore, if the requirement is that when considering how a potential prosecution should be disposed of – whether to instigate court action or to offer a formal caution – the officer making the decision may be required by an Enforcement Policy to take into consideration the character of the defendant, the officer need only consider the character of the defendant when making that decision.

Consideration is subjective – it requires free thought and an independent decision.

What the officer is required to do is to consider what he or she knows about the character of the defendant – good or bad, and to consider how that defendant will respond to whatever method of disposal is preferred.

The officer making the decision should record the reason for the decision being made, showing that he or she has taken the character of the defendant into consideration in its making the requirement.

You must be able to show that you have given it some weight, and that the consideration you gave the issue was true consideration that did impinge on the decision making process, not just a token consideration because you had to do so.

Failure to take account of, consider or have regard to things that you are required to take account of, etc., can cause your decision to be challenged through the mechanism of judicial review.

Any authority that does not consider all of the relevant issues, particularly those it is specifically required to consider, whether they are Codes of Practice, Statutory Guidelines, Planning Policy Guidance or whatever, can be accused of flawed decision making, have its decision overturned by the High Court and returned to it for reconsideration.

Where directory words are used, officers should ensure that what is required is actually done, so that those affected by a decision, whilst perhaps not liking the decision itself, cannot challenge the process by which the decision was made.

Prompt action

Time lapse between offence and bringing case to trial

Summary proceedings must be commenced by the laying of an information within 6 months of the prosecutor having sufficient evidence to make a *prima facie* case, and in either way or indictable matters, they should be laid within 12 months. Failure to comply may lead to a challenge based upon the case being out of time.

Officers must ensure that all proceedings are commenced as expeditiously as possible.

Where there is the possibility of delay, for whatever reason, the reason for the delay and the manner in which the prosecutor has attempted to mitigate it, should be documented in order that the court may be appraised, if necessary. In such scenarios the defence will be to have the prosecution dismissed without consideration of the evidence on the basis of prejudice through delay, and breach of Article 6 of the European Convention on Human Rights.

It is highly unlikely that the court will be sympathetic where lack of progress is attributed to lack of time or resources, pressure of other work. Valid reasons for unavoidable delay will have to be presented.

In some cases commencement dates are prescribed within the legislation, to take account of the fact that covert illegal action may not be discovered for a significant period.

What constitutes commencement? When is a case considered to be started? The date a case is commenced is the date on which the information making the allegation is laid.[1] For all practical purposes, this is the date on which the Informations arrive at the magistrates' court. Where they are taken and delivered by hand that date is clear. Where they are sent by post, certain rules apply to when delivery is assumed to have taken place. Similarly it is now established that where Informations are laid by electronic mail they are laid on being sent, not on being received.[2]

When is an offence discovered?

In some cases it will be very clear – when an officer finds a breach of the legislation. Further, if an officer decides that they will not commence proceedings in respect of the breach but will serve a notice requiring remediation, it is not until they revisit at the end of the compliance period and discover that the requirements of the notice have not been satisfied that time begins to run on an allegation of non-compliance. In both cases the date of discovery is clear and can be pinpointed – the date of the initial visit, or the date of the inspection to ascertain compliance.

The date of the discovery of an offence is the date upon which an investigating officer has reasonable ground to believe that an offence has been committed.

SERVICE OF NOTICES BY LOCAL AUTHORITIES

Generally, notices must be in writing, dated and signed by an authorised officer. A facsimile signature can be used. A notice must be served on the person responsible which is usually defined in the legal provision in each act or regulation. Usually the notice will specify the contravention of the law or a defect which needs to be remedied. It may also specify the works to be done to rectify the contravention or defect.

Addressing a notice on individuals

The notice is addressed to the individual concerned, using the full name, not initials and the envelope addressed likewise. Any failure to comply with the notice can be dealt with against the person named on the notice and the person to whom it is addressed can be prosecuted.

Bodies corporate

Notices can be served on and summonses issued against a number of bodies corporate including limited companies, unincorporated bodies and local authorities. The name on the notice should be that of the company, with its registered office being given as its address. Naming the company ensures that it is the company that is required to comply with the requirements of the notice.

The name and address on the envelope is a different matter. A Company secretary is the legal persona of a company upon whom notices can be served. They can be sent the notice in an envelope and will accept delivery on behalf of the company. Envelopes containing notices should therefore be addressed to the Company secretary with the company name and address. This ensures that somebody with appropriate authority accepts the notice. In the case of local authorities it is usually the Chief Legal Officer who is authorised to accept documents on behalf of the authority, and documents addressed to that person by name or by job title are deemed as having been properly served on the authority concerned.

In the cases of unincorporated bodies, clubs, school governing bodies it is wise to ask on whom service of the notice may be effected. Where the body has a formal constitution this person should be identifiable, where it does not, naming all of the managing committee on the face of the notice and serving each individual ensures that service is properly made and effected.

Some Acts will specify that service shall be effected in a particular way and on a particular party, and failing to comply with that laid down method will mean that effective service will not be achieved.

If service of the document cannot be proved, no prosecution, based on allegation of non-compliance with the requirements of the notice, can proceed.

A notice that has been addressed on its face to the wrong party is a ground of appeal against it. Addressing a notice to the company secretary rather than to the company itself gives rise to that ground of appeal.

Service

This may be effected by one of the following:

(i) delivery to the person;

(ii) in the case of service on an officer of the council, leaving it or sending it in a prepaid letter addressed to him at his office;

(iii) in the case of any other person, leaving it or sending it in a prepaid letter addressed to him at his usual or last known residence;

(iv) for an incorporated company or body, by delivering it to its secretary or clerk at its registered or principal office, or sending it in a prepaid letter addressed to him at that office;

(v) in the case of an owner by virtue of the fact he receives the rackrent as agent for another person, by leaving it or sending it in a prepaid letter addressed to him at his place of business;

(vi) in the case of the owner or occupier of a premises, where it is not practicable after reasonable enquiries to ascertain his name and address or where the premises is unoccupied, by addressing it to the 'owner' or 'occupier' of the premises (naming them) to which the notice relates and delivering it to some person on those premises, or, if there is no one to whom the notice can be delivered, by affixing it or a copy of it, to a conspicuous part of the premises (Public Health Act 1936 section 285).

Proof of receipt and/or service of notices should be obtained.

The Public Health Act 1936 (sections 283 to 285) is the basis of service of notice in licensing law and notices served in any of these ways will be held in later proceedings to have been properly served, but it is possible to serve in some other way and prove service in subsequent proceedings.

The main provision for service of notice is found in the Local Government Act 1972 and, unless otherwise excluded by a particular provision of another Act, is available in addition to methods of service available under other legislation.

1. Subject to subsection (8) below, subsections (2) to (5) below shall have effect in relation to any notice, order or other document required or authorised by or under any enactment to be given to or served on any person by or on behalf of a local authority or by an officer of a local authority.

2. Any such document may be given to or served on the person in question either by delivering it to him, or by leaving it at his proper address, or by sending it by post to him at that address.

3. Any such document may:

   (a) in the case of a body corporate, be given to or served on the secretary or clerk of that body;

   (b) in the case of a partnership, be given to or served on a partner or a person having the control or management of the partnership business.

4. For the purposes of this section and of section 26 of the Interpretation Act 1889 (service of documents by post) in its application to this section, the proper address of any person to or on whom a document is to be given or served shall be his last known address, except that:

   (a) in the case of a body corporate or their secretary or clerk, it shall be the address of the registered or principal office of that body;

   (b) in the case of a partnership or a person having the control or management of the partnership business, it shall be that of the principal officer of the partnership;

   and for the purposes of this subsection the principal office of a company registered outside the United Kingdom or of a partnership carrying on business outside the United Kingdom shall be their principal office within the United Kingdom.

5. If the person to be given or served with any document mentioned in subsection (1) above has specified an address within the United Kingdom other than his proper address within the meaning of subsection (4) above as the one at which he or someone on his behalf will accept documents of the same description as that document, that address shall also be treated for the purposes of this section and section 26 of the Interpretation Act 1889 as his proper address.

6. ...(Repealed)

7. If the name or address of any owner, lessee or occupier of land to or on whom any document mentioned in subsection (1) above is to be given or served cannot after reasonable enquiry be ascertained, the document may be given or served either by leaving it in the hands of a person who is or appears to be resident or employed on the land or by leaving it conspicuously affixed to some building or object on the land.

8. This section shall apply to a document required or authorised by or under any enactment to be given to or served on any person by or on behalf of the chairman of a parish meeting as it applies to a document so required or authorised to be given to or served on any person by or on behalf of a local authority.

9. The foregoing provisions of this section do not apply to a document which is to be given or served in any proceedings in court.

10. Except as aforesaid and subject to any provision of any enactment or instrument excluding the foregoing provisions of this section, the methods of giving or serving documents which are available under those provisions are in addition to the methods which are available under any other enactment or any instrument made under any enactment.

11. In this section 'local authority' includes a joint authority, an economic prosperity board, a combined authority, a joint waste authority, a police and crime commissioner and the Mayor's Office for Policing and Crime.

(Section 233 Local Government Act 1972)

The provisions for the service of notice are included in some other acts and regulations. A list of the sources of these provisions is included in Appendix 2.

Authentication of documents

The following provisions of the Local Government Act 1972 will apply unless a particular piece of legislation has provisions which cover these issues.

1. Any notice, order or other document which a local authority is authorised or required by or under any enactment (including any enactment in this Act) to give, make or issue may be signed on behalf of the authority by the proper officer of the authority.

2. Any document purporting to bear the signature of the proper officer of the authority shall be deemed, until the contrary is proved, to have been duly given, made or issued by the authority of the local authority. In this subsection the word 'signature' includes a facsimile of a signature by whatever process reproduced.

3. Where any enactment or instrument made under an enactment makes, in relation to any document or class of documents, provision with respect to the matters dealt with by one of the two foregoing subsections, that subsection shall not apply in relation to that document or class of documents.

4. In this section 'local authority' includes a joint authority, an economic prosperity board, a combined authority, a joint waste authority, a police and crime commissioner and the Mayor's Office for Policing and Crime.

(Section 234 Local Government Act 1972)

Authorisation of officers

Local authorities must have a system of authorisations and delegations. Licensing officers have power to carry out their functions because that power is given to them by the local authority by whom they are employed.

They are authorised by their employing local authority to carry out a number of functions, and whilst they are carrying out that function they have the right to do whatever the statute says they can and no more. This authorisation should be a Council minute signed by the Leader or Chair, naming the authorised person by full name, stating that they are authorised by the Council, and stating for what purpose the person is authorised.

If a licensing officer is not authorised, anything done by way of enforcement, whether serving a Notice or bringing proceedings is void – it has no lawful standing.

Such actions taken by unauthorised officers open the council and the officer up to claims for trespass, suing for the cost of doing any work required by the officer and any losses to the business or even misrepresentation.

Officers should **always** have a certified copy of the authorisation with them while at work and acting on behalf of the local authority.

The keeping of accurate records of authorisation is vital, not least because the Office of Surveillance Commissioners (OSC) will be expecting to see records of authorisations.

Information regarding ownerships, etc.

The Local Government (Miscellaneous Provisions) Act 1976 makes provision for a local authority to require information to be provided about the ownership of property, etc. Where a local authority, in performing functions under any legislation, requires information regarding interests in any land it may serve a notice, specifying the land and the legal provision containing the particular function under which it is acting, requiring the person concerned to declare, within not less than 14 days, the nature of his interest and the names and addresses of all persons who he believes have interests.

The notice may be served on any of the following:

    (a) the occupier;
    (b) the freeholder, mortgagee or lessee;
    (c) the person directly or indirectly receiving the rent;
    (d) the person authorised to manage the land or arrange for its letting.

Failure to comply with the notice or the making of false statements carries a maximum penalty of level 5 on the standard scale (section 16).

These powers are given to county councils, district councils, London borough councils, the Common Council, the Council of the Isles of Scilly, police authorities, joint authorities, police and crime commissioner, the Mayor's Office for Policing and Crime, an economic prosperity board, a combined authority, joint waste authorities, the London Fire and Emergency Planning Authority and parish and community councils (section 44(1)).

PREPARING A CASE

Collecting evidence

You must first collect evidence of fact from witnesses, identify the reasons/causes of the breach or incident then speak to those who may be responsible for the breach or incident, and finally gather interview information.

Evidence is all the material on which either party seeks to rely on to prove their case, i.e. it is what the court sees or hears, e.g. testimony or oral evidence, documents, objects, videos, photographs.

Gather all documents related to the cases so long as you can provide witnesses who can tell the court about their origin and content. Never write on original correspondence – using sticky notes that can be removed is preferable.

Figure 1.1 shows a method of ensuring the relevant, admissible and fair evidence is collected for a particular offence. Each row should address an individual element of the offence that needs to be proved (first column). The second column shows the details of the offence element that you have collected. The remaining columns to the right are self-evident in that this is the source and references so that the evidence is well organised and can be used in court for efficient reference.

Disclosable documents

Disclosure is the process of informing the defence of unused material which has been recorded or retained by the authority and not disclosed in evidence capable of undermining the prosecution case or assisting the defendant, e.g. notebooks, draft versions of witness statements if different to the final version, interview records, communication with experts, and any material casting doubt on the reliability of the evidence or a witness.

Each party to a case is required to inform the other side of the existence of all disclosable documents that are or have been in their possession or control. Always keep your notebook tidy and neat as the relevant pages of officers' notebooks are disclosable. If the book is untidy and disorganised and the court sees it, it can be used to suggest a less than professional attitude.

Internal office documentation must also be disclosed – file notes or telephone messages written on whatever was to hand when the call was received are part of the case file, and are therefore disclosable.

Anything that relates to a prosecution, or goes onto a case file has the potential to be disclosed, therefore officers should ensure that nothing other than relevant information is included.

The process of disclosure requires the prosecutor to provide to the defence a copy of all relevant material pertaining to a case. The material is divided into:

• material upon which reliance will be placed;
• material which is disclosed but upon which reliance will not be placed; and
• other material which exists, is relevant, but will not be disclosed.

Into this third class goes material for which Legal Professional Privilege is claimed. Advice from inhouse lawyers needs to be kept separate from the case notes so that it attracts legal professional privilege.

Exhibits

Give the court articles of real evidence, in the form of photographs, plans or maps of the location, and real 'things' that they can look at, weigh in their hands and scale against the rest of the scene. This is likely to influence the court more than witness statements.

Figure 1.1 Evidence form – Section XX Act Year

Ref	Offence element (point to prove)	Details of the offence element Case law/definitions/ interpretation	Evidence of the offence element		Page number and paragraph number of witness statement	Page number of interview under caution
			Name of witness giving evidence of the element	Relevant exhibit no. (if any) witness initial and number		
1	Name					
2	Address					
3	Date and time of offences					
4						
5						
6						
7						
8						
9						
10						

Notes

You must ensure that there is sufficient evidence to prove that each element of the offence has been committed.

The evidence must be referred to in a witness statement for it to be admissible in court.

The exhibit(s) must be attached to witness statements for it to be admissible in court.

Taken together the witness statements must cover every element of the offence – so some witnesses may prove some elements and others may prove other elements or reinforce the same elements where necessary.

Any physical evidence taken should be bagged and tagged, and then stored in a lockable evidence cupboard, to which access is restricted.

A register should be kept indicating:

- when the item of evidence was put in the cupboard;
- by whom; and
- if and when the item was taken out of the cupboard, by whom and for what purpose, for example, an item of food removed for it to be inspected by a company representative.

Photographs/videos

The role of the witness is to give evidence as to what was seen or heard; the court makes decisions of fact. Give short factual statements of what you saw to introduce a photograph or video which speaks for itself. It is vital to let the photograph/video speak for itself. The rule is 'Simply exhibit the photograph, and exhibit the photograph simply'.

Identification evidence

The leading case for identification is that of *R* v. *Turnball*.[3] In that case the Court of Appeal laid down guidelines that should be applied where identification evidence is disputed.

Condensed they are:

A Amount of time the suspect was observed by the witness
D Distance between suspect and witness
V Visibility at the time of observation
O Obstructions between witness and suspect
K Known previously? (i.e. does the witness know the suspect?)
A Any significant reason for the witness to remember the suspect?
T Time elapsed since witness saw the suspect and description written down
E Error or discrepancies in the witnesses description.

This is a list of all those points that a witness identifying a suspect **must** cover when making their identification, whether orally in the witness box or in a witness statement made as part of the case if their identification evidence is to be relied on. If one of these is not covered the Judge is obliged to warn the jury about the risks of relying on the identification evidence, and he may still do even if all of the points are covered, and the issue of identity is critical. It is a misdirection should he fail to do so and grounds for appeal.

Enforcement officers should subject their own identification evidence to this test to be sure that it will stand up in court.

Proving competence and authority

It is necessary to prove that the officer taking the case or is the main witness is competent to do so. It needs to be shown that the officer holds a relevant qualification and is sufficiently competent in the particular field of the case. It may be necessary to show knowledge and skills are held and practiced, refresher training done, how knowledge is maintained and that practical skills are up to date.

Where newly qualified officers take a case, knowledge is the most up to date, and skills, once obtained, can be practiced often.

Where officers practice within their field of expertise and keep their skills and knowledge up to date it is very difficult to challenge competency successfully.

It is necessary to show that the officer is duly authorised and was acting within their authority (see page 23).

Witness statements

A witness statement is a document recording the evidence of a person, which is signed by that person to confirm that the contents of the statement are true. A statement should record what the witness saw, heard or felt. In the case of the licensing officer this would include all of the exhibits that he collected in the course of the inquiry. You should rely on notes made at the time of the incident when making their witness statement. Lawyers expect to find a very close connection between the notes and the text of the witness statement.

Officers should consider producing a witness statement at timely intervals. Breaking the case up into manageable amounts makes the witness statement likely to be more reliable and therefore less vulnerable to challenge. Make sure in your witness statement you make it clear that you are familiar with the leading case, the current Code of Practice, the relevant guidance, i.e. the key legal background to the case. You should check all facts written in the statement are correct with the correct date and supported by evidence. Never add or remove things from your statement. Describe behaviour and observable facts in the statement and avoid professional terms and jargon, using everyday language. Ensure any opinion stated is within your expertise and be aware of any inconsistencies.

Witness statements can be made under Section 9 of the Criminal Justice Act 1967 which states that the contents of a written statement will be admissible, without the witness attending court to give oral evidence.

Witness statements should be on prescribed forms. They should describe the sequence of events in a logical order, usually chronological. They should be certified by the witness that the evidence is true. They should be short and the information in it should be relevant to the offence.

Witness statements from lay people

It is important to impress on witnesses that they should only say what they know to be true, and thereafter to leave it to the court to determine effect.

The first step is to make sure that witnesses know that they should not discuss the matter with each other. Quite clearly you cannot stop this happening. All you can do is make sure you advise them, in simple and understandable terms not to talk to anyone about their evidence. Stress the importance of it being their evidence and remaining uncontaminated. Make sure they understand your advice. The second step is to impress on a witness about to provide a witness statement the need to ensure that they can give their own evidence of what they are about to write by highlighting that their evidence will not be accepted per se, but will be challenged in cross examination. If the witness comes under pressure in the witness box and concedes that they have chatted with other witnesses about their evidence, on their head be it. They will also have to concede that you told them not to, either in cross examination or under re-examination, so at the very least you can demonstrate you did try to prevent evidence contamination.

Can you write the statement for the witness? You can write down the statement, at the dictation of the witness, but you must only record what you are told. You may not add bits, and you should not sanitise the use of English to make the statement more readable. A defence lawyer can usually tell whether a witness is using his own words or words are being used for him, and any suspicion that a statement has been doctored or sanitised will lead to the witness who made it being called and cross examined on it.

The first thing you can do is talk the witness through their evidence. Get them to go through what they saw, heard, etc. in chronological order. You cannot impress on them what is important and what is not, because to do so is to affect their judgement but at least they will have had an opportunity to rehearse what they will say before committing it to paper. If the witness does not remember something vital you can push very gently – 'Did you not say to me … that may be worth recording', etc. You cannot remind them, only they can remember.

Where a witness writes their own statement, the rule is that you can guide the pen but you may not push it. It is perfectly acceptable to suggest to the witness that he writes his statement in chronological order, because that makes sense to record the story as it unfolded. Given that you will have talked it through as suggested above, this will ensure that the witness has a clear picture in his mind, recalls as much as possible and records it in admissible form. When the witness has finished writing, you can read over the statement and point out that he may have forgotten to mention something he has told you orally, giving him the chance to include it, provided the exchange to which reference is made actually took place.

How much assistance can you give to a witness who is about to appear and give oral evidence for the authority? Coaching is not allowed. Officers must be very vigilant to ensure that nothing that they do or say can be construed as witness coaching.

Sub judice

Information that is relevant to a case that has not yet come before the court is *sub judice*, literally '*in the course of trial*', and must not be discussed in the public domain. It must be preserved intact, to be considered in court by a jury or lay bench who are hearing its contents for the first time, and can form a view as to its validity in context of the hearing and alongside all other relevant evidence.

The reason for the rule of *sub judice* is simple – it allows arbitrators to consider evidence on the basis of what they hear and see in court. Where information has been in the public domain it is virtually impossible for an individual not to form a view about it themselves and for their own judgement not to be coloured by what they have heard from others.

Expert witnesses

Experts, and indeed licensing officers, who are considering offering themselves as expert witnesses must be aware of the rules governing the conduct of expert witnesses and must abide by them. Failure to do so may damage the credibility of the expert witness in the eyes of the court, with their client or to others who may have considered engaging him. Expert witnesses hold a very privileged position in litigation, and with this comes responsibilities.

An expert witness may give the court evidence of his or her opinion, unlike a witness to fact, who is restricted to purely factual evidence. In order to give opinion evidence, the expert must know what each witness has said, and in order to facilitate him having this knowledge he is allowed to sit in the well of the court and listen to the evidence of the witnesses as to fact. He may then comment on that evidence when he gives evidence, by for example, indicating which scenario presented in the evidence is more likely, in his view, to be the case.

The expert witness has more freedom in court than the lay witness.

It is incumbent both on those who act as expert witnesses, and those who chose to offer expert witnesses in support of their cases, to ensure that the restrictions of the role are understood.

Where an expert is asked a question or considers a proposition that takes him outside the core of his field of expertise he should advise the court that he cannot give expert evidence on the point.

What is an expert witness?

An individual must actually be an expert in a particular and relevant field. Clear evidence of expertise in the form of qualifications, papers published, etc. must be provided to satisfy the court of expertise. The court would look for evidence that the expert was so regarded by his peers. The court will also consider whether her expertise is up to date – an individual who has not updated their knowledge will not be regarded as an expert. Equally the court will want to know how much experience an expert has.

However, it is in the discretion of the court as to whether an expert witness is required.

The purpose of an expert witness is to help the court by providing an expert opinion based on the facts as presented by the lay witnesses, and if the court feels that it (or the jury) is capable of making a decision without such assistance, it will refuse to allow an expert witness to be called. If it does agree, the court also has to be persuaded that the person tendered as an expert witness is in fact someone who can be regarded as an expert.

Selecting an expert witness

Strictly define that area in which the assistance of expert evidence is required.

It is necessary to be clear about the exact areas of evidence upon which expert opinion is required and to ensure that the expert chosen has sufficient expertise to consider all of them.

It should be remembered that a particular expert's field of expertise may not necessarily be as wide as the area upon which expertise is required, and in some cases more than one expert may be required.

Ensure that the court will accept the expert chosen.

- Is his expertise up to date?
- Is his expertise mainly theoretical or does he have field experience?
- Is he mainstream and speaks as the majority in this field or not?

Consider whether you can understand your expert.

- What your expert has to say has to be comprehensible to the court or the jury.
- Can your expert speak so that lay jurors or magistrates understand him and will be persuaded by what he says?

Take the advice of your expert.

- Does he agree with your evidence and your methods?
- Are there more tests he would recommend?
- Does he suggest another approach?
- Ensure his opinion does not conflict with your evidence.
- Does your expert know what is expected of him?

Expert rules

An expert is there to assist the court and his overriding duty is to the court, not to the party calling him. He must stay within his area of expertise and must inform the court if he speaks on matters outside that, when his evidence will not be treated as expert evidence.

It is incumbent on him to make it clear to the court when he speaks outside his area of expertise, since that is within his knowledge more than anyone else's.

The expert is required to refer to what facts and assumptions he has used and made when coming to his conclusions and also to state what material facts do not support his conclusions.

Where the expert has not been able to research the issue fully, because for example insufficient data are available or because the theories relied upon may not be widely accepted he should note that his view is provisional.

If, after exchange of statements, the expert changes his mind on a material matter, he should tell the other side and the court as soon as possible.

Photographs, plans surveys and other documents relied on by the experts in coming to their stated views must be exchanged at the same time as the reports, so that each side can come to a fully advised view of the other's position.

There is an exception where the law says the relevant officer can act on an opinion. Licensing officers are lay witnesses and can only give evidence of fact unless the law says they can.

Importantly, the licensing officer will give opinion evidence on only one point, whether in his or her opinion and based upon all of the factual evidence he or she gives as a lay witness something is prejudicial to health or is a nuisance, for example.

However, licensing officers can be regarded as expert witnesses but only if they have specialised in a particular field, and have reached a very high level in it.

Professional witnesses

In cases where it is perceived that there may be some difficulty persuading lay persons to give evidence, local authorities can consider the use of professional witnesses.

Engaging professional witnesses, usually from an investigation agency, undertake monitoring, as well as ensuring that the evidence gathered will be admissible and available. In such cases the cost of engaging the witnesses can legitimately be claimed as investigation costs when the matter goes to court.

Provided the witnesses are properly briefed as to what the investigation requires, and can be demonstrated to have no interest in the outcome of it, they are a more certain source of evidence than lay persons.

Interviews under caution (IUCs)

Before carrying out interviews you should prepare well. Prepare your questions in advance. Ensure you have the ability to record the questions and answers in a notebook, interview record form or tape recorder.

The right to legal advice – An interview under caution is evidence and is admissible for the truth of its content in court.

When a person attends an interview voluntarily he must be informed about his continuing right to consult privately with a solicitor and that free independent legal advice is available.

The person to be interviewed has the right to speak with a solicitor.

It is essential he is properly and appropriately advised, by someone who not only understands the law but understands the way in which it works in the context of an interview under caution.

Why do you carry out an interview under caution? – IUCs are governed by the requirements of PACE Code of Practice B.

A prosecutor carries out an IUC when he has *prima facie* evidence to suggest that an offence has been committed and that the interviewee is either the sole accused, or has played a part in a multi handed offence.

Interviews under cautions are not necessary if you already have enough evidence to establish a *prima facie* case against the proposed defendant. Conducted properly an IUC should fill gaps in the inquiry by establishing the role of the accused, providing the interviewer with details not known at the time of the offence, or, in some cases allowing the interviewee to clear himself and eliminate himself from the inquiry. An IUC should not be done until the obvious ways of obtaining evidence have been exhausted.

Ending an interview under caution – The interview must cease when 'the officer in charge of the investigation … reasonably believes there is sufficient evidence to provide a realistic prospect of conviction for that offence'.

Interviewing officers conducting IUCs need to carefully consider every answer that they get and think about the cumulative effect of the answer. If that answer gives them enough evidence to believe that they have a reasonable prospect of a successful prosecution, they must stop the interview. There is no discretion to do otherwise and the case will not be assisted if they do.

At the end of the interview invite the interviewee to sign the notes or the seal on the tape. Record the time the interview started and ended, any breaks taken and all persons present.

Evidence of previous condition and failure to comply

To introduce the previous nature of the premises in a prosecution is to open a witness statement with a sentence such as '*XYZ premises is rated as Risk Category 1 because of previous issues with ... etc.*'. Risk assessment based inspection is standard practice and there should be no problem in introducing these facts.

You can also include a potted inspection history to the statement as an exhibit. For example, '*This visit was the third within a period of x months. The inspection history of the premises is attached as Exhibit AB1*' but no further detailed comment on the inspection history should be made.

Claiming costs

'Where ... any person convicted of an offence before the Crown Court, the Court may make such order as to the costs to be paid by the accused to the prosecutor as it considers just and reasonable' – Prosecution of Offenders Act 1985 s18(1)c.

Provided a local authority can show that its claim for costs is just and reasonable there is no *prima facie* reason for the costs not being awarded. Costs start to run when the licensing officer enters the premises on a routine or investigatory visit. It is important that when collating a costs application to accompany a prosecution that everything that can justifiably be claimed as costs of the investigation is included.

A complete Schedule of Costs should include:

- costs of the time spent by the licensing officer in the investigation;
- administrative staff costs in typing Notices, making copies and transcribing interview notes;
- costs of photo processing and analysts cost;
- costs incurred in service of documents, by e.g. personal service or by recorded delivery post;
- any conferences or meetings with the Council's legal team should be costed, with the costs of both licensing officer and lawyer being included in the claim;
- telephone calls relating to the matter, whether internally, with other officers of the Council or with the defendant or his legal representative;
- expenses that are incurred in attending on witnesses and taking statements, etc.; and
- costs of licensing officer, lawyers and witnesses in attending at the hearing.

Try, so far as you can, to estimate what the defendant may claim and cover it off.

Include the time spent advising and assisting the defendant. This has a dual purpose, it undermines a claim that he didn't know and couldn't be expected to know what was required and further rebuts any claim that the council has been unhelpful or uncooperative.

If you can show that he has made substantial profits from his activity, advise the prosecutor to give details of your calculation to the court – it will have to be either accepted or denied and if denied will have to be explained.

Give your lawyer as much background information as you can and make sure potential 'areas of dispute' are addressed either in the facts of the case or during cross examination.

If the defence, having been sent a copy of the prosecution's application for costs, do not tell the prosecutor or the court that they propose to object to some or all of them before the verdict is delivered, they may not raise objection to any part of the costs application.

Local authorities are required by rules deriving from the *Associated Octel* judgement to serve their Schedule of Costs on the defendant at the earliest opportunity, in order that the defence may decide what parts, if any, they wish to challenge as unjust or unreasonable.

(*R* v. *Associated Octel Co Ltd*[4])

Victim Personal Statements

Victims Personal Statements (VPS) give victims an opportunity to say how the crime has affected them, which information the judge or magistrates can take into account when imposing sentence. Details of their use can be found in the Consolidated Criminal Practice Direction – 8th July 2002.

They can be of considerable assistance to the court when considering sentence where there has been a workplace death or serious injury in outlining the human effects of what appears to be a 'mere' breach of a technical regulation or requirement.

There are a number of rules regarding collation of a VPS.

1. The victim must be told about the scheme and offered the opportunity to make a VPS before the final disposal of the case.
2. The victim cannot be compelled to give a VPS, victims must wish to give such a statement. VPSs are often intensely personal statements and not every victim wishes to go through the process of quantifying the effect of the incident on their lives and their thoughts for the future at what is a very traumatic time for them.
3. Should the victim wish to make VPS it must be served in the proper form, being either in the form of a witness statement (section 9 Criminal Justice Act 1967) or must be contained in an experts report, for example where the witness has been receiving counselling or psychiatric help consequent to the incident.

The VPS, once made, is served on the defendants or his solicitor and on the court prior to sentence. The court may then take account of the effects of the incident on the victim when considering sentence. Local authorities proposing to take a VPS should consider seeking professional assistance either from the police or from trained counsellors to ensure that the statement is taken properly and is a full reflection of the victim's position, but also to ensure that the process is no more traumatic for the victim or the person taking the statement that it needs to be.

Evidence suggests, however, that victims who do produce a VPS feel that their views are taken into account, and that the sentence reflects this. Where a VPS would be appropriate local authorities should assist the court and the victim by obtaining and using them.

VPS can also apply to environmental crime, for example, on costs of clear up, effect on the local community, short and long term effects on the environment and to those affected by the damage. The Environment Agency can give guidance on costs and timescales of recovery, Also considered should be local tourist boards (in cases of oil spills onto holiday beaches or areas of outstanding natural beauty, the wider tourist industry in an area may suffer), chamber of trade and local interest groups.

The prosecutor should speak on behalf of the environment as a victim and should put before the court details of all of the damages caused by environmental crime and all of the impacts short and long term on the environment and to all of those who are affected by damage to the environment.

Sentencing

It will help the court greatly, and shorten hearings, if a case summary can be supplied by the prosecutor at the time of the issue of proceedings. This document should set out the facts on which the prosecution rely and contain aggravating and mitigating features. These are known as Friskies Schedules which have their origin in a prosecution brought by the Health and Safety Executive.[5] In this case, the Court of Appeal made strong recommendations about presentation of health and safety cases, which should assist the judge to understand the aggravating and mitigating features of the case and to pass an appropriately tailored sentence.

The rules are:

1. The prosecutor should list, in writing, the facts of the case and the aggravating features that he says are present.
2. This list should be served on the defendant and the court.
3. In the event of the defendant pleading guilty, the defence should submit a similar document, but in their cases outlining those mitigating features they say are present.
4. If the prosecution and the defence agree the aggravating and mitigating features they should commit this to writing and put the agreement to the court. Where there is no agreement, the judge will hear argument on the disputed points, form a view and advise the parties of his view, which would be relevant in any appeal against the sentence handed down.

The Court of Appeal has also declared what constitutes aggravating and mitigating factors in health and safety cases, in the case of Howe and Son.[6]

The court stepped outside the facts of the case to list other features that might be relevant to sentence, but noted that depending on the circumstances of the case it would be for the court to decide whether they were aggravating or mitigating, and to what extent. These included degree of risk and extent of danger (i.e. whether death or injury was foreseeable), gravity and extent of breach, whether it was an isolated incident or continued over a long period, and the defendant's resources and the effect of a fine on their business, noting that this list is not exhaustive.

Whilst both the cited cases were brought under health and safety legislation, it will assist the court to take a similar approach irrespective of the legislation under which proceedings are brought. Applying the same principle by providing information that will assist the magistrates or judge when considering sentence should ensure that the sentence is tailored to reflect the degree of criminality involved in the offence, which is in the interests of prosecution, defendant and the general public's belief in the criminal justice system.

Laying informations

A prosecutor who wants the court to issue a summons must serve an information in writing on the court officer or unless other legislation prohibits this, present an information orally to the court, with a written record of the allegation that it contains (Criminal Procedure Rules 2011 7.2).

An information is a document which states in clear and unambiguous language the offence with which the accused is charged. There should be only one offence in the information. You should accurately describe the name and address of the accused and the specific offence with which the accused is charged in ordinary language, avoiding technical terms. It should include the appropriate section of the act or regulation contravened. The information should also include the date of offence where possible. If the date is not accurately known then you can use the phrases 'on or about' or 'between date 1 and date 2'. Finally, refer to the section where the penalty for the offence is contained.

Briefing solicitors or Counsel

Counsel is required to keep a distance from the client – he or she can only be contacted via the instructing solicitor. Solicitors also need to be briefed before they come to court. It is therefore important that the brief contains all of the necessary information, and copies of all of the relevant documents, as well as clear and unequivocal advice as to what the client wants.

Solicitors or Counsel may have no knowledge at all of the case or the facts surrounding it, they have to be told everything, particularly those things that, as professionals, licensing officers consider to be taken as read, or common knowledge, since for someone outside the field of environmental health, they are not.

If you want Counsel to attend at court and represent the local authority it is necessary to be clear and to say so. It is also necessary to underline what else you want, for example, that an application for costs (to include Counsel's reasonable fees) should be made, that a Confiscation Order should be sought, etc. and to advise what would be acceptable by way of plea bargain, should it be offered.

Where Counsel is asked to advise, it is essential to be clear about what advice you actually want.

It is important to include some practical information which solicitors or Counsel would be keen to have, such as the fact that the lay witness is reluctant and very nervous, or that the defendant has a violent temper. This allows them to be ready for whatever may happen, and to plan accordingly.

Proving the case

In a court case, all that is necessary is to know what you have to prove and to show, by way of evidence that you can do so.

It is very important to present all the relevant facts to the court. The courts want a presentation of all of the relevant facts, for example, in the case of environmental damage, the court should be made properly aware of environmental damage and understand how long the environment would take to recover from a particular incident, or the true cost of remediation. A plea of guilty should not prevent all of the relevant facts coming before the court, with sufficient detail so that a bench of lay magistrates can understand the complexity of the damage caused by the incident and the scale and cost of remediation.

The information laid before the court has to be comprehensive, understandable and, for example, in the case of environmental damage should take account of the following features which demonstrate aggravating damage:

- adverse effects on human health, flora and fauna;
- noxious and persistent damage by pollutants;
- the need for extensive clean up operations; and
- the effect of the incident on lawful activities.

Previous convictions of a like nature are also aggravating features.

Any financial gain to the defendant as a result of the incident should be demonstrated as it should be reflected in the fine, such that polluters cannot make enough from their activities to pay any fine and still be in pocket as the result of their activities.

The court wants:

- projections of clean up costs and times particularly since the cost of clean up all too often falls to publicly funded bodies;
- evidence of the polluting incident, e.g. colour photographs or video footage can have a significant effect on those who do not have regular contact with such incidents; and
- estimates of the financial benefit to the defendant, which are very valuable.

Prosecution file

The following list is an example of what could be contained in a prosecution file.

- File front sheet should include:
  - defendant's name and address, date of birth, gender, occupation, nationality;
  - defendant's previous convictions;

- alleged offences with details of offence and relevant legislation;
- date of alleged offence;
- date of laying information;
- officer's authorisation details;
- case summary;
- evidence – witness statements, photographs and physical evidence;
- witness reliability;
- whether evidence is admissible, substantial and reliable;
- whether there is a reasonable prospect of conviction; and
- whether it is in the public interest to prosecute.
- Initial witness assessment (whether witnesses are vulnerable and require support)
- Witness non-availability
- Witness statements
- Exhibit list including documents
- Record of interviews
- Unused material likely to undermine the case
- Schedule of prosecution costs
- Officer's contemporaneous notes.

If there is a not guilty plea add:

- Schedule of sensitive material
- Schedule of non-sensitive unused material
- Disclosure report
- Update on witness non-availability.

COURT PROCEDURE

The magistrates' court is the usual court in which local authorities take prosecutions and where appeals are heard. There are three lay magistrates sitting in court and they are assisted by the clerk of the court. The clerk of the court sits in front of the magistrates and gives them legal and procedural advice. The clerk also deals with general administrative work.

When in court, before the session starts, stand up when the magistrates or judge enter the court. The magistrates will normally bow and you should bow immediately afterwards. Sit down after the magistrates sit down.

The clerk will read the charge to the defendant and the defendant will plead guilty or not guilty.

Where the defendant pleads guilty the prosecuting lawyer will read a statement of facts relating to the case. This includes submissions of the seriousness of the case, notice of the offence, description of the circumstances giving aggravating and mitigating circumstances, costs and compensation claims. The defendant's solicitor will then plead mitigation. The magistrates will then normally impose a sentence but they may ask for more information before passing sentence.

Where a not guilty plea is made the prosecution will present its case, calling witnesses and the defendant's lawyer will cross-examine those witnesses. Then the defendant's lawyer presents this defence including witnesses. The prosecution lawyer will cross-examine the defendant's witness.

An EHP who has given evidence and sits in court may find the evidence of the defendant incredulous, or the comments of the defence solicitor in mitigation farcical, and may feel frustrated in the extreme that he or she cannot counter it. The temptation is to rolling your eyes or shaking your head in disbelief is considerable, as it may be seen as the only way in which the evidence being advanced can be challenged. The temptation must be resisted, because behaviour in court that may influence the magistrates or jury can form the ground for appeal. Be impassive to the antics of the defence.

Giving evidence

The prosecuting lawyer will ask the main witness, probably the licensing officer taking the case, questions about themselves and the details of the offence or offences based on the witness statement. When answering the lawyer, face the bench or the jury to ensure they understand what you are saying as they will be making the decision on the case. In the Crown Court make eye contact with one or two of the jury. If they think the witness is looking at them, they concentrate. Crisp sharp evidence engages juries.

If you are asked questions by the magistrates address them collectively as 'your worships' and individually as 'Sir' or 'Ma'am'. Crown Court judges should be addressed as 'My lord' or 'My lady'. Appear serious, caring, flexible, well informed, fair and reasonable in giving evidence.

When giving oral evidence it is absolutely critical that the witness says enough to establish those things they are required to prove are indeed demonstrated, and is done sufficiently robustly to rebut any challenge that may be put in cross-examination.

When giving evidence in the magistrates' court, watch the clerk. If he or she is writing they consider the point sufficiently important to record it. If they are not, they do not, so stop. Slow down if the clerk seems to writing too fast and may be missing some important points. The same applies in the Crown Court and in other forms of administrative tribunal – if the judge or the inspector is writing, the point is important, carry on, slowly. If they are not, stop.

Watch your own lawyer. If they are nodding encouragingly, keep going. If they look as if they want to speak, let them. The rule is to keep it short, keep it simple and keep it relevant.

When it comes to language in court, the rule is no acronyms, no jargon and no unexplained use of technical terminology.

Juries like clear, unfussy information in understandable chunks and not embellished with unnecessary details. The witness needs to get the message across as one who speaks their language.

Opinion

Opinion is a judgement or belief not founded on certainty or proof or an evaluation or judgement given by an expert. Opinion is the view you hold – the judgement you form which need not be certain or proved to be true. If an officer can show he is properly informed and has considered his view, his opinion is valid, and he is entitled to act on it. See expert witnesses and opinion (page 30).

Role of the lawyer

The role of the local authority lawyer is to take the local authority's instructions and act so as to give effect to those instructions.

The lawyer must always advise the local authority of the likely outcome of the proposed action, and the likely outcome of any alternatives that he may propose, but at the end of the day it is always for the client (local authority/ licensing officer) to decide which action should be taken.

The instructing department must be free to exercise their professional judgement as to the disposal of matters. The role of the legal department is to follow their clients' instructions. The role of the instructing department is to take legal advice, consider it and then make a decision taking account of all of the relevant considerations.

Appeals

The person or body that appeals is the 'plaintiff'. If the licensing authority is defending against one of its decisions it will be the 'respondent'. They may both be separately represented by lawyers who are likely to wish to cross-examine the merits of the case. The court is deciding on an administrative matter so the strict rules of evidence do not apply, witnesses can be called in support of the case and hearsay evidence can be used.

The person appealing would normally open the case and call his or her witnesses. However, in licensing cases, the court may invite the licensing authority to speak first, if everyone agrees. This is to help the court understand how the licensing authority came to the licensing decision in the first place. The onus is on the person appealing to persuade the court that the licensing authority should not have exercised its discretion in the way that it did.

Whatever decision the court reaches it will give reasons for its decision in writing.

APPENDIX 1

Local authority definitions

The definition of a local authority is very similar in all Acts of Parliament and regulations. The references for their definition are below.

- Animal Boarding Establishments Act – Section 5(2);
- Breeding of Dogs Act 1973 – Section 5(2);
- Caravan Sites and Control of Development Act 1960 – Section 29(1);
- Clean Air Act 1993 – Section 55 and 64;
- Control of Pollution Act 1974 – Section 73(1);

- Criminal Justice and Public Order Act 1994 – Section 77(6);
- Dangerous Wild Animals Act 19 – Section 7(4);
- Environment Act 1995 – Section 91, 82 and 105;
- Environmental Damage (Prevention and Remediation) Regulations 2009 – Regulation 2;
- Environmental Permitting (England and Wales) Regulations 2010 – Regulation 6;
- Environmental Protection Act 1990 – Section 78A(9) – Contaminated land;
- Environmental Protection Act 1990 – Section 4 and (Section 30 (Waste authorities));
- Food Safety Act 1990 – Section 5 and Food Hygiene (England) Regulations 2006 – Regulation 2;
- Housing Act 2004 – Section 261;
- Licensing Act 2003 – Section 3;
- Local Government (Miscellaneous Provisions) Act 1982 – Section 13(11) 20(1) and 44(1) 29(4);
- Pet Animals Act – Section 7(3);
- Public Health Act 1936 – Section 1(2) and 91;
- Riding Establishments Act – Section 6(4);
- Fire Safety and Safety of Places of Sport Act 1987 – Section 41;
- Metropolitan Public Carriage Act 1869 – Section 2;
- Local Government (Miscellaneous Provisions) Act 1976 (as amended) – Section 44;
- Scrap Metal Dealers Act 2013 – Section 22;
- Marriages and Civil Partnerships (Approved Premises) Regulations 2005 – Regulation 2;
- Marriage Act 1949 – Section 46A;
- Civil Partnerships Act 2004 – Section 28;
- Weights and Measures Act 1985 – Section 69;
- Animal Welfare Act 2006 – Section 62;
- Animal Health Act 1981 – Section 50;
- Performing Animals (Regulation) Act 1925 – Section 5;
- Children And Young Persons Act 1933 and 1963 – Section 579 Education Act 1996;
- Fireworks Regulations 2004 – regulation 3;
- Explosive Regulations 2014 – regulation 2;
- Highways Act 1980 – Section 329;
- New Roads And Street Works Act 1991 – section 329 Highways Act 1980; and
- Poisons Act 1972 – Section 11.

APPENDIX 2

Service of notices references

- Control of Pollution Act 1974 – Section 105;
- Environment Act 1995 – Section 123 and 124;

- Environmental Permitting (England and Wales) Regulations 2010 – Regulation 10;
- Environmental Protection Act 1990 – Section 160;
- Housing Act 2004 – Section 246;
- Licensing Act 2003 – Section 113;
- Local Government Act 1972 – Sections 233 and 234;
- Public Health Act 1936 – Section 283 to 285;
- Fire Safety and Safety of Places of Sport Act 1987 – Section 38;
- Private Hire Vehicles (London) Act 1998 – Section 34;
- Marriages and Civil Partnerships (Approved Premises) Regulations 2005 – Regulation 13;
- Animal Health Act 1981 – Section 83;
- Highways Act 1980 – Section 332;
- New Roads And Street Works Act 1991 – Section 97.

NOTES

1. *Beardsley* v. *Giddings* (1904) 1KBD847.
2. *R* v. *Pontypridd Juvenile Court* (unrep).
3. 1976 63 Cr Ap R 132 [1977]QB 224(CA).
4. *R* v. *Associated Octel Co Ltd* (Costs) TLR Nov 15 1996 646.
5. *R* v. *Friskies Petcare Ltd* (2000) 2 CAR(S) 401.
6. *R* v. *Howe and Son (Engineering) Ltd* (1999) 2 All ER 249.

Chapter 2

PUBLIC HEALTH

ANIMAL BOARDING ESTABLISHMENTS

References

Animal Boarding Establishments Act 1963 as amended by the Animal Welfare
Act 2006.
CIEH Model Licence Conditions and Guidance for Cat Boarding Establishments
2013.
CIEH Model Licence Conditions and Guidance for Dog Breeding Establishments.

Extent

This licensing procedure applies in England, Wales and Scotland but not in
Northern Ireland (section 7(3)).

Scope

A person keeping an Animal Boarding Establishment for cats and/or dogs is
required to hold a licence issued by the local authorities (section 1(1)).
   The Act defines the keeping of an Animal Boarding Establishment as:

> ...to the carrying on by him at premises of any nature (including a private
> dwelling) of a business of providing accommodation for other people's
> animals:

Provided that:

(a) a person shall not be deemed to keep a boarding establishment for
    animals by reason only of his providing accommodation for other peo-
    ple's animals in connection with a business of which the provision of
    such accommodation is not the main activity; and
(b) nothing in this Act shall apply to the keeping of an animal at any
    premises in pursuance of a requirement imposed under, or having effect
    by virtue of, the Animal Health Act 1981 (section 5(1)).

FC 2.1 Animal boarding establishments – Animal Boarding Establishments Act 1963

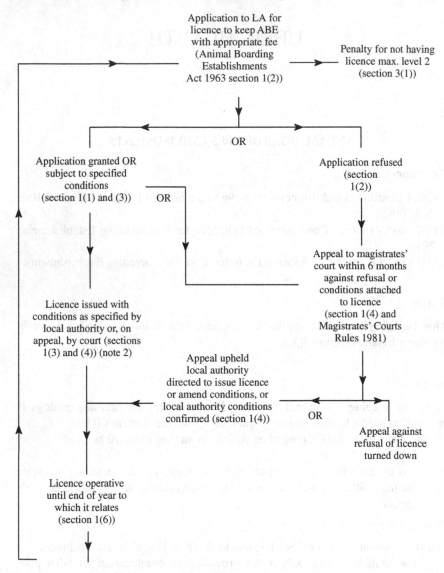

Notes
1. This procedure applies in Scotland but not in Northern Ireland (section 7(3)).
2. Penalty for contravening licence conditions is max. level 2 fine and up to 3 months' imprisonment or both (section 3(1)).
3. Conditions may only be varied on the renewal of licence.

Applications (FC 2.1)

Applications made to the local authority must specify the premises concerned and a fee as determined by the local authority must be paid before any licence is granted (section 1(2)). The factors which may be taken into account in deciding whether or not to issue a licence are listed under 'Conditions' below but these are without prejudice to the local authority's discretion to refuse a licence on any other grounds (section 1(3)).

Licences

A licence normally remains in force until the end of the year to which it relates in which it is granted, however, the applicant determines whether it comes into force on the date it is granted or the first day of the following calendar year (section 1(5) and (6)).

In the event of the death of a licence holder, the licence passes to his personal representative and operates for a period of 3 months from the date of death before expiring. The local authority may, on application, agree to extension of that period (section 1(7)).

Conditions

The local authority is required to specify such conditions in the licence as appears to it necessary or expedient for securing all or any of the following objectives:

(a) that animals will at all times be kept in accommodation suitable as respects construction, size of quarters, number of occupants, exercising facilities, temperature, lighting, ventilation and cleanliness;

(b) that animals will be adequately supplied with suitable food, drink and bedding material, adequately exercised, and (so far as necessary) visited at suitable intervals;

(c) that all reasonable precautions will be taken to prevent and control the spread among animals of infectious or contagious diseases, including the provision of adequate isolation facilities;

(d) that appropriate steps will be taken for the protection of the animals in case of fire or other emergency;

(e) that a register be kept containing a description of any animals received into the establishment, date of arrival and departure, and the name and address of the owner, such register to be available for inspection at all times by an officer of the local authority, veterinary surgeon or veterinary practitioner authorised under section 2(1) of this Act (section 1(3)).

Guidance on conditions relating to these licences is given in two booklets published by the Chartered Institute of Environmental Health in November 1995, *Guide to Dog Boarding Establishments* www.cieh.org/policy/dog_guidance.html and *Guide to Cat Boarding Establishments* www.cieh.org/WorkArea/showcontent.aspx?id= 49634. The Institute encourages local authorities to discuss with establishment owners appropriate standards and timescales based on the documents.

Power of entry

Local authority officers, veterinary surgeons or practitioners authorised in writing by the local authority for this purpose, may, upon producing their authority, if required, inspect a licenced Animal Boarding Establishment and any animals found there at all reasonable times. Persons wilfully obstructing or delaying authorised officers are subject on conviction to a maximum penalty at level 2 on the standard scale (sections 2 and 3(2)).

There is no power of entry to unlicensed premises.

Disqualifications and cancellations

In making a conviction under the Animal Boarding Establishment Act 1963 or under the Protection of Animals Act 1911, the Protection of Animals (Scotland) Act 1912, the Pet Animals Act 1951 and the Animal Welfare Act 2006, the court may cancel any Animal Boarding Establishment licence held by the person and may disqualify him from holding such a licence, whether or not he currently holds one, for any specified period. The cancellation or disqualification may be suspended by the court pending an appeal (sections 3(3) and (4)).

Local authorities must refuse applications for Animal Boarding Establishments licences from persons disqualified under:

(a) the Animal Boarding Establishment Act 1963;
(b) the Pet Animals Act 1951 from keeping a pet shop;
(c) the Protection of Animals (Amendment) Act 1954 from having the custody of animals (section 1(2));
(d) the Animal Welfare Act 2006.

Definition

animal means any cat or dog (section 5(2)).

LICENCE TO HOLD AN ANIMAL GATHERING

References

Animal Health Act 1981.
The Animal Gatherings Order 2010.
The Animal Gatherings (Wales) Order 2010.

Scope

This procedure applies to cattle (excluding bison and yak), sheep, pigs, deer and goats.

The Order is enforced by a local authority.

Extent

This procedure applies in England and similar provisions apply to Wales.

Introduction

If a person wants to hold an animal gathering, the premises must have an Animal Gatherings Order (AGO) licence from Animal and Plant Health Agency (APHA).

The procedures apply to cattle, sheep, pigs, goats and deer.

This includes when animals are brought together to be sold, to be sent elsewhere (for example, for slaughter), for show or exhibition and to be inspected for breed characteristics (for example, to assess pedigree status).

It is an offence to hold an animal gathering without a licence.

The licence application must include a disease control contingency plan outlining how to respond to a notifiable disease at the event, licence implementation plan outlining how the licence conditions will be met and a site plan outlining the animal area and the extent of the site to be licensed (this is agreed with APHA).

This licence is currently free.

At least 14 days before an animal gathering is held the local authority must be informed about the times when the licensed premises will be open to receive animals and the purpose of the animal gathering.

Notices and licences (FC 2.2)

A notice, licence or authorisation issued by a veterinary inspector under this Order:

(a) must be in writing;
(b) may be amended, suspended or revoked; and
(c) may be subject to such conditions and requirements as the veterinary inspector considers necessary to control the introduction into or spread of disease within or from the licensed premises.

A licence is not needed to hold an animal gathering if all the animals brought to the holding are owned by the same person, all the animals come from one premises or it takes place on premises owned by the animal owner (article 4 Animal Gatherings Order 2010).

The use of premises for animal gatherings

No person may use any premises for an animal gathering unless those premises are licensed by a veterinary inspector. An application for a licence must be in writing and must include the name of the licensee; the premises on which the animal gathering may take place; and particulars that identify the animal area (article 5).

Restriction for 27 days between animal gatherings

An animal gathering cannot take place on licensed premises until 27 days has elapsed from the day on which the last animal has left the licensed premises; and all equipment to which animals had access has been cleansed of visible contamination, after the last animal has left the licensed premises (article 6).

FC 2.2 Licence to hold an animal gathering – The Animal Gatherings Order 2010

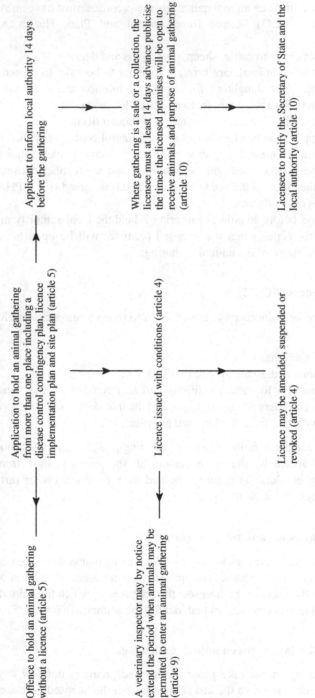

Offence to hold an animal gathering without a licence (article 5)

Application to hold an animal gathering from more than one place including a disease control contingency plan, licence implementation plan and site plan (article 5)

Applicant to inform local authority 14 days before the gathering

Where gathering is a sale or a collection, the licensee must at least 14 days advance publicise the times the licensed premises will be open to receive animals and purpose of animal gathering (article 10)

Licensee to notify the Secretary of State and the local authority (article 10)

A veterinary inspector may by notice extend the period when animals may be permitted to enter an animal gathering (article 9)

Licence issued with conditions (article 4)

Licence may be amended, suspended or revoked (article 4)

Notes

1. An animal gathering cannot take place on licensed premises until 27 days has elapsed from the day on which the last animal has left the licensed premises; and all equipment to which animals had access has been cleansed of visible contamination, after the last animal has left the licensed premises (article 6 and 12). Exceptions to this in article 7.

2. This order applies to England and similar provisions apply to Wales.

This restriction does not apply where all animal areas on the licensed premises are paved, and the areas and any equipment are scraped clean, swept, cleansed by washing and then disinfected with an approved disinfectant before the next animal gathering. A veterinary inspector can issue an authorisation to a licensee to hold an exempt animal gathering subject to other requirements (article 7).

Destruction, treatment or disposal of feeding stuffs and other materials

The licensee must ensure that all feeding stuffs to which animals had access, and all bedding, animal excreta, other material of animal origin and other contaminants derived from animals in the animal area are destroyed, treated so as to remove the risk of transmission of disease, or disposed of so that animals do not have access to them (article 8).

Admittance of animals only for a period of up to 48 hours

No person can bring or permit any animal onto premises used for an animal gathering after the gathering has been open to receive animals for a period of 48 hours. A veterinary inspector may by notice extend the period when animals may be permitted to enter an animal gathering (article 9).

Requirement to notify and publicise the times for bringing animals to animal gatherings

Where the animal gathering is a sale or a collection, the licensee must at least 14 days in advance publicise the times when the licensed premises will be open to receive animals and the purpose of the animal gathering; and notify the Secretary of State and the local authority (article 10).

Dedicated slaughter sales and dedicated slaughter collections

A dedicated slaughter sale or a dedicated slaughter collection must be on a paved animal area and cannot be part of an animal gathering held for any other purpose (article 11).

Restrictions

Once the last animal at an animal gathering has left the licensed premises animals are not allowed onto that premises until all equipment to which animals had access during an animal gathering has been cleansed of visible contamination.

After the animal gathering equipment can only be removed if it has been scraped clean, swept, cleansed by washing and then disinfected with an approved disinfectant; or a period of 27 days has elapsed since the last animal has left the licensed premises.

If there is no appropriate facility to properly cleanse and disinfect the equipment on the licensed premises the equipment may be moved to the nearest appropriate facility where it must be cleansed and disinfected (article 12).

Enforcement

The Secretary of State may discharge the duties under the order directly and delegate prosecutions under section 73 of the Animal Health Act 1981 to the Director of Public Prosecutions (article 13).

Licence conditions

Licence conditions may include:

- a statement of compliance with the conditions in the licence and implementation plan;
- a statement of compliance with the disease control contingency plan when dealing with notifiable diseases;
- a statement that all reasonable steps to prevent the spread of disease during animal gatherings (biosecurity measures) will be taken;
- a biosecurity officer is appointed to make sure licence conditions are upheld (this can be the licence holder);
- animals must only enter or leave the animal area on a vehicle (they must only be loaded onto and off vehicles in the animal area);
- there is a fence around the licensed area to prevent animals escaping;
- records are kept to allow tracing if notifiable disease is suspected; and
- a statement of compliance with animal welfare rules.

Records

The following records should be kept:

- the origin of the animals;
- the destination of the animals (if not available, details about the buyer);
- details about any vehicles used to transport the animals so that they can be traced;
- names and addresses of all staff working in the animal area; and
- whether staff have contact with livestock anywhere else.

Records must be up to date and must be kept for 6 months.

Offences

Issue of false licences, etc. and in blank

A person is guilty of an offence:

    (a) who grants or issues a licence, certificate or instrument made or issued, or purporting to be made or issued under this Act, order of the Minister, or a regulation of a local authority, which is false, unless he shows to the court's satisfaction that he did not know of the falsity, and that he could not with reasonable diligence have obtained knowledge of it; or

(b) who grants or issues such a licence, certificate or instrument not having, and knowing that he has no lawful authority to grant or issue it (section 67 Animal Health Act 1981).

Similar provisions apply to licences issued in blank, i.e. not being before its issue filled up to specify any particular animal or thing (section 68 Animal Health Act 1981).

Falsely obtaining licences, etc.

A person is guilty of an offence:

(a) who for the purpose of obtaining a licence, certificate or instrument makes a declaration or statement false in any material particular; or

(b) who obtains or endeavours to obtain a licence, certificate or instrument by means of a false pretence, unless he shows to the court's satisfaction that he did not know of that falsity, and that he could not with reasonable diligence have obtained knowledge of it (section 69 Animal Health Act 1981).

Alteration of licences, etc.

A person is guilty of an offence against this Act, who, with intent unlawfully to evade this Act, or an order of the Minister, or a regulation of a local authority:

(a) alters, or falsely makes, or ante-dates, or counterfeits a licence, declaration, certificate or instrument made or issued, or purporting to be made or issued, under the Act, order or regulation; or

(b) offers or utters such a licence, declaration, certificate or instrument knowing it to be altered, or falsely made, or ante-dated or counterfeited (section 70 Animal Health Act 1981).

Other offences as to licences

A person is guilty of an offence:

(a) who, with intent unlawfully to evade this Act, order of the Minister, or a regulation of a local authority, does anything for which a licence is required under the Act, order or regulation, without having obtained a licence; or

(b) who, where the licence has expired does the thing licensed; or

(c) who uses or offers or attempts to use a licence:

  (i) an instrument not being a complete licence; or

  (ii) an instrument untruly purporting or appearing to be a licence,

unless he shows to the court's satisfaction that he did not know of that incompleteness or untruth, and that he could not with reasonable diligence have obtained knowledge of it (section 71 Animal Health Act 1981).

Holding an animal gathering without a licence or failure to comply with the licence conditions can result in prosecution. A licence can be suspended or withdrawn and the licensee may have to comply with extra monitoring (section 73 Animal Health Act 1981).

Definitions

animals means cattle (excluding bison and yak), deer, goats, sheep and pigs.
animal gathering means an occasion at which animals are gathered for one or more of the following purposes:

(a) a sale, show or exhibition;
(b) collection for onward consignment within Great Britain;
(c) inspection to confirm the animals possess specific breed characteristics (Article 2).

dedicated slaughter collection means an animal gathering for the purpose of onward consignment direct to slaughter in Great Britain.
dedicated slaughter sale means an animal gathering for the purpose of a sale before onward consignment direct to slaughter in Great Britain.

DANGEROUS WILD ANIMALS

References

Dangerous Wild Animals Act 1976 as amended by the Animal Welfare Act 2006 and Dangerous Wild Animals Act 1976 (Modification) (No. 2) Order 2007.
DEFRA Circular 1/2002 The Keeping of Wild Animals: An Introductory Guide to: The Protection of Animals Act 1911, The Performing Animals (Regulation) Act 1925, The Pet Animals Act 1951, The Dangerous Wild Animals Act 1976, The Zoo Licensing Act 1981.

Extent

This licensing procedure applies in England, Wales and Scotland but not in Northern Ireland (section 10(3)).

Scope

Any person keeping any dangerous wild animal is required to hold a licence from a local authority (section 1(1)).

The animals covered by this provision are listed in the Schedule to the Act and this has been amended from time to time. The latest modification was in 2008 (Scotland) and the animals requiring their keeper to be licensed are given in paragraph 1 of schedule 1 of the act and paragraph 1 of the schedule in the Dangerous Wild Animals Act 1976 (Modification) (No. 2) Order 2007. The common names of the animals in the schedules are reproduced in the list below.

This licensing procedure does not apply to animals kept in:

(a) a zoo within the meaning of the Zoo Licensing Act 1981;
(b) a circus;
(c) pet shops; and
(d) places which are designed establishments under the Animals (Scientific Procedures) Act 1986 (section 5).

A person is held to be the keeper of the animal if he has it in his possession and the assumption of possession continues even if the animal escapes or it is being transported, etc. This removes the need for carriers or veterinary surgeons to be licenced (section 7(1)).

List of common names of the animals in the schedules

The Tasmanian devil

Grey kangaroos, the euro, the wallaroo and the red kangaroo

New-world monkeys (including capuchin, howler, saki, spider, uakari and woolly monkeys

Old-world monkeys (including baboons, the drill, colobus monkeys, the gelada, guenons, langurs, leaf monkeys, macaques, the mandrill, mangabeys, the patas and proboscis monkeys and the talapoin)

Leaping lemurs (including the indri, sifakas)

Large lemurs (the broad-nosed gentle lemur and the grey gentle lemur are excepted)

Anthropoid apes (including chimpanzees, gibbons, the gorilla and the orang-utan)

The giant armadillo

The giant anteater

Crested porcupines

The giant panda and the red panda

Jackals, wild dogs, wolves and the coyote (foxes, the raccoon dog and the domestic dog are excepted)

The bobcat, caracal, cheetah, jaguar, lion, lynx, ocelot, puma, serval, tiger and all other cats (the domestic cat is excepted)

Hyaenas (except the aardwolf)

Badgers (except the Eurasian badger), otters (except the European otter), and the tayra, wolverine, fisher and ratel (otherwise known as the honey badger)

The African, large-spotted, Malay and large Indian civets, and the fossa

The walrus, eared seals, sealions and earless seals (the common and grey seals are excepted)

Elephants

Asses, horses and zebras (the donkey, domestic horse and domestic hybrids are excepted)

Rhinoceroses

Tapirs

The aardvark

The pronghorn

Antelopes, bison, buffalo, cattle, gazelles, goats and sheep (domestic cattle, goats and sheep are excepted)

FC 2.3 Dangerous wild animals – Dangerous Wild Animals Act 1976

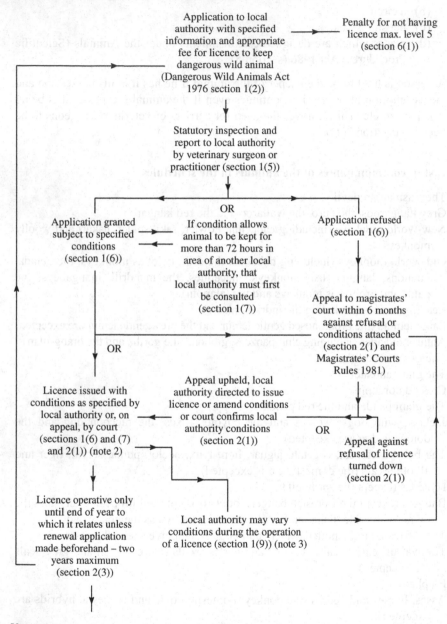

Application to local authority with specified information and appropriate fee for licence to keep dangerous wild animal (Dangerous Wild Animals Act 1976 section 1(2))

Penalty for not having licence max. level 5 (section 6(1))

Statutory inspection and report to local authority by veterinary surgeon or practitioner (section 1(5))

OR

Application granted subject to specified conditions (section 1(6))

If condition allows animal to be kept for more than 72 hours in area of another local authority, that local authority must first be consulted (section 1(7))

Application refused (section 1(6))

Appeal to magistrates' court within 6 months against refusal or conditions attached (section 2(1) and Magistrates' Courts Rules 1981)

OR

Licence issued with conditions as specified by local authority or, on appeal, by court (sections 1(6) and (7) and 2(1)) (note 2)

Appeal upheld, local authority directed to issue licence or amend conditions or court confirms local authority conditions (section 2(1))

OR

Appeal against refusal of licence turned down (section 2(1))

Licence operative only until end of year to which it relates unless renewal application made beforehand – two years maximum (section 2(3))

Local authority may vary conditions during the operation of a licence (section 1(9)) (note 3)

Notes
1. This procedure applies in England, Wales and Scotland but not in Northern Ireland (section 10(3)).
2. Penalty for contravening licence condition is maximum level 5 (section 6(1)).
3. Unless the variation was requested by the licence holder, the local authority must notify him of the variation and allow a reasonable time for compliance.

Camels

The moose or elk and the caribou or reindeer (the domestic reindeer is excepted)

The giraffe and the okapi

The hippopotamus and the pygmy hippopotamus

Old-world pigs (including the wild boar and the wart hog) (the domestic pig is excepted)

New-world pigs (otherwise known as peccaries)

Mammalian hybrids with a parent or (parents) of a specified kind

Cassowaries

The ostrich

Alligators and caimans

Crocodiles and the false gharial. The gharial (otherwise known as the gavial)

Mole vipers and certain rear-fanged venomous snakes (including the moila and montpellier snakes, twig snakes), the boomslang, the red-necked keelback and the yamakagashi (otherwise known as the Japanese tiger-snake)

Certain front-fanged venomous snakes (including cobras, coral snakes, the desert black snake, kraits, mambas, sea snakes and all Australian poisonous snakes (including the death adders)). The gila monster and the (Mexican) beaded lizard

Certain front-fanged venomous snakes (including adders, the barba amarilla, the bushmaster, the copperhead, the fer-de-lance, moccasins, rattlesnakes and vipers)

Wandering spiders

The Sydney funnel-web spider and its close relatives

Brown recluse spiders (otherwise known as violin spiders)

The black widow spider (otherwise known as red-back spider) and its close relatives

Buthid scorpions.

Added in 2007

The Argentine black-headed snake, the Peruvian racer, the South American green racer, the Amazon false viper, the Middle Eastern thin-tailed scorpion and the dingo.

Applications for licence (FC 2.3)

Any application made to a local authority for a licence must be made (unless in exceptional circumstances) by the person who proposes to own and possess the animal and must:

(a) specify the species and number of animals to be kept;

(b) specify the premises where the animals will normally be kept;

(c) be made to the local authority for those premises;

(d) be made by a person 18 years of age or over and not disqualified from holding a licence under the Act; and

(e) be accompanied by a fee stipulated by the local authority at a level sufficient to meet the direct and indirect costs involved.

Applications not complying with these requirements may not be granted (section 1(2) and (4)).

Reports

Before granting any licence the local authority is required to consider a report of an inspection of the premises by a veterinary surgeon or practitioner authorised by it (section 1(5)).

Matters for consideration

The local authority may not grant a licence unless:

(a) it will not be contrary to the public interest on grounds of safety, nuisance or otherwise to issue a licence;
(b) the applicant is suitable;
(c) animals will:
  (i) be held in secure accommodation suitable in size for the animals kept and which is suitable as regards construction, temperature, lighting, ventilation, drainage and cleanliness; and
  (ii) have adequate and suitable food, drink and bedding and be visited at regular intervals;
(d) be appropriately protected in case of fire or other emergency;
(e) be subject to precautions to control infectious diseases;
(f) be provided with adequate exercise facilities (section 1(3)).

Licences

According to the wishes of the applicant, a licence comes into force on either the day on which it is granted, in which case it expires on 31 December of that same year, or on 1 January next in which case it expires on 31 December of that next year. If an application for renewal is made before the date of expiration, the licence continues until the application is determined.

On the death of a licence holder, the licence continues in the name of the personal representative for 28 days only and then expires unless application is made for a new licence within that time, in which case it continues until the new application is determined (sections 2(2), (3) and (4)).

Conditions

The local authority is required to specify conditions which:

(a) require the animals to be kept only by persons specified in the licence;
(b) require the animals to be normally held at the premises specified in the licence;
(c) require the animals not to be moved from those premises unless in circumstances allowed for in the licence;
(d) require the licence holder and person keeping the animals to be insured against liability for damage caused by the animals to the satisfaction of the local authority;
(e) restrict the species and numbers of animals;

(f) require a copy of the licence to be made available by the licence holder to persons entitled to keep the animals; and

(g) any other conditions necessary or desirable to secure the objectives specified in paragraphs (c)–(f) listed under 'Matters for consideration' above (section 1(6) and (7)).

The local authority may attach any other conditions which it thinks fit but if it is to permit the animal to be taken into another local authority area for more than 72 hours, it must consult that local authority (section 1(8)).

Conditions not required by the Act to be attached to the licence may be revoked or modified by the local authority or new conditions may be added. These variations come into effect immediately if they were requested by the licence holder but otherwise the local authority must notify him and allow a reasonable time for compliance (section 1(10)).

Disqualifications and cancellations

Where a person is convicted of an offence under the Dangerous Wild Animals Act 1976 or under:

(a) Protection of Animals Act 1911;

(b) Protection of Animals (Scotland) Act 1912 to 64;

(c) Pet Animals Act 1951;

(d) Performing Animals (Regulation) Act 1925;

(e) Animals (Cruel Poisons) Act 1962;

(f) Animal Boarding Establishments Act 1963;

(g) Riding Establishments Act 1964 and 70;

(h) Breeding of Dogs Act 1973;

(i) Animals Welfare Act 2006 sections 4 to 9 and 11

the court may cancel any licence he may hold to keep a dangerous wild animal and disqualify him, whether or not he is a current holder, from holding such a licence for such period as the court thinks fit. The cancellation or disqualification may be suspended by the court in the event of an appeal (section 6(2) and (3)).

Power of entry

Local authorities may authorise competent persons to enter premises either licensed under the Act or specified in application for a licence, at all reasonable times, producing if required their authority, and the authorised officers may inspect these premises and an animal in them. The local authority may charge the person making the application, for the inspection.

The penalty for wilfully obstructing or delaying an authorised officer is a maximum fine at level 5 (sections 3 and 6(1)).

Seizure of animals

If a dangerous wild animal is being kept without the authority of a licence or in contravention of a licence condition, the local authority may seize the animal and retain it, destroy it or otherwise dispose of it. The local authority is not liable to compensation and may recover costs from the keeper of the animal at the time of this seizure (section 4).

Definitions

circus includes any place where animals are kept or introduced wholly or mainly for the purpose of performing tricks or manœuvres.
premises includes any place (section 7(4)).

DOG BREEDING ESTABLISHMENTS

References

Breeding of Dogs Acts 1973 (as amended by the Animal Welfare Act 2006) and 1991.
Breeding and Sale of Dogs (Welfare) Act 1999.
The Breeding of Dogs (Licensing Records) Regulations 1999.
Home Office Circular 53/1999.

Extent

This licensing procedure applies in England, Wales and, with modifications, Scotland but not in Northern Ireland (section 7(2)).

Scope (FC 2.4)

A licence is required to carry out a business of the breeding of dogs for sale. The Act indicates that the keeping of bitches who give birth to five or more litters in any period of 12 months is presumed to constitute such a business and required to be licensed although local authorities may conclude in the circumstances of a particular case that licensing is required for a lesser number. Where there is no sale of any puppies during the 12 months being considered licensing is not required.

Bitches count towards the qualifying total if they are kept at any time during the 12 month period by the applicant/licence holder or by their relatives at the premises, by the applicant/licence holder elsewhere or by someone else as part of a breeding agreement. This is intended to prevent evasion of licensing by distributing bitches amongst different people and/or premises (section 4A).

Inspection

Upon receipt of an application for a licence for the first time, the local authority must arrange for an inspection by a veterinary surgeon/practitioner **and** by an officer of the local authority. Either or both may carry out inspections relating to subsequent applications (section 1(2A)).

FC 2.4 Dog breeding establishments – Breeding of Dogs Act 1973 as amended

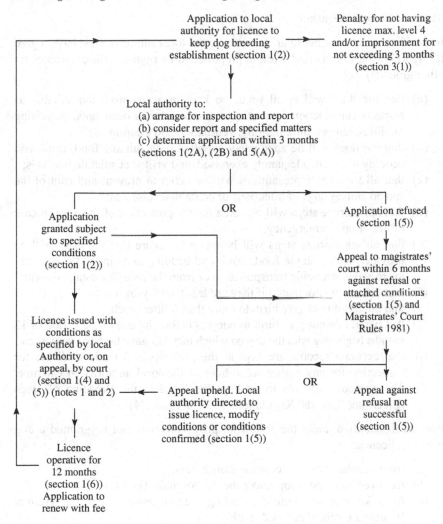

Notes
1. Penalty for non-compliance with conditions is max. fine of level 4 and/or imprisonment for not exceeding 3 months (section 3(1)).
2. Conditions may only be varied on the renewal of a licence.

Report

The local authority must arrange for a report of inspection to be prepared and considered before making its decision (section 1(2B)).

Matters for consideration

In deciding whether or not to grant a licence the local authority must have regard to the following matters (but without prejudice to its right to refuse a licence on other grounds):

(a) that the dogs will at all times be kept in accommodation suitable as respects construction, size of quarters, number of occupants, exercising facilities, temperature, lighting, ventilation and cleanliness;

(b) that the dogs will be adequately supplied with suitable food, drink and bedding material, adequately exercised, and visited at suitable intervals;

(c) that all reasonable precautions will be taken to prevent and control the spread among dogs of infectious or contagious diseases;

(d) that appropriate steps will be taken for the protection of the dogs in case of fire or other emergency;

(e) that all appropriate steps will be taken to secure that the dogs will be provided with suitable food, drink and bedding material and adequately exercised when being transported to or from the breeding establishment;

(f) that bitches are not mated if they are less than 1 year old;

(g) that bitches do not give birth to more than 6 litters each;

(h) that bitches do not give birth to puppies before the end of the period of 12 months beginning with the day on which they last gave birth to puppies; and

(i) that accurate records are kept at the premises and made available for inspection for any authorised officer of the local authority to examine. The particular records to be kept are listed in the Breeding of Dogs (Licensing Records) Regulations 1999 (section 1(4)).

Persons disqualified under the following provisions may not be granted a dog breeding licence:

(a) from keeping a dog breeding establishment;

(b) from keeping a pet shop under the Pet Animals Act 1951;

(c) from keeping an animal boarding establishment under the Animal Boarding Establishments Act 1963;

(d) from having the custody of animals under the Protection of Animals (Amendment) Act 1954;

(e) under section 34(2), (3) or (4) of the Animal Welfare Act 2006 (section 1(2) and (4));

(f) under subsection (1) of section 40 of the Animal Health and Welfare (Scotland) Act 2006.

Licences

The local authority must determine the application within 3 months (section 1(5A)). Licences run for 12 months from the start date being either the date requested by the applicant or the date of issue, whichever is the later (section 1(6) and (7)). On the death of a licence holder, licences pass to his personal representative for a period of 3 months from the day of his death and then expire. The local authority may extend that 3-month period at its discretion (section 1(8)).

Conditions

On granting a licence the local authority must specify such conditions as appear necessary or expedient to achieve the objectives set out in (a)–(i) in the paragraph headed 'Matters for consideration' above (section 1(4)).

Disqualifications and cancellations

In making a conviction under this Act, the court may cancel any dog breeding licence held by the convicted person, disqualify him from holding a licence for such period as the court thinks fit and disqualify him from having custody of any dog for such period as specified. The cancellation or disqualification may be suspended by the court pending appeal (section 3(3) and (4)).

Fees

Local authorities may charge fees for both the applications for licences and for the related inspections. The level of fees may reflect the reasonable administrative and enforcement costs incurred and may be varied to take account of different circumstances (section 3A).

Appeals

Any aggrieved person may appeal to a magistrates' court (sheriff court in Scotland) against a refusal to grant a licence and against the imposition of any condition. The court may direct in either case, as it thinks proper (section 1(5)).

Offences/penalties

Persons operating without a licence or breaching licence conditions are guilty of an offence and on conviction liable to a penalty of either up to 3 months' imprisonment or a fine not exceeding level 4 or to both (sections 1(9) and 3(1)).

Power of entry

Local authority officers, veterinary surgeons or practitioners authorised in writing by the local authority may, upon producing if required their authority, enter any premises licenced as a dog breeding establishment at any reasonable time and inspect it and any animals found there (section 2(1)). Persons wilfully obstructing or delaying authorised officers are liable on conviction to a maximum fine of level 3 (section 2(2) and 3(2)).

Authorised Officers are also given powers of entry into premises other than private dwellings although entry into garages, outhouses, and other structures forming part of the private dwelling is provided for where it is suspected that a person is operating without a licence. Before entry is requested a warrant must be obtained from a Justice of the Peace and must be effected at reasonable times. The penalty for obstruction against these powers is also a maximum fine of level 3 (1991 Act section 1).

PERFORMING ANIMALS REGISTRATION

Reference

Performing Animals (Regulation) Act 1925.

Extent

This procedure applies in England, Wales or Scotland.

Scope

If a person keeps or trains animals for public performance – for example, in a circus, television or film production, or a theatre performance – they will need to register with the local authority.

Registration

If a person exhibits, uses or trains performing animals, they must be registered with a local authority (section 1(1)).

The local authority must keep a register of any person who exhibits or trains animals. Applications must be made on a prescribed form and on payment of a fee he will be registered (section 1(2)).

The application for registration must have particulars about the animals and the general nature of the performances in which the animals are to be exhibited or for which they are to be trained (section 1(3)).

The registered person must be given a certificate of registration in the prescribed form containing entry in the register (section 1(4)).

The local authority must make the register available for inspection by any person (section 1(5)).

Details on the register may be varied by the applicant on application (section 1(6)).

Copies of every certificate of registration must be sent to the Secretary of State (section 1(7)).

Fees may be charged for inspection of the register, for taking copies, making extracts, or for inspection of copies of certificates of registration issued by them (section 1(8)).

FC 2.5 Registration of performing animals – Performing Animals (Regulation) Act 1925

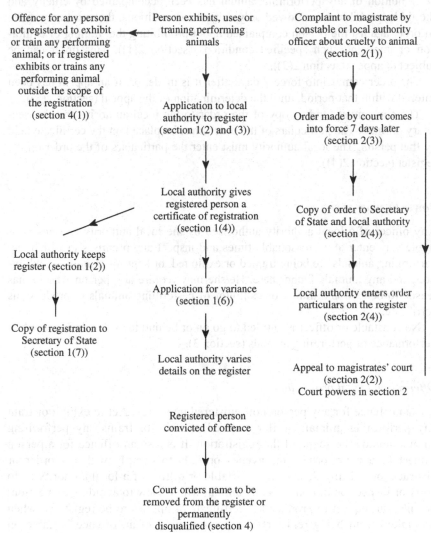

Offence for any person not registered to exhibit or train any performing animal; or if registered exhibits or trains any performing animal outside the scope of the registration (section 4(1))

Person exhibits, uses or training performing animals

Complaint to magistrate by constable or local authority officer about cruelty to animal (section 2(1))

Application to local authority to register (section 1(2) and (3))

Order made by court comes into force 7 days later (section 2(3))

Local authority gives registered person a certificate of registration (section 1(4))

Copy of order to Secretary of State and local authority (section 2(4))

Local authority keeps register (section 1(2))

Application for variance (section 1(6))

Local authority enters order particulars on the register (section 2(4))

Copy of registration to Secretary of State (section 1(7))

Local authority varies details on the register

Appeal to magistrates' court (section 2(2))
Court powers in section 2

Registered person convicted of offence

Court orders name to be removed from the register or permanently disqualified (section 4)

Note
This procedure applies in England, Wales or Scotland.

Court powers

Where it is proved to the satisfaction of a court of summary jurisdiction on a complaint made by a constable or an officer of a local authority that the training or exhibition of any performing animal has been accompanied by cruelty and should be prohibited or allowed only subject to conditions, the court may make an order against the person complained about prohibiting the training or exhibition or imposing such specified conditions (section 2(1)). This order can be subject to appeal (section 2(2)).

The order comes into force 7 days after it is made, or, if an appeal has been entered within that period, until the determination of the appeal (section 2(3)).

The court must send a copy of the order to the local authority and the Secretary of State. The particulars of the order must be placed on the certificate held by that person. The local authority must enter the particulars of the order on the register (section 2(4)).

Power to enter premises

Any officer of a local authority authorised by the local authority and any constable may enter at all reasonable times and inspect any premises in which any performing animals are being trained or exhibited, or kept for training or exhibition, and any animals found there. He/she may require any person who he has reason to believe is a trainer or exhibitor of performing animals to produce his certificate.

No constable or officer is entitled to go on or behind the stage during a public performance of performing animals (section 3).

Offences and legal proceedings

It is an offence for any person not registered under this Act to exhibit or train any performing animal; or if registered exhibits or trains any performing animal outside the scope of the registration. It is also an offence for a person subject to a court order contravenes or fails to comply with the order or obstructs or wilfully delays any constable or officer of a local authority as to entry or inspection or conceals any animal with a view to avoiding inspection; or fails, on request to produce his certificate or applies to be registered when prohibited from being registered. A person guilty of an offence is liable on summary conviction upon a complaint made by a constable or an officer of a local authority to a fine not exceeding level 3 on the standard scale (section 4(1)).

Where a person is convicted of an offence against this Act, or against the Protection of Animals Act 1911 or the Animal Welfare Act 2006 the court may in addition to or in lieu of imposing any other penalty, if registered under this Act, order that his name be removed from the register or order that the person be either permanently or for such time as may be specified in the order be disqualified for being registered (section 4).

Exceptions

The Act does not apply to the training of animals for bona fide military, police, agricultural or sporting purposes, or the exhibition of any trained animals (section 7).

Definitions

exhibit means exhibit at any entertainment to which the public are admitted, whether on payment of money or otherwise.
train means train for the purpose of any such exhibition.

PET SHOPS

References

Pet Animals Act 1951 as amended by the Pet Animals Act 1951 (Amendment) Act 1983 and the Animal Welfare Act 2006.
DEFRA Circular 1/2002.

Extent

This licensing procedure applies in England, Wales and Scotland but not in Northern Ireland (section 8).

Scope (FC 2.6)

A person keeping a pet shop requires a licence from the local authority (section 1(1)). The Act defines 'the keeping of a pet shop' as:

...the carrying on at premises of any nature (including a private dwelling) of a business of selling animals as pets, and as including references to the keeping of animals in any such premises ... with a view to their being sold in the course of such a business, whether by the keeper thereof or by any other person: Provided that:

(a) a person shall not be deemed to keep a pet shop by reason only of his keeping or selling pedigree animals bred by him, or the offspring of an animal kept by him as a pet;

(b) where a person carries on a business of selling animals as pets in conjunction with a business of breeding pedigree animals, and the local authority are satisfied that the animals so sold by him (in so far as they are not pedigree animals bred by him) are animals which were acquired by him with a view to being used, if suitable, for breeding or show purposes but have subsequently been found by him not to be suitable or required for such use, the local authority may if they think fit direct that the said person shall not be deemed to keep a pet shop by reason only of his carrying on the first-mentioned business.

FC 2.6 Pet shops – Pet Animals Act 1951

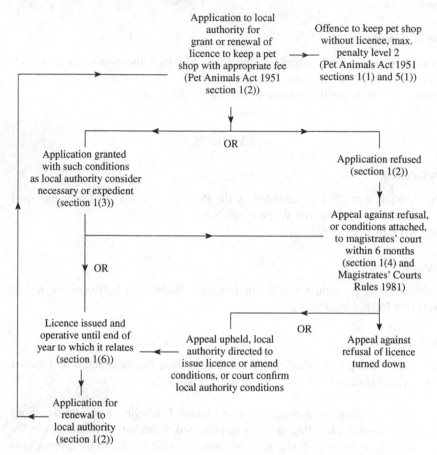

Application to local authority for grant or renewal of licence to keep a pet shop with appropriate fee (Pet Animals Act 1951 section 1(2))

Offence to keep pet shop without licence, max. penalty level 2 (Pet Animals Act 1951 sections 1(1) and 5(1))

OR

Application granted with such conditions as local authority consider necessary or expedient (section 1(3))

Application refused (section 1(2))

Appeal against refusal, or conditions attached, to magistrates' court within 6 months (section 1(4) and Magistrates' Courts Rules 1981)

OR

Licence issued and operative until end of year to which it relates (section 1(6))

Appeal upheld, local authority directed to issue licence or amend conditions, or court confirm local authority conditions

OR

Appeal against refusal of licence turned down

Application for renewal to local authority (section 1(2))

Notes
1. Penalty for contravening licence conditions maximum level 2 (section 5(1)).
2. This procedure applies in Scotland and Wales but not in Northern Ireland (section 8).
3. Conditions may only be varied on the renewal of a licence.

References in this Act to the selling or keeping of animals as pets shall be construed in accordance with the following provisions, that is to say:

(a) as respects cats and dogs, such references shall be construed as including references to selling or keeping, as the case may be, wholly or mainly for domestic purposes; and

(b) as respects any animal, such references shall be construed as including references to selling or keeping, as the case may be, for ornamental purposes (section 7(1) and (2)).

Applications

Applications made to the local authority must specify the premises concerned and a fee as determined by the local authority must be paid before a licence is granted (section 1(2)).

The factors which may be taken into account in deciding whether or not to issue a licence are listed under 'Conditions' below but these are without prejudice to the local authority's discretion to refuse a licence on any other grounds (section 1(3)).

Licences

Licences remain in force until the end of the year to which they relate and the latter is determined by the applicant as being either the year in which it is granted or the following year. In the first case, i.e. the year in which it is granted, the licence comes into force on the day it is granted and expires on 31 December of that year, in the second case, i.e. the year following that in which it is granted, it comes into force on 1 January of that year and expires on 31 December of that year (section 1(5) and (6)).

Conditions

The local authority must attach any conditions which it considers to be necessary or expedient for securing all or any of the following objectives:

(a) that animals will at all times be kept in accommodation suitable as respects size, temperature, lighting, ventilation and cleanliness;

(b) that animals will be adequately supplied with suitable food and drink and (so far as necessary) visited at suitable intervals;

(c) that animals, being mammals, will not be sold at too early an age;

(d) that all reasonable precautions will be taken to prevent the spread among animals of infectious diseases;

(e) that appropriate steps will be taken in case of fire or other emergency (section 1(3)).

No conditions may be specified which relate to a matter dealt with by the Regulatory Reform (Fire Safety) Order 2005.

Power of entry

Local authority officers, veterinary surgeons or practitioners authorised in writing by the local authority for this purpose, may, upon producing their authority if required, inspect a licensed pet shop and any animals there at all reasonable times. Persons wilfully obstructing authorised officers are subject on conviction to a maximum fine of level 2 (sections 4 and 5(1)).

There does not appear to be a power of entry for premises which do not hold a pet shop licence, e.g. which may be suspected of operating an unlicensed pet shop or in respect of which an application has been made. In the latter case refusal of the licence application would be the obvious remedy.

Disqualifications and cancellations

In making a conviction under the Pet Animals Act 1951, the Protection of Animals Act 1911, the Protection of Animals (Scotland) Act 1912 or the Animal Welfare Act 2006 sections 4 to 9 and 11, the court may cancel any pet shop licence held by the person and may disqualify him from holding such a licence, whether or not he currently holds one, for any specified period. The cancellation or disqualification may be suspended by the court pending an appeal (section 5(3) and (4)).

Where proceedings for cruelty or neglect are brought under the Protection of Animals Act 1911, the Protection of Animals (Amendment) Act 2000 enables courts to make orders regarding the care, disposal or slaughter of animals kept for sale in pet shops.

Local authorities must refuse licence applications from persons currently disqualified by a court from holding a pet shop licence (section 1(2)). This provision differs from that relating to Animal Boarding Establishments where disqualifications under other Acts are also relevant.

Definitions

animal includes any description of vertebrate.

pedigree animal means an animal of any description which is by its breeding eligible for registration with a recognised club or society keeping a register of animals of that description (section 7(3)).

There is no definition of the word 'premises', this having been removed by the Amendment Act of 1983, but the sale of animals as pets as a business is prohibited in any part of a street or public place or at a stall or barrow in a market (section 2 as amended).

RIDING ESTABLISHMENTS

Reference

Riding Establishments Acts 1964 and 1970 as amended by the Animal Welfare Act 2006.

Extent

This licensing procedure applies in England, Wales and Scotland but not in Northern Ireland (section 9(2)).

Scope

A person keeping a riding establishment is required to hold a licence issued by the local authority (section 1(1)).

The 1964 Act defines the keeping of a riding establishment as:

> ...the carrying on of a business of keeping horses for either or both of the following purposes, that is to say, the purpose of their being let out on hire for riding or the purpose of their being used in providing, in return for payment, instruction in riding, but as not including a reference to the carrying on of such a business:
>
> (a) in a case where the premises where the horses employed for the purposes of the business are kept are occupied by or under the management of the Secretary of Defence; or
> (b) solely for police purposes; or
> (c) by the Zoological Society of London; or
> (d) by the Royal Zoological Society of Scotland (section 6(1)).

Horses kept by a university providing veterinary courses are also exempt and the place at which a riding establishment is run is to be taken as the place at which the horses are kept (section 6(2) and (3)).

Applications (FC 2.7)

Applications made to the local authority must specify the premises concerned and a fee as determined by the local authority must be paid before any licence is granted (section 1(2)).

Reports

The local authority is required to receive a report by a listed veterinary surgeon or practitioner (see 'Power of entry' below) before making a decision and the report must be based on an inspection made not more than 12 months before the application was received (section 1(3)).

Matters for consideration

In deciding whether or not to grant a licence, or provisional licence, the local authority must, without prejudice to its right to refuse a licence on other grounds, have regard to the following matters:

(a) whether that person appears to them to be suitable and qualified, either by experience in the management of horses or by being the holder of an

FC 2.7 Riding establishments – Riding Establishments Act 1964 as amended

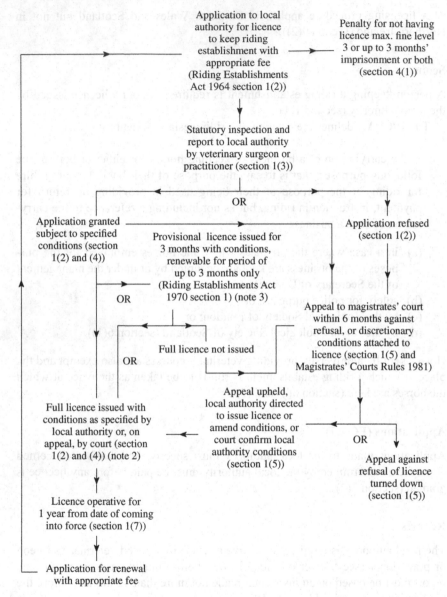

Notes

1. This procedure applies in England, Wales and Scotland but not in Northern Ireland (section 9(2)).
2. Penalty for contravening licence condition is maximum level 3 and up to 3 months' imprisonment or both (section 4(1) as amended).
3. There is no appeal against the decision to issue a provisional licence only or against the conditions attached to it.
4. Conditions may only be varied at the renewal of the licence.

approved certificate or by employing in the management of the riding establishment a person so qualified, to be the holder of such a licence; and

(b) the need for securing:

   (i) that paramount consideration will be given to the condition of the horses and that they will be maintained in good health, and in all respects physically fit and that, in the case of a horse kept for the purpose of its being let out on hire for riding or a horse kept for the purpose of its being used in providing instruction in riding, the horse will be suitable for the purpose for which it is kept;

   (ii) that the feet of all animals are properly trimmed and that, if shod, their shoes are properly fitted and in good condition;

   (iii) that there will be available at all times, accommodation for horses suitable as respects construction, size, number of occupants, lighting, ventilation, drainage and cleanliness and that these requirements be complied with not only in the case of new buildings but also in the case of buildings converted for use as stabling;

   (iv) that in the case of horses maintained at grass there will be available for them at all times during which they are so maintained adequate pasture and shelter and water and that supplementary feeds will be provided as and when required;

   (v) that horses will be adequately supplied with suitable food, drink and (except in the case of horses maintained at grass, so long as they are so maintained) bedding material, and will be adequately exercised, groomed and rested and visited at suitable intervals;

   (vi) that all reasonable precautions will be taken to prevent and control the spread among horses of infectious or contagious disease and that veterinary first-aid equipment and medicines shall be provided and maintained in the premises;

   (vii) that appropriate steps will be taken for the protection and extrication of horses in case of fire and, in particular, that the name, address and telephone number of the licence holder or some other responsible person will be kept displayed in a prominent position on the outside of the premises and that instructions as to action to be taken in the event of fire, with particular regard to the extrication of horses, will be kept displayed in a prominent position on the outside of the premises;

   (viii) that adequate accommodation will be provided for forage, bedding, stable equipment and saddlery (section 1(4)).

Persons under 18 years old or persons or bodies corporate disqualified under the following provisions may not be given a licence:

(a) from keeping a riding establishment under the Riding Establishment Act 1964;

(b) from keeping a pet shop under the Pet Animals Act 1951;

(c) from having custody of animals under the Protection of Animals (Amendment) Act 1954;

(d) from keeping an animal boarding establishment under the Animal Boarding Establishment Act 1963;
(e) under section 30 (2), (3) or (4) of the Animal Welfare Act 2006.
(f) under subsection (1) of section 40 of the Animal Health and Welfare (Scotland) Act 2006 (asp 11) (section 1(2)).

Licences

Full licences continue for 1 year beginning on the day on which they came into force and then expire. The date of operation, depending upon the wishes of the applicant, is either the day on which it is granted or 1 January next (section 1(6) and (7)).

Provisional licences operate for 3 months from the day on which they are granted and are used where the local authority is satisfied that it would not be justified in issuing a full licence. The 3 months' period of operation may, on application before the expiration of the 3 months, be extended for a further period of not exceeding 3 months so long as this would not exceed a 6-month period in 1 year (Riding Establishments Act 1970 section 1).

On the death of a licence holder, licences pass to his personal representative for a period of 3 months and then expire. The 3-month period may be extended at the local authority's discretion (section 1(8)).

Any person guilty of an offence under any provision of this Act other than section 2(4) is liable to a fine not exceeding level 3 or 3 months' imprisonment or both (section 4(1)).

Conditions

On granting a licence the local authority is required to specify conditions as appear necessary or expedient to achieve all the objectives set out in (b)(i)–(viii) in the paragraph headed 'Matters for consideration' above. In addition the following conditions are required by the Act, whether specified in the licence or not:

(a) a horse found on inspection of the premises by an authorised officer to be in need of veterinary attention shall not be returned to work until the holder of the licence has obtained at his own expense and has lodged with the local authority a veterinary certificate that the horse is fit for work;
(b) no horse will be let out on hire for riding or used for providing instruction in riding without supervision by a responsible person of the age of 16 years or over unless (in the case of a horse let out for hire for riding) the holder of the licence is satisfied that the hirer of the horse is competent to ride without supervision;
(c) the carrying on of the business of a riding establishment shall at no time be left in the charge of any person under 16 years of age;
(d) the licence holder shall hold a current insurance policy which insures him against liability for any injury sustained by those who hire a horse from him for riding and those who use a horse in the course of receiving from him in return for payment, instruction in riding and arising out of the hire or use of

a horse and which also insures such persons in respect of any liability which may be incurred by them in respect of injury to any person caused by, or arising out of, the hire or use of a horse;

(e) a register shall be kept by the licence holder of all horses in his possession aged 3 years and under and usually kept on the premises, which shall be available for inspection by an authorised officer at all reasonable times (section 1(4A)).

Disqualifications and cancellations

In making a conviction under this Act, under the Protection of Animals Act 1911, the Protection of Animals (Scotland) Act 1912, the Pet Animals Act 1951, the Animal Boarding Establishment Act 1963 or sections 4 to 9 and 11 of the Animal Welfare Act 2006, the court may cancel any riding establishment licence held by the convicted person and may disqualify him from holding such a licence, whether or not he is a current holder, for such period as the court thinks fit. The cancellation or disqualification may be suspended by the court pending an appeal (section 4(3) and (4)).

Power of entry

Local authority officers and veterinary surgeons and practitioners, authorised in writing by the local authority, may, upon producing their authority if required, enter and inspect the following premises at all reasonable times:

(a) licensed riding establishments;
(b) premises subject to an application for a licence to run a riding establishment; and
(c) unlicensed premises suspected of being used as a riding establishment.

This power extends to the inspection of any horses found on the premises (section 2(1) to (3)).

Persons wilfully obstructing or delaying authorised officers are subject on conviction to a maximum fine not exceeding level 2 (sections 2(4) and 4(1)).

Definitions

approved certificate means:

(a) any one of the following certificates issued by the British Horse Society, namely, Assistant Instructor's Certificate, Instructor's Certificate and Fellowship;
(b) Fellowship of the Institute of Horse; or
(c) any other Certificate for the time being prescribed by order of the Secretary of State.

horse includes any mare, gelding, pony, foal, colt, filly or stallion and also any ass, mule or jennet.

premises includes land (section 6(4)).

LICENCE FOR A TRAVELLING CIRCUS WITH WILD ANIMALS

References

Animal Welfare Act 2006.
The Welfare of Wild Animals in Travelling Circuses (England) Regulations 2012.
Guidance on the welfare of wild animals in travelling circuses (England) regulations 2012 (DEFRA).

Scope

This procedure applies to wild animals in a travelling circus and those performing animals.

These regulations cease to have effect 7 years from the day these regulations came into force, that is 20/3/2020.

Extent

These procedures apply in England.

Introduction

A licence is required for a travelling circus with wild animals that give performances or displays. Licence applications should be sent to the Animal and Plant Health Agency (APHA) circus licensing team. APHA will arrange an inspection, which must take place before a licence can be issued. A licensing panel considers the application. If the application is approved, APHA will issue the licence.

Licensing or registration of activities involving animals (FC 2.8)

A licence is required for an activity which involves animals and this includes a travelling circus (section 13(1) Animal Welfare Act 2006 and regulation 3).

The Welfare of Wild Animals in Travelling Circuses (England) Regulations 2012

Grant and renewal of licence

The Secretary of State must, having received an application in writing for a licence grant a licence to an operator, or renew a licence, if satisfied that the licensing conditions are or will be met. The Secretary of State may not grant or renew a licence unless the travelling circus has been inspected by an inspector. He/she must take account of the results of any inspection before deciding whether to grant or renew a licence, and may grant or renew a licence for any period of up to 3 years (regulation 4(1)). The applicant's conduct as the operator of the travelling circus may be taken into account when deciding to grant the licence (regulation 4(2)).

FC 2.8 Travelling circus with wild animals – the Welfare of Wild Animals in Travelling Circuses (England) Regulations 2012 and Animal Welfare Act 2006

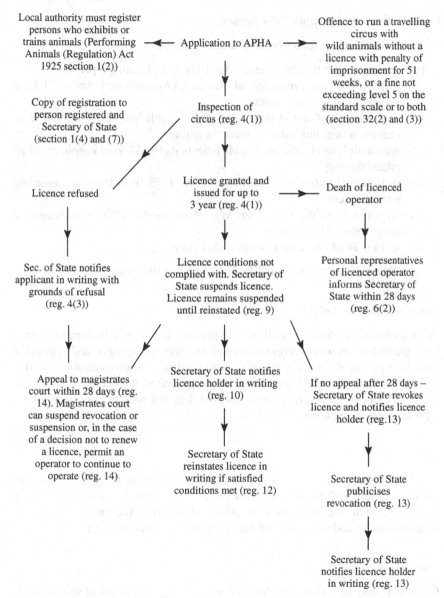

Notes
1. If operator is disqualified, licence is automatically revoked if appeal time lapses.
2. These procedures apply in England.

A decision by the Secretary of State not to grant or renew a licence must be notified to the applicant in writing and that notice must include a statement of the grounds of the decision; set out the right of appeal and time period for an appeal to a magistrates' court (regulation 4(3)).

People who may not apply for a licence

A person cannot apply for a licence if disqualified under:

(a) section 5(3) of the Pet Animals Act 1951 from keeping a pet shop;

(b) section 1 of the Protection of Animals (Amendment) Act 1954 from having custody of an animal;

(c) section 3(3) of the Animal Boarding Establishments Act 1963 from keeping a boarding establishment for animals;

(d) section 4(3) of the Riding Establishments Act 1964 from keeping a riding establishment;

(e) section 3(3) of the Breeding of Dogs Act 1973 from keeping a breeding establishment;

(f) section 6(2) of the Dangerous Wild Animals Act 1976 from keeping a dangerous wild animal; or

(g) section 34 of the Animal Welfare Act 2006,

and any licence issued to a disqualified person is invalid (regulation 5).

Death of a licence holder

In the event of the death of an licenced operator, that licence is deemed to have been granted to personal representatives of the licensed operator and remains in force for a period of 3 months beginning with the date of death (regulation 6 (1)).

Within 28 days of the death of the licensed operator, the personal representatives must notify the Secretary of State that they are now the operators of the travelling circus (regulation 6 (2)).

Fees

The fee for the licence application is £389.36 (regulation 8(1)). An Inspection fee is payable at £72.53 for each hour or part hour spent by an inspector on an inspection, including time spent travelling to or from the site of inspection, accommodation and on associated administration (regulation 8(3)).

Conditions

Conditions may be included in the licence involving acquisition of wild animals, notification of tour itinerary, wild animal records, care plans, persons with access to licensed animals, veterinary surgeons, responsibility of the operator to promote the welfare of licensed animals, specific welfare requirements for display, training and performance, specific welfare requirements for environment and specific welfare requirements for transportation.

Information to be included with a licence application

1. Care plan for each animal, agreed with your lead vet showing compliance with licensing conditions and welfare standards.
2. List of animals covered by the licence (with details of their unique identification or microchip number).
3. Locations of all animals covered by the licence.
4. Individual record for each animal.
5. Tour itinerary with dates at each site (this must be given at least 30 days before the tour starts).
6. List of people authorised to have contact with and care for the animals.
7. Copy of any licence or authorisation from a home country (if not from England) (Schedule and guidance).

Conditions of the licence

Operators must meet the following licensing conditions:

- implement care plans for all animals including requirements on:
  - dict
  - performance
  - environment
  - breeding
- arrange a visit by a vet at least every 3 months;
- prove that animals are taken good care of;
- make sure a suitably qualified and experienced person is in overall charge of the animals;
- make sure that there are enough staff to maintain public and animal safety;
- only allow suitably qualified and experienced people unsupervised access to the animals;
- keep a journey plan when the circus moves;
- have clear procedures for staff working with animals (for example, contingency plans for fires or if an animal escapes); and
- prove that the circus is safe for people and animals (for example, having emergency drills for fires or escapes) (Schedule and guidance).

Records to be kept

Operators must keep up to date written records of:

- tour itineraries
- licensed animals
- locations of licensed animals
- individual records for all animals
- care plans, including:
  - assessment of welfare risks
  - behavioural and environmental enrichment
  - breeding policy

- dietary requirements
- preventative medicine programme (for example, vaccinations and worming treatments)
- daily welfare records
- retirement, when animals are no longer able to take part in shows
- list of people authorised to care for the animals
- list of authorised people on duty (this must be visible to staff where animals are kept)
- journey plans (including details of the plan and the actual journey).

Operators may need to send copies of the records to APHA at any time as evidence that they are complying with licence conditions (Schedule and guidance).

Give notice of any changes

Operators must tell APHA as soon as possible if:

- there is a change in the tour itinerary;
- there is a change in the list of animals;
- a new wild animal for use in the travelling circus is obtained (this must be done at least 14 days before receiving the animal); or
- the licence holder dies and another party wishes to operate under the existing licence.

Suspension of a licence

The Secretary of State may suspend a licence at any time if satisfied that the licensing conditions have not been complied with (regulation 9).

If a licence is suspended the Secretary of State must notify the holder in writing without undue delay by way of a notice of suspension which has immediate effect and continues in effect unless the licence is reinstated (regulation 10).

A notice of suspension must state the grounds for being satisfied that the licensing conditions are not being complied with; specify the measures that the operator must take in order to secure compliance and set out the right of appeal and time period for appeal to a magistrates' court (regulation 11).

Reinstatement of licence

The Secretary of State must reinstate a suspended licence by notice in writing once satisfied that the licensing conditions have been or will be complied with. Where a licence is reinstated the period for which it is granted may be varied (regulation 12).

Revocation of licence

The Secretary of State may revoke a licence that has been suspended for more than 28 days unless there is an outstanding appeal to the magistrates' court and may publicise a revocation. When revoking a licence the Secretary of State must notify the holder in writing and revocation takes effect from the time of notification. Where an operator is disqualified, the operator's licence is automatically revoked when the time limit for any appeal against that disqualification expires or, if an appeal is brought, when that appeal is dismissed (regulation 13).

Appeals

A person who is aggrieved by a decision not to grant or renew a licence or the decision to suspend or revoke a licence may appeal to a magistrates' court.

The procedure on an appeal to a magistrates' court is complaint and the Magistrates' Courts Act 1980 applies to the proceedings. The appeal period is 28 days beginning with the day following the date on which the decision is notified.

A court may suspend a suspension or revocation or, in the case of a decision not to renew a licence, permit an operator to continue to operate a travelling circus (regulation 14).

Penalties

It is an offence to run a travelling circus with wild animals without a licence and could face imprisonment for a term not exceeding 51 weeks, or a fine not exceeding level 5 on the standard scale or to both (section 32(2) and (3)).

Section 1 of the Performing Animals (Regulation) Act 1925

Restriction on exhibition and training of performing animals

No person shall exhibit or train any performing animal unless he is registered (section 1(1)).

Every local authority must keep a register of any person who exhibits or trains animals after making an application to the local authority of the district in which he resides, or if he has no fixed place of residence in Great Britain, to the local authority of such one of the prescribed districts as he may choose on payment of the relevant fee (section 1(2)).

The application for registration must contain particulars of the animals and the general nature of the performances in which the animals are to be exhibited or for which they are to be trained and those particulars must be entered in the register and the register should be open at all reasonable times (section 1(3) and (5)).

The local authority must give a certificate of registration to the person whose name appears on the register kept with the particulars entered in the register (section 1(4)).

The registered person can requested a variance to the conditions on the register, and where the particulars are varied the existing certificate must be cancelled and a new certificate issued (section 1(6)).

A copy of every certificate of registration must be sent to the Secretary of State (section 1(7)).

A local authority may charge fees appropriate for inspection of the register, for taking copies or making extracts from it or for inspection of copies of certificates of registration (section 1(8)).

Definitions

licence means a licence to operate a travelling circus.
licensing conditions means the conditions set out in the Schedule.
operator means a person responsible for the operation of a travelling circus.
travelling circus means a circus which travels from place to place for the purpose of giving performances, displays or exhibitions, and as part of which wild animals are kept or introduced (whether for the purpose of performance, display or otherwise); and any place where a wild animal associated with such a circus is kept.

ZOO LICENCES

References

Zoo Licensing Act 1981.
DEFRA circular 02/2003.

Extent

This procedure applies in England, Scotland and Wales.

Scope

This procedure applies to the licensing of zoos.

Licensing of zoos by local authorities (FC 2.9)

It is unlawful to operate a zoo without a licence issued by the local authority for the area within which the whole or the major part of the zoo is situated (section 1(1)). This applies to any zoo to which members of the public have access, with or without charge for admission, on 7 days or more in any period of 12 consecutive months (section 1(2A)). It also applies even if members of the public do not have such access if a licence is in force (section 1(2B)).

Application for licence

At least 2 months before making an application to the local authority for a zoo licence, the applicant must give notice in writing to the local authority of his intention to make the application. He must publish a notice of that intention in a local newspaper and a newspaper with a national circulation and exhibit a copy of the notice at the site and the notice must state that the notice to the local authority can be inspected (section 2(1)).

The notice must identify the situation of the zoo and specify:

(a) the kinds of animals listed in taxonomic category of order and approximate number of each group kept or to be kept for exhibition on the premises and the arrangements for their accommodation, maintenance and well-being;

(b) the approximate numbers and categories of staff employed or to be employed in the zoo;

(c) the approximate number of visitors and motor vehicles for which accommodation is or is to be provided;

(d) the approximate number and position of the means of access provided or to be provided to the premises (section 2(2));

(e) how the conservation measures are being or will be implemented at the zoo (section 2(2A).

Any notice given to the local authority must be kept available by the authority at their offices for public inspection free of charge at reasonable hours (section 2(3)).

Consideration of application

When considering an application for a licence the local authority must take into account any representations made by or on behalf of any of the following persons:

(a) the applicant;

(b) the chief officer of police (or in Scotland the chief constable) for any area in which the whole or any part of the zoo is situated;

(c) the relevant fire and rescue authority;

(d) the governing body of any national institution concerned with the operation of zoos;

(e) where part of the zoo is not situated in the area of the local authority with power to grant the licence:
 (i) a planning authority for the area in which the part is situated (other than a county planning authority);
 (ii) if the part is situated in Wales, the local planning authority for the area in which it is situated;

(f) any person alleging that the establishment or continuance of the zoo would injuriously affect the health or safety of persons living in the neighbourhood of the zoo;

(g) any other person whose representations might, in the opinion of the local authority, show grounds on which the authority has a power or duty to refuse to grant a licence (section 3).

Grant or refusal of licence

Before granting or refusing to grant a licence for a zoo, the local authority must:

(a) consider zoo inspectors' reports;

FC 2.9 Zoo licence – Zoo Licensing Act 1981

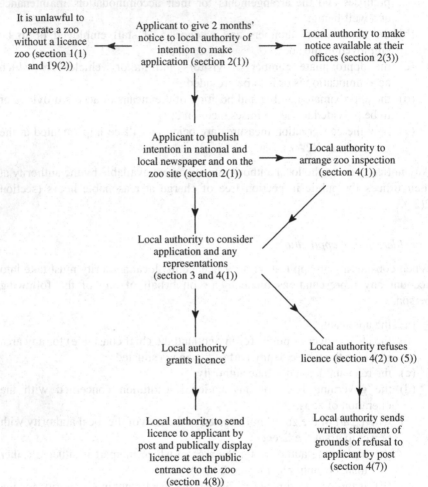

It is unlawful to operate a zoo without a licence zoo (section 1(1) and 19(2))

Applicant to give 2 months' notice to local authority of intention to make application (section 2(1))

Local authority to make notice available at their offices (section 2(3))

Applicant to publish intention in national and local newspaper and on the zoo site (section 2(1))

Local authority to arrange zoo inspection (section 4(1))

Local authority to consider application and any representations (section 3 and 4(1))

Local authority grants licence

Local authority refuses licence (section 4(2) to (5))

Local authority to send licence to applicant by post and publically display licence at each public entrance to the zoo (section 4(8))

Local authority sends written statement of grounds of refusal to applicant by post (section 4(7))

Notes
1. Secretary of State may consult local authority on conditions he may direct to be attached to the licence (section 4(7)).
2. A person aggrieved by the refusal of an application or a condition in the licence may appeal to the magistrates' court (section 18(1)).

(b) consult the applicant about the conditions they propose would be attached to the licence, if one were granted; and

(c) make arrangements for an inspection to be carried out (section 4(1)).

The local authority must refuse to grant a licence for a zoo if they are satisfied that the establishment or continuance of the zoo would injuriously affect the health or safety of persons living in the neighbourhood of the zoo, or seriously affect the preservation of law and order or if they are not satisfied that the conservation measures will be implemented in a satisfactory manner at the zoo (section 4(2)).

The local authority may refuse to grant a licence for a zoo if the above does not apply and they are not satisfied that the standards of accommodation, staffing or management are adequate for the proper care and well-being of the animals or any of them or otherwise for the proper conduct of the zoo. They may also refuse to grant a licence if the applicant, or where the applicant is a body corporate, the body or any director, manager, secretary or other similar officer of the body, or any person employed as a keeper in the zoo has been convicted of an offence under the Act or under any of the following acts involving the ill-treatment of animals (section 4(3) to (5)):

- the Protection of Animals Act 1911;
- the Protection of Animals (Scotland) Acts 1912 to 1964;
- the Performing Animals (Regulation) Act 1925;
- the Pet Animals Act 1951;
- the Animals (Cruel Poisons) Act 1962;
- the Animal Boarding Establishments Act 1963;
- the Riding Establishments Acts 1964 and 1970;
- the Breeding of Dogs Act 1973;
- the Dangerous Wild Animals Act 1976;
- the Endangered Species (Import and Export) Act 1976;
- Part 1 of the Wildlife and Countryside Act 1981;
- sections 4, 5, 6(1) and (2), 7 to 9 and 11 of the Animal Welfare Act 2006;
- sections 28C or 28F(16) of the Animal Health Act 1981 (c.22);
- sections 19 to 24, 25(7), 29 or 40(11) of the Animal Health and Welfare (Scotland) Act 2006 (asp 11).

If the local authority are not satisfied that any planning permission for the establishment or continuance of the zoo during the period for which the licence would be in force, has been, or is deemed to be, granted, they must either refuse to grant the licence or grant the licence but suspend its operation until the local planning authority have notified the local authority that any such planning permission has been or is deemed to be granted (section 4(6)).

If the local authority refuse to grant the licence they must send a written statement of the grounds of their refusal to the applicant by post (section 4(7)).

When a licence is granted the local authority must send it to the applicant by post and the licence or a copy of it must be publicly displayed at each public entrance to the zoo (section 4(8)).

Period and conditions of licence

An original licence is granted for 4 years beginning with the date on the licence (section 5(1)). A fresh licence to the holder of an existing licence is granted for 6 years beginning with the end of the period of the existing licence (section 5(2)).

A licence is granted subject to conditions requiring the conservation measures to be implemented at the zoo and other conditions as the local authority think necessary or desirable for ensuring the proper conduct of the zoo during the period of the licence, including insurance against liability for damage caused by animals (section 5(2A) and (3)).

In deciding what conditions to attach to a licence, a local authority must have regard to any standards specified by the Secretary of State and sent by him to the authority (section 5(4)).

The Secretary of State may, after consulting the authority, direct them to attach one or more conditions to a licence, and the authority must attach them (section 5(5)). But he cannot direct the authority to attach a condition which is inconsistent with the implementation at the zoo of the conservation measures (section 5(5A)) and the authority cannot attach any condition inconsistent with one they are directed to attach (section 5(6)). The authority cannot attach a condition which relates only or primarily to the health, safety or welfare of persons working in the zoo (section 5(7)).

Renewal of licence (FC 2.10)

Where application for the renewal of an existing licence is made within 6 months before the end of the period of the licence or a shorter time the local authority has in special circumstances allowed, the local authority may either extend the period of the existing licence or direct the applicant to apply for a fresh licence (section 6(1)).

Before extending the period of an existing licence the authority must make arrangements for an inspection to be carried out and consider the report made about the inspection (section 6(1A)).

Where an application for a fresh licence is made by the holder of an existing licence, the existing licence continues in force until the application is disposed of or withdrawn (section 6(2)).

Any extension of the period of an existing licence must be granted for 6 years beginning with the end of the period of the existing licence; and the local authority must ensure the holder of the licence is notified in writing of the extension (section 6(3)).

The local authority must give notice to the holder of any licence granted by the authority, within 9 months before the end of the period of the licence, of the latest date on which application for renewal may be made (section 6(4)).

Transfer, transmission and surrender of licence

A licence for a zoo may, with the approval of the local authority, be transferred to another person, and in that case the transferee shall become the licence holder

FC 2.10 Zoo licence renewal and transfer

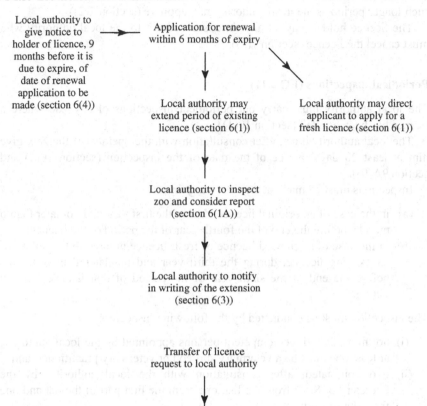

Local authority to give notice to holder of licence, 9 months before it is due to expire, of date of renewal application to be made (section 6(4))

Application for renewal within 6 months of expiry

Local authority may extend period of existing licence (section 6(1))

Local authority may direct applicant to apply for a fresh licence (section 6(1))

Local authority to inspect zoo and consider report (section 6(1A))

Local authority to notify in writing of the extension (section 6(3))

Transfer of licence request to local authority

Local authority to notify transferor and transferee of the accepted transfer (section 7(1))

Notes

1. On the death of the licence holder, the personal representatives of the deceased are deemed to be the licence holders for 3 months after the death or a longer period the local authority may approve (section 7(2)).
2. The licence holder may, at any time, surrender it to the local authority who must cancel the licence (section 7(3)).
3. Where an application has been made for renewal of a zoo's licence and the authority propose to extend the period of the licence; and one or more inspections of the zoo are required by section 16, the authority may combine those inspections with the inspection which is required by section 6 (section 9A(3)).
4. A person aggrieved by a refusal to transfer a licence may appeal to the magistrates' court (section 18(1)).

from the date specified by the authority and notified by them to the transferor and transferee (section 7(1)).

On the death of the licence holder, the personal representatives of the deceased are deemed to be the licence holders for 3 months after the death or such longer period as the local authority may approve (section 7(2)).

The licence holder may, at any time, surrender it to the local authority who must cancel the licence (section 7(3)).

Periodical inspections (FC 2.11)

The local authority must carry out periodical inspections of any zoo where a licence has been granted (section 10(1)).

The local authority must, after consultation with the operator of the zoo, give him at least 28 days' notice of the date of the inspection (section 10(2) and section 9A(9)).

Inspections must be made at the following times:

(a) in the case of an original licence, during the first year and not later than 6 months before the end of the fourth year of the period of the licence;

(b) in the case of a renewed licence or fresh licence granted to the holder of an existing licence, during the third year and not later than 6 months before the end of the sixth year of the period of that licence (section 10(3)).

The inspection must be conducted by the following inspectors:

(i) not more than three competent persons appointed by the local authority, at least one must be a veterinary surgeon or veterinary practitioner; and

(ii) two nominated, after consultation with the local authority, by the Secretary of State from the list, one from the first part of the list and one from the second;

and the names of all persons inspecting must be notified to the operator of the zoo (section 10(4(a)).

The operator may give notice of objection to the local authority about any one or more of the inspectors, and the local authority or the Secretary of State may if they think fit give effect to the objection (section 10 (4(b)) and section 9A).

Up to three representatives of the operator may accompany the inspectors on the inspection and the inspectors may require the attendance of any veterinary surgeon or practitioner employed by the zoo (section 10(4(c))).

The inspection must extend to all features of the zoo directly or indirectly relevant to the health, welfare and safety of the public and the animals, including measures for the prevention of the escape of animals (section 10(4(d))).

The inspectors must require the production of all records kept by the operator about the conditions of the and the operator must produce the records (section 10(4(e))).

The inspectors must send their report to the local authority, and the report may include advice on the keeping of records and recommendations for any practicable improvements designed to bring any features of the zoo up to the normal standards of modern zoo practice and the inspectors must have regard to any standards specified by the Secretary of State (section 10(5)).

Any disagreement between the inspectors over recommendations to be made in their report relating to the welfare of the animals may be referred to the Secretary of State, who may, after consultation with persons as he thinks fit, give guidance he thinks proper (section 10(6)).

The local authority must send a copy of the report to the operator of the zoo within 1 month and give him an opportunity to comment on it (section 10(7)).

Procedure relating to inspections before the grant, refusal, renewal or significant alteration of licences

Where an application has been made for renewal of a zoo's licence and the authority propose to extend the period of the licence; and one or more inspections of the zoo are required by section 16, the authority may combine those inspections with the inspection which is required by section 6 (section 9A(3)).

Where more than one inspection is required under section 16, the authority may combine the inspections (section 9A(4)).

Where in the course of an inspection it becomes apparent to the inspectors that a significant alteration to the licence is likely to be needed, they shall consult the licence holder about that alteration, consider whether any new conditions will be needed to secure that alteration are likely to be met if the licence is altered and include their findings and recommendations in a report (section 9A(5)). This does not apply where the inspection is an inspection before a significant alteration (section 9A(6)).

Except in the case of an inspection carried out before the grant or refusal of an original licence, the inspector shall consider whether the conditions attached to the licence are met (section 9A(10)).

In the case of an inspection carried out before the grant or refusal of a licence, the inspector shall consider whether the conditions proposed by the authority are likely to be met if the licence is granted (section 9A(11)).

In the case of an inspection carried out before the period of an existing licence is extended, the inspector shall consider whether the conditions attached to the licence are likely to be met if the period of the licence is extended (section 9A(12)).

In the case of an inspection carried out before the significant alteration of a licence, the inspector shall consider whether any new or varied conditions proposed by the authority or by the Secretary of State are likely to be met if the licence is altered and whether the conservation measures will be implemented at the zoo if the licence is altered (section 9A(13)).

Except in the case of an inspection carried out before the grant or refusal of an original licence, the inspector may require the production of all records kept by the holder of the licence relating to the conditions requiring the conservation measures to be implemented at the zoo, and the holder must produce the records (section 9A(14)).

The inspector must send his report to the authority, and within 1 month after receiving the report and the authority must send a copy to the applicant or operator and give him an opportunity to comment on it (section 9A(15)).

Special inspections

The local authority may at any time carry out a special inspection of a zoo for which a licence is in force if they consider it appropriate to do so having regard to any report on the zoo, or any representations made to them on behalf of a properly constituted body concerned with any aspect of the management of zoos or the welfare of animals, or any other circumstances which in their opinion call for investigation (section 11(1)).

A special inspection must be conducted by persons who appear to the local authority to be competent for the purpose and who are authorised by the authority to conduct the inspection (section 11(2)).

Where the purpose of the inspection relates to the health of animals, the inspectors must be or include a veterinary surgeon or veterinary practitioner with experience of animals of kinds kept in the zoo (section 11(3)). On appointing such persons the authority must inform them and the operator of the zoo the purpose and scope of the inspection (section 11(4)).

The powers to require production of records, etc. apply to special inspections (section 11(5)).

Special inspections of closed zoos

A local authority may at any time carry out a special inspection of a zoo or a section of a zoo if they consider it appropriate to do so having regard to supervising the implementation of a plan or ensuring the welfare of animals following the closure of a zoo (section 11(A)(1)).

The inspectors must send their report to the authority, and, except where the operator of the zoo cannot after reasonable enquiries have been made be found, the authority shall send a copy to the operator and give him an opportunity to comment on it (section 11(A)(4)).

Informal inspections

A local authority must make such arrangements as they think fit to ensure that any licensed zoo is inspected informally by an inspector once in any calendar year if there are no formal inspections required. The inspector must be appointed by the authority and must be a competent person (section 12).

Local authority zoos

When a local authority is the owner of a zoo as soon as practicable after granting a licence for the zoo, or extending the period of a licence, or receiving an inspectors' report the authority must send a copy of the licence, or notification in writing of the extension, or a copy of the report (with any local authority comments) to the Secretary of State (section 13(2) and (3)).

Alterations can be made by the Secretary of State as if for a private zoo (section 13(4) and (5)). The Secretary of State has similar powers to the local authority powers in the case of the closure of a local authority owned zoo, for the welfare of animals and inspection of the zoo (section 13(6) to (11)).

FC 2.11 Zoo licences – inspections

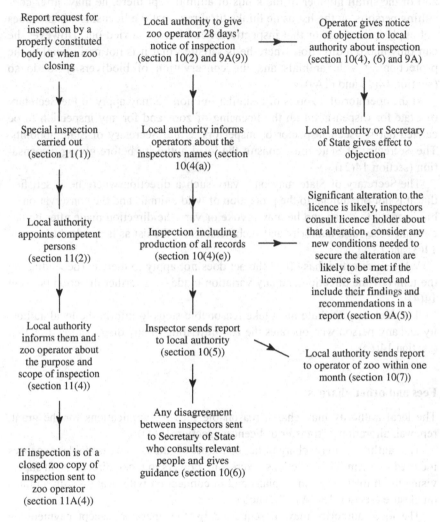

Report request for
inspection by a
properly constituted
body or when zoo
closing

Local authority to give
zoo operator 28 days'
notice of inspection
(section 10(2) and 9A(9))

Operator gives notice
of objection to local
authority about inspection
(section 10(4), (6) and 9A)

Special inspection
carried out
(section 11(1))

Local authority informs
operators about the
inspectors names (section
10(4(a))

Local authority or Secretary
of State gives effect to
objection

Local authority
appoints competent
persons
(section 11(2))

Inspection including
production of all records
(section 10(4)(e))

Significant alteration to the
licence is likely, inspectors
consult licence holder about
that alteration, consider any
new conditions needed to
secure the alteration are
likely to be met if the
licence is altered and
include their findings and
recommendations in a
report (section 9A(5))

Local authority
informs them and
zoo operator about
the purpose and
scope of inspection
(section 11(4))

Inspector sends report
to local authority
(section 10(5))

Local authority sends report
to operator of zoo within one
month (section 10(7))

If inspection is of a
closed zoo copy of
inspection sent to
zoo operator
(section 11A(4))

Any disagreement
between inspectors sent
to Secretary of State
who consults relevant
people and gives
guidance (section 10(6))

Dispensation for particular zoos (FC 2.12)

If the local authority inform the Secretary of State that in their opinion a dispensation should be made because of the small number of animals kept in the zoo or the small number of the kinds of animal kept there, he may, after consulting persons on the list as he thinks fit, direct that the licensing of zoos does not apply to that zoo or that inspections need not be carried out. However, he can only make a direction where he is satisfied that it is not prejudicial to the protection of wild animals and the conservation of biodiversity to do so (section 14(1) and (1A)).

If the operator of a zoo is of a similar opinion he may apply to the Secretary of State for dispensation on the licensing of zoos and for any inspection to be carried out by such inspector or inspectors as the Secretary of State appoints. The Secretary of State must consult the local authority before giving dispensation (section 14(2)).

The Secretary of State may only vary such a direction where he is satisfied that it is not prejudicial to the protection of wild animals and the conservation of biodiversity to do so and he may revoke or vary the direction made after he has consulted the local authority and such persons on the list as he thinks fit (section 14(3) and (3A).

Where a direction has effect the act does not apply to the zoo (depending on the terms of the direction and any variation made by a further direction (section 14(4)).

The Secretary of State must take reasonable steps to inform the local authority and any person who operates the zoo in writing of any dispensation direction (section 14(6)).

Fees and other charges

The local authority may charge reasonable fees for applications for the grant, renewal, alteration or transfer of licences (section 15(1)).

The authority may charge the operator of the zoo reasonable expenses incurred by them for inspections, exercising powers to make directions, in supervising the implementation of plans and in connection with dealing with animals on closure (section 15(2A), (2B) and (5)).

The local authority may, if requested by the operator, accept payment by instalments (section 15(3)). Any fee or other charge payable by any person shall be recoverable by the local authority as a debt (section 15(4)).

Power to alter licences (FC 2.13)

A granted licence can be altered by the local authority if in their opinion it is necessary or desirable to do so for ensuring the proper conduct of the zoo during the period of the licence (section 16(1)).

Where the authority have made a direction relating to conditions, the period specified in that direction has expired and the authority are satisfied that a

FC 2.12 Dispensation for particular zoos

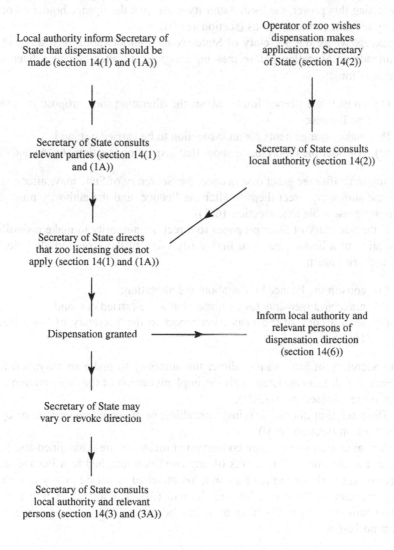

Local authority inform Secretary of State that dispensation should be made (section 14(1) and (1A))

Operator of zoo wishes dispensation makes application to Secretary of State (section 14(2))

Secretary of State consults relevant parties (section 14(1) and (1A))

Secretary of State consults local authority (section 14(2))

Secretary of State directs that zoo licensing does not apply (section 14(1) and (1A))

Dispensation granted

Inform local authority and relevant persons of dispensation direction (section 14(6))

Secretary of State may vary or revoke direction

Secretary of State consults local authority and relevant persons (section 14(3) and (3A))

condition specified in that direction requiring any conservation measure to be implemented at the zoo is not met in relation to any section of the zoo, the authority must make alterations to the licence to ensure that the relevant section of the zoo is closed permanently to the public (section 16(1) and (1B)). Before exercising this power, the local authority must give the licence holder an opportunity to make representations (section 16(2)).

Except where the Secretary of State requires an alteration or if a report recommends an alteration before making a significant alteration to a licence the authority must:

(a) consult the licence holder about the alteration they propose to make to the licence;
(b) make arrangements for an inspection to be carried out; and
(c) consider the report made about that inspection (section 16(2A) and (2B)).

At any time after the grant of a licence, the Secretary of State may, after consulting the authority, direct them to alter the licence, and the authority must do so within a reasonable time (section 16(3)).

If the Secretary of State proposes to direct the authority to make a significant alteration to a licence, he must first notify them of the proposed alteration and the authority must:

(a) consult the licence holder about the alteration;
(b) make arrangements for an inspection to be carried out; and
(c) send a copy of the inspection report to the Secretary of State (section 16(3A)).

The Secretary of State cannot direct the authority to make an alteration to the licence which is inconsistent with the implementation of the conservation measures at the zoo (section 16(3B)).

The alteration may be varying, cancelling or attaching conditions or by any combination (section 16(4)).

Any alteration must ensure conservation measures are maintained and implemented at the zoo and the terms of any condition attached to a licence are not inconsistent with the terms of a condition attached or varied in any direction of the Secretary of State (section 16(4A) and (5)). The alteration will not have effect until written notification of it has been received by the licence holder (section 16(6)).

Enforcement of licence conditions

Where the local authority, after giving the licence holder an opportunity to be heard, are not satisfied that a condition attached to a licence is met in relation to the zoo or a section of it, the authority must make a direction specifying:

(a) the licence condition which they are not satisfied is met;
(b) whether they are not satisfied that that condition is met in relation to the zoo or a section of the zoo;

FC 2.13 Zoo licence – alteration

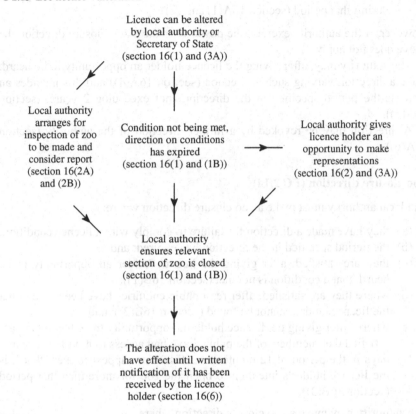

Licence can be altered
by local authority or
Secretary of State
(section 16(1) and (3A))

Local authority
arranges for
inspection of zoo
to be made and
consider report
(section 16(2A)
and (2B))

Condition not being met,
direction on conditions
has expired
(section 16(1) and (1B))

Local authority gives
licence holder an
opportunity to make
representations
(section 16(2) and (3A))

Local authority
ensures relevant
section of zoo is closed
(section 16(1) and (1B))

The alteration does not
have effect until written
notification of it has been
received by the licence
holder (section 16(6))

Notes
1. The Secretary of State may, after consulting the authority, direct them to alter the licence, and the authority must do so within a reasonable time (section 16 (3)).
2. Where Secretary of State requires alteration local authority must send a copy of the report to him (section 16 (3A)).
3. A person aggrieved by the alteration of licence issued by the local authority may appeal to the magistrates' court (section 18(1)).

(c) steps to be taken by the licence holder to ensure that that condition is met within a specified period, in less than 2 years from the date of the direction; and

(d) whether the zoo or a section of it is required to be closed to the public during that period (section 16A(1) and (2)).

However, if the authority exercise the power to make a zoo closure direction the above does not apply.

The authority may, after giving the licence holder an opportunity to be heard, make a direction varying such a direction (section 16A(4)) and this includes an increase the period specified in the direction (not exceeding 2 years) (section 16A(5)).

A direction may be revoked by a further direction of the authority (section 16A(6)).

Zoo closure direction (FC 2.14)

The local authority must make a zoo closure direction where:

(a) they have made a direction for failure to comply with a licence condition;

(b) the period specified in the direction has expired; and

(c) they are satisfied, after giving the licence holder an opportunity to be heard, that a condition is not met (section 16B(1));

(d) where they are satisfied, after reasonable enquiries have been made, that the licence holder cannot be found (section 16B(2)); and

(e) where, after giving the licence holder an opportunity to be heard they are satisfied that members of the public have had access to it on fewer than 7 days in the period of 12 months and it does not appear to them that it is the licence holder's intention to open the zoo for more than that period (section 16B(3)).

The authority may make a zoo closure direction where:

(a) they have made a direction for failure to comply with a licence condition;

(b) the period specified in the direction has expired; and

(c) they are satisfied, after giving the licence holder an opportunity to be heard, that a condition specified in that direction, other than one which requires any conservation measure, is not met (section 16B(4)).

The authority may, after giving the licence holder an opportunity to be heard, make a zoo closure direction if:

(a) any reasonable requirements relating to the premises or conduct of the zoo notified by them to the licence holder in consequence of the report of any inspection are not complied with within a time as is reasonable in the circumstances;

(b) they are satisfied that the zoo has been conducted in a disorderly manner or so as to cause a nuisance;

(c) the licence holder (or, where the licence holder is a body corporate, the body or any director, manager, secretary or other similar officer of the body) is convicted of any offence under the act; or

(d) any person who, to the knowledge of the licence holder, has been convicted is employed as a keeper in the zoo (section 16B(5)).

But the authority cannot make a zoo closure direction if a direction is in force in respect of the zoo and:

(a) when that direction was made there were grounds upon which the authority could have made a zoo closure direction, but they chose not to do so; and

(b) the grounds upon which they would make a zoo closure direction are the same as any of those upon which they could have made one when they made the direction instead (section 16B(6)).

No zoo closure direction may be made on grounds involving the care or treatment of animals unless the authority have first consulted such persons on the list as the Secretary of State may nominate (section 16B(7)).

Where the authority make a zoo closure direction, the zoo's licence is revoked from the date on which the direction has effect (section 16B(8)).

Zoo closure direction for zoos without licences

Where a zoo is being operated without a licence or no direction has effect and it appears to the local authority to have been operated for more than 7 days in 12 months (section 16C(1)) and the authority are satisfied, after reasonable enquiries have been made, that the operator of a zoo cannot be found, they must make a zoo closure direction (section 16C(2)).

However, if the authority informs the Secretary of State that in their opinion a direction should be made and he makes a direction, the authority must give the operator at least 35 days' notice in writing of their intention to make a zoo closure direction and give the operator an opportunity to be heard (section 16C(3)). If, after the 35 day period and after giving the operator an opportunity to be heard, it does not appear to the authority that the view they reached was incorrect, they must make a zoo closure direction unless notice has been given of intention to make an application for a licence for the zoo (section 16C(4)). Where notice is given to the authority of intention to make an application for a licence for the zoo, but an application for a licence for the zoo is not made within 3 months beginning on the date on which the notice was given or the application for a licence for the zoo is refused and no appeal is brought against the refusal within time, or if an appeal is brought against the refusal within the time, it is abandoned, or the court confirms the decision to refuse the application, the authority must make a zoo closure direction (section 16C(5)).

Directions

The local authority must take reasonable steps to notify the operator of the zoo in writing about any direction under the Act (section 19A(1)). But where the operator of the zoo cannot, after reasonable enquiries have been made, be found, the authority must take reasonable steps to notify any person appearing to them

FC 2.14 Zoo licence closure

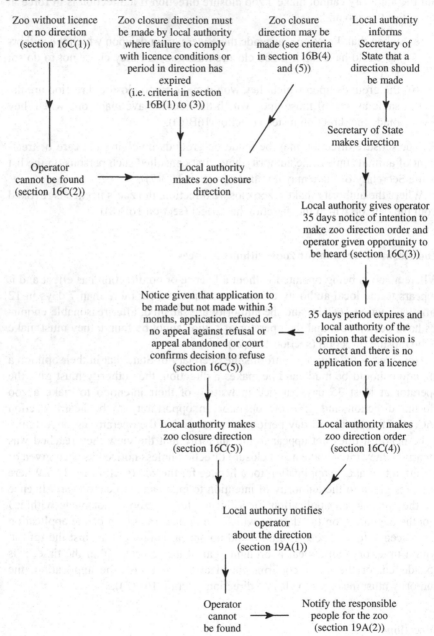

Zoo without licence or no direction (section 16C(1))

Zoo closure direction must be made by local authority where failure to comply with licence conditions or period in direction has expired (i.e. criteria in section 16B(1) to (3))

Zoo closure direction may be made (see criteria in section 16B(4) and (5))

Local authority informs Secretary of State that a direction should be made

Secretary of State makes direction

Operator cannot be found (section 16C(2))

Local authority makes zoo closure direction

Local authority gives operator 35 days notice of intention to make zoo direction order and operator given opportunity to be heard (section 16C(3))

Notice given that application to be made but not made within 3 months, application refused or no appeal against refusal or appeal abandoned or court confirms decision to refuse (section 16C(5))

35 days period expires and local authority of the opinion that decision is correct and there is no application for a licence

Local authority makes zoo closure direction (section 16C(5))

Local authority makes zoo direction order (section 16C(4))

Local authority notifies operator about the direction (section 19A(1))

Operator cannot be found

Notify the responsible people for the zoo (section 19A(2))

Notes

1. If there is poor care or treatment of animals the local authority consult persons nominated by the Secretary of State (section 16B(7)).
2. A person aggrieved by a zoo closure direction or refusal to approve a plan by the local authority may appeal to the magistrates' court (section 18(1)).

to be responsible for the zoo in writing of the direction (section 19A(2)). The operator of a zoo or other person notified in writing of a direction must comply with a direction (section 19A(3)).

Application of section 16E

The powers to deal with the welfare of animals following closure of zoo (section 16E) apply where a zoo closure direction has been made, from the date on which the direction has effect and a zoo whose licence has expired or been surrendered, from the date of its expiration or surrender (section 16D(1)). These powers apply to a section of a zoo which is closed permanently to the public due to alterations to the zoo's licence from the date on which those alterations have effect (section 16D(3)). If a section of a zoo which was closed permanently to the public due to alterations to the zoo's licence reopens after further alterations to the licence, the powers cease to apply to that section (section 16D(4)).

The Act ceases to apply to a zoo when the local authority have notified the operator, or, where the operator cannot be found, any person appearing to them to be responsible for the zoo, in writing that they are satisfied that all animals which are to be disposed of have been disposed of; and satisfactory arrangements for the care of any animals kept in the zoo which are not to be disposed are in place, and there are reasonable grounds for believing that satisfactory arrangements will continue to be maintained (section 16D(2)).

Welfare of animals following closure of zoo (FC 2.15)

Where a zoo closure direction has been made as soon as reasonably practicable the operator must give the authority a plan of the arrangements he proposes to make in relation to the animals kept in the zoo for their future care; or for their disposal and for their care until they are disposed of (section 16E(2)).

The operator must supply the authority with any information they request about the care or disposal of animals kept in the zoo (section 16E(3)). Where the authority notify the operator that they approve a plan he shall implement it under the supervision of the authority (section 16E(4)).

Except with the agreement of the authority, the operator must not dispose of any animal kept in the zoo before a plan has been approved by the authority; or dispose of any animal kept in the zoo not in accordance with the plan (section 16E(5)).

Where the authority are not satisfied with the prepared plan; the authority are not satisfied with the way in which the plan is being implemented; the operator of the zoo has not prepared a plan within a reasonable period or the authority consider that urgent steps need to be taken by the operator to safeguard the welfare of animals kept in the zoo, the authority may, after giving the operator an opportunity to be heard, make a direction as to the future care of the animals kept in the zoo, or for their disposal and for their care until they are disposed of (section 16E(6)).

Where the zoo operator has not complied with a direction to the satisfaction of the authority; or the authority consider that urgent steps need to be taken by them to safeguard the welfare of animals kept in the zoo, the authority must, after giving the operator an opportunity to be heard, make arrangements for the future care of the animals kept in the zoo, or for their disposal and for their care until they are disposed of (section 16E(7)).

Where the local authority are satisfied, after reasonable enquiries have been made, that the operator of the zoo cannot be found the above provisions do not apply, i.e. section 16E(2) to (7) (section 16E(1)).

Where the authority are satisfied, after reasonable enquiries have been made, that the operator of the zoo cannot be found, they shall make arrangements for the future care of the animals kept in the zoo, or for their disposal and for their care until they are disposed of (section 16E(8)).The authority may care for any animal on the premises of the zoo; or remove any animal found on the premises of the zoo and either retain it in the authority's possession or dispose of it (section 16E(9)).

The authority may make a direction varying or revoking any other direction but, unless they are satisfied, after reasonable enquiries have been made, that the operator cannot be found, they cannot do so without first giving him an opportunity to be heard (section 16E(11)).

Power of authority to dispose of animals

The authority may sell or otherwise dispose of any animal if:

(a) after making reasonable inquiries they are satisfied that the animal is owned by the operator of the zoo;

(b) after making reasonable inquiries they are unable to identify or unable to find the animal's owner;

(c) they have obtained the consent of the owner of the animal;

(d) the owner of the animal has been asked for his consent before a date specified in the request, but that date has passed and the authority have not received it, and the owner has not arranged to take possession of the animal or arranged for possession to be taken by another person; or

(e) the owner has arranged to take possession of the animal or for its possession to be taken by another person, but the date for implementation of the arrangements has passed and they remain unimplemented (section 16F(2)).

Where an animal is sold or given away any person to whom the animal is sold or given shall have a good title to it (section 16F(3)).

Where the authority have sold all the animals, the authority shall pay the operator of the zoo a sum equal to the total proceeds of the sales of animals less any part of the charge which the authority are entitled to make which has not been paid (section 16F(5)).

Where the authority have identified a person other than the operator whom they are satisfied was the owner of an animal immediately before its sale they must pay that person a sum equal to the proceeds of the sale of that animal, less the costs incurred by them in connection with the sale and in caring for the animal before the sale (section 16F(7)).

If the person to whom the authority are required to make a payment cannot be found within 4 months beginning with the date of the sale of the last animal the proceeds of sale will be the authority's (section 16F(8)) and any remaining proceeds of the sales will be the authority's (section 16F(9)).

An authority must make any payment between 1 and 4 months beginning with the date of the sale of the last animal (section 16F(10)).

Nothing in this section can prevent an authority from making arrangements, on the advice of a veterinary surgeon or practitioner, for an animal to be put down without delay where it is necessary or expedient to do so in the interests of its welfare (section 16F(11)).

Powers of entry

A person duly authorised by the authority may, on producing his authority, enter the premises of the zoo for the purposes of:

(a) inspecting any animal found there;
(b) inspecting the accommodation of any such animal;
(c) caring for any such animal; or
(d) removing any such animal (section 16G(1)).

This does not apply to a private dwelling (section 16G(2)).

However, a justice of the peace can issue a warrant for entry to a private dwelling or where admission has been refused (section 16G(3)).

The warrant must specify the length of time for which it is valid; and the times at which entry can be obtained, and may contain restrictions as the justice thinks fit (section 16G(4)).

A person authorised must, if required, produce his authority and warrant before entering the premises of the zoo or part of the premises to which the warrant relates (section 16G(5)).

Appeals

A person aggrieved by the refusal to grant a licence; any condition attached to a licence; any variation or cancellation of a condition; the refusal to approve the transfer of a licence; a direction or any variation of such a direction; a zoo closure direction; the refusal to approve a plan; or any arrangements for welfare of animals on closure may appeal to a magistrates' court (section 18(1)).

The appeal by way of a complaint for an order must be made within 28 days from the date on which the person wishing to appeal receives written notification of the authority's relevant decision (section 18(2)).

A magistrates' court can confirm, vary or reverse the local authority's decision and generally give such directions as it thinks proper (section 18(3)).

In Scotland the sheriff has the same power but can award expenses as he thinks fit (section 18(4)). The decision of the sheriff on an appeal is final (section 18(6)).

Where the dispute is about a condition attached to a licence, or the variation of a condition, and it imposes a requirement on the licence holder to carry out

FC 2.15 Zoo licence – dealing with animals after closure

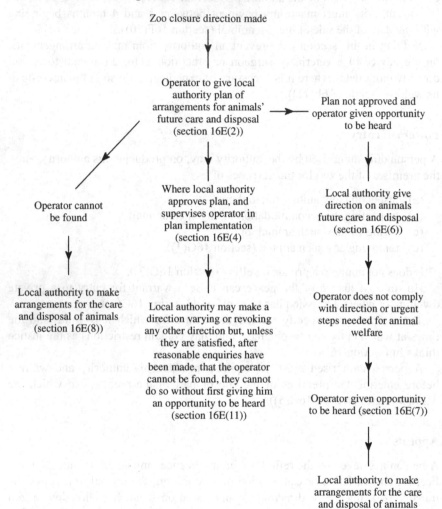

Zoo closure direction made

Operator to give local
authority plan of
arrangements for animals'
future care and disposal
(section 16E(2))

Plan not approved and
operator given opportunity
to be heard

Operator cannot
be found

Where local authority
approves plan, and
supervises operator in
plan implementation
(section 16E(4)

Local authority give
direction on animals
future care and disposal
(section 16E(6))

Local authority to make
arrangements for the care
and disposal of animals
(section 16E(8))

Local authority may make a
direction varying or revoking
any other direction but, unless
they are satisfied, after
reasonable enquiries have
been made, that the operator
cannot be found, they cannot
do so without first giving him
an opportunity to be heard
(section 16E(11))

Operator does not comply
with direction or urgent
steps needed for animal
welfare

Operator given opportunity
to be heard (section 16E(7))

Local authority to make
arrangements for the care
and disposal of animals
(section 16E(7))

Notes
1. Local authority can vary or revoke the zoo closure direction but operator must be given an opportunity to be heard if he can be found (section 16E(11)).
2. A person aggrieved by a direction of the local authority may appeal to the magistrates' court (section 18(1)).

works, the condition or the variation does not have effect during the appeal period or during the period before the appeal is determined or abandoned (section 18(7)).

An alteration to a licence has effect if the authority's decision is confirmed or varied, or if subsequently abandoned on the day following the day on which the appeal is determined, or on such other day as the court directs or if no appeal is brought in time on the expiration of that time (section 18(7A)).

A direction requiring the zoo or a section of it to be closed to the public; or which requires the operator of the zoo to carry out works he would not otherwise be required to carry out; or which requires disposal of any animals or any variation of such a direction does not have effect during the appeal period or during the period before the appeal is determined or abandoned (section 18(9)).

A zoo closure direction has effect if an appeal is brought within time, and the authority's decision is confirmed or varied, or subsequently abandoned on the day following the day on which the appeal is determined, or on such other day as the court directs or if no appeal is brought within time on the expiration of that time (section 18(10)).

Offences and penalties

It is an offence for the operator to operate a zoo without a licence (section 19(1)). It is also an offence for the operator of a zoo to fail, without reasonable excuse, to comply with any condition attached to a licence (section 19(2)).

It is an offence without reasonable excuse:

- to intentionally obstruct an inspector acting under this Act (section 19(3) and (3A));
- to fail to comply with a requirement in a direction to close the zoo or a section of it to the public in accordance with the direction (section 19(3B));
- for any person notified in writing of a zoo closure direction to fail to comply with that direction (section 19(3C));
- if the operator of a zoo fails to comply with a local authority's request for information (section 19(3D));
- if the operator of a zoo without the agreement of the authority disposes of any animal kept in the zoo before a plan has been approved by the authority or otherwise than in accordance with a plan which has been approved by the authority (section 19(3E));
- if the operator of a zoo fails to comply with a direction about the welfare of animals of which he is notified in writing (section 19(3F)); or
- the holder of a licence for a zoo fails to display the zoo licence or a copy of it publicly at each public entrance to the zoo (section 19(3G)).

A person guilty of an offence is liable on summary conviction to a fine not exceeding level 3 or 4 on the standard scale depending on the offence (section 19(4)).

If a body corporate commits an offence and proof can be demonstrated about consent or connivance of, or to have been attributed to any neglect on the part of, any director, manager, secretary or any other similar officer of the body corporate, or any person who was purporting to act in any such capacity, he, as well as the body corporate commits an offence (section 19(5)).

Typical local authority procedure

At least 2 months before making an application for a licence, the applicant must give notice in writing of their intention to make the application. The notice must identify:

- the zoo's location;
- the types of animals and approximate number of each group kept for exhibition on the premises and the arrangements for their accommodation, maintenance and well-being;
- the approximate numbers and categories of staff to be employed in the zoo;
- the approximate number of visitors and motor vehicles for which accommodation is to be provided;
- the approximate number and position of access to be provided to the premises; and
- how required conservation measures will be implemented at the zoo.

At least 2 months before making the application, the applicant must also publish notice of that intention in one local newspaper and one national newspaper and exhibit a copy of that notice. The notice must identify the location of the zoo and state that the application notice to the local authority is available to be inspected at the council offices.

Application evaluation process

When considering an application, a local authority will take into account any representations made by or on behalf of:

- the applicant;
- the chief officer of police (or in Scotland the chief constable) in the relevant area;
- the appropriate authority – this is either the enforcing authority or relevant authority in whose area the zoo will be situated;
- the governing body of any national institution concerned with the operation of zoos;
- where part of the zoo is not situated in the area of the local authority with power to grant the licence, a planning authority for the relevant area (other than a county planning authority) or, if the part is situated in Wales, the local planning authority for the area in which it is situated;
- any person alleging that the zoo would affect the health or safety of people living in the neighbourhood;
- anyone stating that the zoo would affect the health or safety of anyone living near it; or
- any other person whose representations might show grounds on which the authority has a power or duty to refuse to grant a licence.

Before granting or refusing to grant the licence, the local authority will consider any inspectors' reports based on their inspection of the zoo, consult the applicant about any conditions they propose should be attached to the licence and make arrangements for an inspection to be carried out. At least 28 days' notice of the inspection will be provided by the local authority.

The local authority will not grant the licence if they feel that the zoo would adversely affect the health or safety of people living in near it, or seriously affect the preservation of law and order or if they are not satisfied that appropriate conservation measures would be satisfactorily implemented.

An application may also be refused if:

- the local authority are not satisfied that accommodation, staffing or management standards are suitable for the proper care and well-being of the animals or for the proper conduct of the zoo; or
- the applicant, or if the applicant is an incorporated company, the company or any of the company's directors, managers, secretaries or other similar officers, or a keeper in the zoo, has been convicted of any offence involving the ill-treatment of animals.

Applications to renew a licence will be considered no later than 6 months before the expiry of the existing licence, unless a shorter time period is allowed by the local authority.

The Secretary of State, after consulting the local authority, may direct them to attach one or more conditions to a licence.

The local authority may advise the Secretary of State that, because of the small number of animals kept in the zoo or the small number of the kinds of animal kept there, a direction should be made that a licence is not required.

Appeal

If the applicant is refused a licence, they may appeal to a magistrates' court, or in Scotland, to the sheriff within 28 days from the date on which the applicant receives written notification of the refusal.

A licence holder may appeal to a magistrates' court against:

- any condition attached to a licence or any variation or cancellation of a condition;
- the refusal to approve the transfer of a licence;
- a zoo closure direction; and
- enforcement steps relating to any unmet condition.

The appeal must be brought within 28 days from the date on which the licence holder receives written notification of the authority's decision as to the relevant matter.

Any person who wishes to appeal against a decision to close a zoo may apply to the local magistrates' court or, in Scotland, to the sheriff. Appeals must be made within 28 days of the notice of the local authority decision.

Definitions

animals means animals of the classes Mammalia, Aves, Reptilia, Amphibia, Pisces and Insecta and any other multi cellular organism that is not a plant or a fungus.

circus means a place where animals are kept or introduced wholly or mainly for the purpose of performing tricks or manoeuvres at that place.

section of a zoo means: a particular part of the zoo premises; animals of a particular description in the zoo; or animals of a particular description which are kept in a particular part of the zoo premises.

References to the **closure of a section of a zoo to the public** mean: the closure to the public of a particular part of the zoo premises; ceasing to exhibit animals of a particular description to the public; or ceasing to exhibit animals of a particular description to the public in a particular part of the zoo premises.

wild animals means animals not normally domesticated in Great Britain (section 21).

zoo means an establishment where wild animals are kept for exhibition to the public otherwise than for purposes of a circus or in a pet shop (section 1(2)).

REGISTRATION OF FOOD BUSINESS ESTABLISHMENTS

References

Food Safety Act 1990.
EC Regulation 852/2004 on the Hygiene of Foodstuffs, 29 April 2004.
Official Feed and Food Controls (England) Regulations 2009.
The Food Hygiene (England) Regulations 2013.
Food Law Code of Practice (England) April 2014, DEFRA.

Extent

These provisions apply in England only but there are similar provisions in Wales.

Scope

Regulation 852/2004 requires that each Food Business Operator (FBO) must supply the competent authority, in this case the Food Authority, with details to enable each establishment used for any of the stages of production, processing and distribution of food to be registered (A6.2). An establishment is defined as including any unit of a food business and will therefore include those on land and water and those which are mobile (A2, paragraph 1(c)). The purpose is to allow the Food Authority to perform official controls efficiently (Recital paragraph (19)).

Each separate unit must be registered except those:

(a) which require to be approved – see FC 2.16;

(b) which are not to be regarded as food businesses for the purpose of this procedure. These are:
  (i) primary production for private domestic use;
  (ii) domestic preparation, handling or storage for private domestic consumption;
  (iii) the direct supply by the producer of small quantities of primary products to the final consumer or to a local retail establishment directly supplying the final consumer;
  (iv) collection centres and tanneries.
(c) Ocean going ships, aircraft, trains and long distance coaches will not be expected to register for practical reasons but shore based vessels, e.g. floating restaurants, and boats plying on inland waterways will have to register.
(d) Individual vending machines although distribution centres at which food is stored or transported to vending machines should be registered.
(e) In relation to markets, vehicles and stalls used for transporting or preparing food or the sale of food within the market are to be registered even if owned by the market operator. Operators working from permanent stalls within a market must register.
(f) Moveable establishments other than those which form part of a market or operate within the area of a market, e.g. ice cream vans, hot dog vendors, etc. should be registered by the food business operator with the Food Authority in the area in which they are ordinarily kept (Code of Practice 1.5.1,2 and 8).

Registration (FC 2.16)

Operators must register their premises with the Food Authority at least 28 days before food operations commence and it is an offence not to do so. The Food Authority is required to provide a registration form for this to be undertaken and a model form is contained in Annex 8 (Code of Practice 1.5.3.2 and 3).

On receiving the form the local authority must record the date of receipt on it and, if there are any activities noted that are outside its enforcement role, send it to the appropriate authority. If there are any parts not completed, the Food Authority should make contact with the operator or, if the missing information is substantial, return it to him for final completion (Code of Practice 1.5.5).

Registration is then made by entering the details on a database and keeping the form in the file for those premises. Certificates of registration are not to be issued to the operator but the Food Authority can confirm the receipt of the form and the entry of the premises in the database (Code of Practice 1.5.6). In two tiered areas, the Food Authority is required to send a copy of the information to the county council within 28 days (Code of Practice 1.1.5).

Food Authorities are asked to consider an inspection of the premises when the form is received but registration is not dependent upon the results of that inspection. The registration process is informatory (Code of Practice 1.5.5).

FC 2.16 Registration of food business establishments – EC Regulation 852/2004 on the Hygiene of Foodstuffs

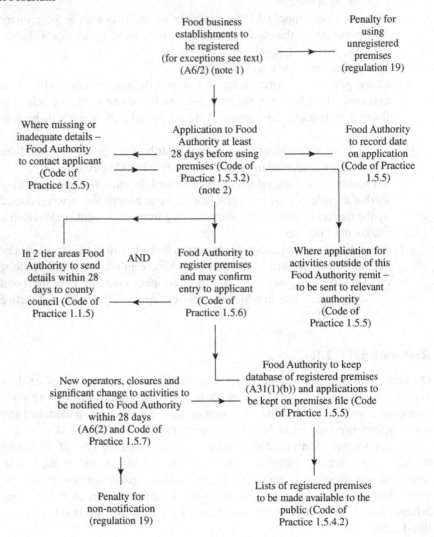

Food business
establishments to
be registered
(for exceptions see text)
(A6/2) (note 1)

Penalty for
using
unregistered
premises
(regulation 19)

Where missing or
inadequate details –
Food Authority
to contact applicant
(Code of
Practice 1.5.5)

Application to Food
Authority at least
28 days before using
premises (Code of
Practice 1.5.3.2)
(note 2)

Food Authority
to record date
on application
(Code of Practice
1.5.5)

In 2 tier areas Food
Authority to send
details within 28
days to county
council (Code of
Practice 1.1.5)

AND

Food Authority to
register premises
and may confirm
entry to applicant
(Code of
Practice 1.5.6)

Where application for
activities outside of this
Food Authority remit –
to be sent to relevant
authority
(Code of
Practice 1.5.5)

New operators, closures and
significant change to activities to
be notified to Food Authority
within 28 days
(A6(2) and Code of
Practice 1.5.7)

Food Authority to keep
database of registered premises
(A31(1)(b)) and applications to
be kept on premises file (Code
of Practice 1.5.5)

Penalty for
non-notification
(regulation 19)

Lists of registered premises
to be made available to the
public (Code of
Practice 1.5.4.2)

Notes
1. 'A' references are to EC Regulation 852/2004, regulation refers to the 2013 England regulations and Code of Practice to the 2014 Code of Practice.
2. A model form is at Annex 8 of the Code of Practice and should be made available by the Food Authority.

Database

Food Authorities must draw up a list of food establishments registered with them and will include within it premises previously registered under the 1991 and 2004 regulations (EC Reg. 882/2004 A31 (1)(b)). The list is to be made available to the general public at all reasonable times and should contain the following details for each business:

(i) name of the food business;
(ii) address;
(iii) particulars and nature of the business.

Details of food premises records are to be made available to the Health Protection Agency and the Consultant in Communicable Disease Control to facilitate the investigation of outbreaks of disease (Code of Practice 1.4.3). These records are subject to both the Data Protection Act 1998 and the Freedom of Information Act 2000.

Changes to activities, etc.

Food Business Operators are required to ensure that the Food Authority always has up to date information on establishments by notifying them of any significant changes in activities or of any closure within 21 days of the changes happening (A6.2 of 852/2004 and Code of Practice 1.5.7). The new operator should make notification of a change of operator. The database should be updated accordingly and the details placed on the premises file (Code of Practice 1.5.7).

SKIN PIERCING AND OTHER SPECIAL TREATMENTS

References

Local Government (Miscellaneous Provisions) Act 1982 sections 13–17 as amended by the Local Government Act 2003 (Section 120).
Body Art – Skin piercing, cosmetic therapies and other special treatments – CIEH Good Practice Guidelines 2001.
Local Government Act 2003: Regulation of cosmetic piercing and skin colouring businesses – Guidance on section 120 and schedule 6. Department of Health – 26/2/04.

Extent

The provisions apply only in England and Wales.

In London similar powers also exist through the London Local Authorities Act 1991.

Scope

The activities which may be controlled by registration and subsequent application of bye-laws are:

    (a) acupuncture

    (b) tattooing

    (c) semi-permanent skin colouring

    (d) cosmetic piercing

    (e) electrolysis (sections 14–15).

Businesses carried on under the supervision of a registered medical practitioner do not require registration under these provisions (sections 14(8) and 15(8)) and they do not apply to skin piercing on animals (section 16(12)).

The only one of these activities which is defined in the Act is 'semi-permanent skin-colouring' which is defined as 'the insertion of semi-permanent colouring into a person's skin' (section 15(9)).

Adoption (FC 2.17)

These provisions do not operate unless they are adopted by a local authority in respect of its own area by the procedure shown in FC 2.17. The adoption may relate to all or any of the activities mentioned below and different dates of operation may be applied to different activities (section 13).

Registration

Both the practitioner and the premises at which the activity is to be carried out are required to be registered (sections 14(1) and (2) and 15(1) and (2)). Where a registered practitioner sometimes visits people to give treatment, the premises where the treatment takes place is not required to be registered (sections 14(2) and 15(2)).

Applications are to be accompanied by such particulars as the local authority may reasonably require and these include details of:

    (a) the premises where the applicant desires to practice; and

    (b) any convictions under section 16 of this Act,

but the local authority cannot require information about people to whom treatment services have been given (sections 14(5) and 15(5)).

Registration is required to be effected by the local authority unless the applicant has had a previous registration cancelled by a court, in which case the consent of that court is required (sections 14(3), 15(3) and 16(8)(b)).

Any person who fails to register themselves and the premises they operate from is guilty of an offence and liable on summary conviction to a fine not exceeding level 3 and if they contravene bye-laws the court can suspend or cancel their registration.

Fees

The local authority may charge reasonable fees for registration at their discretion (sections 14(6) and 15(6)).

FC 2.17 Skin piercing and other special treatments – Sections 13 to 17 Local Government (Miscellaneous Provisions) Act 1982

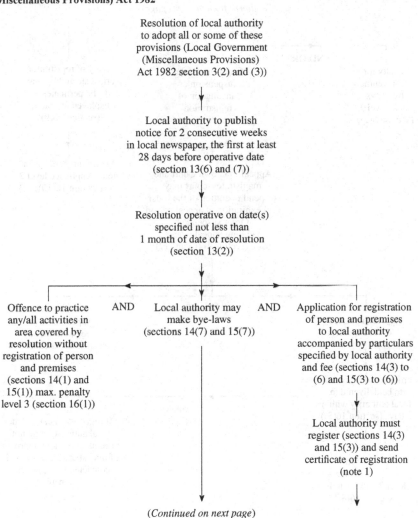

Resolution of local authority
to adopt all or some of these
provisions (Local Government
(Miscellaneous Provisions)
Act 1982 section 3(2) and (3))

Local authority to publish
notice for 2 consecutive weeks
in local newspaper, the first at least
28 days before operative date
(section 13(6) and (7))

Resolution operative on date(s)
specified not less than
1 month of date of resolution
(section 13(2))

Offence to practice
any/all activities in
area covered by
resolution without
registration of person
and premises
(sections 14(1) and
15(1)) max. penalty
level 3 (section 16(1))

AND

Local authority may
make bye-laws
(sections 14(7) and 15(7))

AND

Application for registration
of person and premises
to local authority
accompanied by particulars
specified by local authority
and fee (sections 14(3) to
(6) and 15(3) to (6))

Local authority must
register (sections 14(3)
and 15(3)) and send
certificate of registration
(note 1)

(Continued on next page)

Note
1. Where a previous registration has been cancelled by that local authority, future registrations require the consent of the magistrates' court which convicted the applicant (section 16(8)(b)).

FC 2.17 Continued

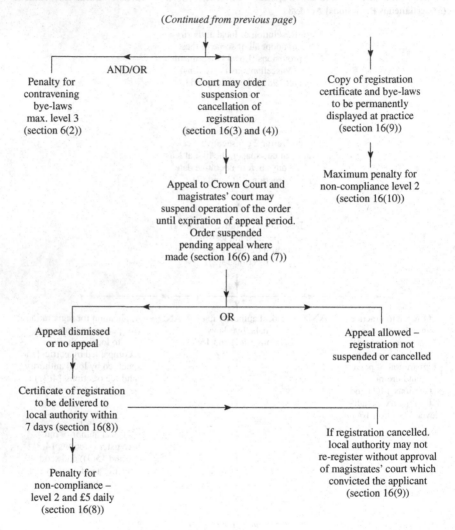

(Continued from previous page)

AND/OR

Penalty for
contravening
bye-laws
max. level 3
(section 6(2))

Court may order
suspension or
cancellation of
registration
(section 16(3) and (4))

Copy of registration
certificate and bye-laws
to be permanently
displayed at practice
(section 16(9))

Appeal to Crown Court and
magistrates' court may
suspend operation of the order
until expiration of appeal period.
Order suspended
pending appeal where
made (section 16(6) and (7))

Maximum penalty for
non-compliance level 2
(section 16(10))

OR

Appeal dismissed
or no appeal

Appeal allowed –
registration not
suspended or cancelled

Certificate of registration
to be delivered to
local authority within
7 days (section 16(8))

If registration cancelled.
local authority may not
re-register without approval
of magistrates' court which
convicted the applicant
(section 16(9))

Penalty for
non-compliance –
level 2 and £5 daily
(section 16(8))

Bye-laws

The local authority operating these procedures may (but does not need to) make
bye-laws which may cover:

(a) cleanliness of premises and fittings;
(b) cleanliness of persons;
(c) cleansing and sterilisation of instruments, materials and equipment
     (sections 14(7) and 15(7)).

Power of entry

This is only available through a warrant by a Justice of the Peace and may be
granted if the Justice of the Peace is satisfied that there is reasonable ground for
entry on suspicion that an offence is being committed and that:

(a) admission has been refused; or
(b) admission is apprehended; or
(c) the case is one of urgency; or
(d) application for admission would defeat the object of entry.

Unless the situation falls under (c) or (d) above, notice of intention to apply for
the warrant must be given to the occupier.

The warrant is operative for 7 days or until entry is secured, whichever is the
shorter period. There is no mention in these provisions of entry by force and
authorised officers effecting entry may be required to show their authority.

The maximum penalty for refusing to permit entry to an authorised officer
acting under warrant is a fine not exceeding level 3 (section 17).

Chapter 3

PUBLIC SAFETY

ALCOHOL AND ENTERTAINMENT LICENCES: LICENSING ACT 2003

References

Licensing Act 2003.

The Anti-social Behaviour Act 2003 sections 40–41.

The Licensing Act 2003 (Permitted Temporary Activities (Notices) Regulations 2005 as amended.

The Licensing Act 2003 (Personal Licences) Regulations 2005 as amended.

The Licensing Act 2003 (Premises Licences and Club Premises Certificates) Regulations 2005 as amended.

The Licensing Act 2003 (Fees) Regulations 2005 as amended.

The Licensing Act 2003 (Hearings) Regulations 2005 as amended.

The Licensing Act 2003 (Licensing authority's register) (other information) Regulations 2005.

The Licensing Act 2003 (Summary Review of Premises Licences) Regulations 2007.

Licensing Act 2003 (Mandatory Conditions) Order 2014.

Guidance to local authorities and police officers was issued by the Secretary of State under section 182 of the Licensing Act 2003 and approved by Parliament in October 2012 (The Guidance).

Background

The 2003 Act amalgamated six previous licensing regimes (alcohol, public entertainment, cinemas, theatres, late night refreshment houses and night cafes) into a single integrated scheme and the separate licensing procedures for each have been repealed. The general framework of the Act is shown in FC 3.1.

Extent

The provisions of the Licensing Act 2003 apply in England only.

The Anti-social Behaviour Act 2003 applies in England, Wales and Northern Ireland (section 181).

Scope and definitions

The Licensing Act 2003 applies to 'licensable activities' which are defined as:

(a) the sale by retail of alcohol;
(b) the supply of alcohol by clubs;
(c) the provision of regulated entertainment; and
(d) the provision of late night refreshment (section 1).

'Regulated entertainment' is fully defined in schedule 1. The term includes 'entertainment' and to come within this definition the first test is that either must be provided for:

• the public or sections of the public;
• exclusively for members of a qualifying club and their guests; or
• if neither of these applies, for consideration and a view to profit.

The second test is that the premises on which the entertainment are provided are made available for the purpose (schedule 1 paragraphs 1–3).
    Entertainment means the following:

1. performance of a play;
2. exhibition of a film;
3. indoor sporting events;
4. boxing or wrestling entertainment;
5. performance of live music;
6. playing of recorded music;
7. performance of dance; and
8. events which have the purpose of entertaining an audience or spectators (schedule 1 paragraph 2).

'Plays', 'film exhibitions', 'indoor sporting events' and 'boxing or wrestling entertainment' are defined in part 3 of schedule 1.

Exempted regulated entertainment

The following are exempted from controls on regulated entertainment:

(a) film exhibitions for advertisement, information and exhibition;
(b) film exhibitions at museums and art galleries;
(c) use of TV or radio receivers;
(d) religious services and places of worship;
(e) garden fetes;
(f) morris dancing;
(g) vehicles in motion (schedule 1 part 2).

Such licensable activity may only be carried on under and in accordance with a premises licence (PL), a temporary event notice (TEN) or, where appropriate, a club premises certificate (CPC) and to operate in a non-authorised premises constitutes an offence (section 2).
    In addition for those selling or supplying alcohol in licensed premises, except qualifying clubs, a personal licence is required.

Late night refreshment

This is defined as taking place where:

(i) a person at any time between 11p.m. and 5a.m. supplies hot food or hot drink to the public, on or from any premises and whether for consumption on or off the premises; or

(ii) between those hours when members of the public are admitted to any premises the person supplies or is willing to supply hot food or hot drink for consumption on or off the premises.

There are exemptions from this definition in relation to supplying certain clubs, hotels and employees and to premises licensed under the Greater London Council (General Powers) Act 1966 and the London Local Authorities Act 1995 (schedule 2).

Licensing objectives

The licensing objectives that must be promoted by the licensing authority and licence holders and to which they must have regard to in their policy framework document are:

- the prevention of crime and disorder;
- public safety;
- the prevention of public nuisance;
- the protection of children from harm (section 4).

Statements of licensing policy (FC 3.1)

Each local authority must prepare and publish a statement of licensing policy every 5 years and in doing so must consult widely including the police, fire and rescue authority, primary care trust or local health board, representatives of holders of licences issued under the Act, representatives of local businesses and the public (section 5). The statement must have regard to the licensing objectives of the Act (section 4) – see above – and to the Guidance issued under section 182.

Licensing committee

Each local authority is required to establish a licensing committee of between 10 and 15 members of the authority to discharge the council's licensing functions except that of the preparation and publication of the statement of policy which must be approved by the whole council. The committee may delegate functions to sub-committees and/or officers, except in the latter case where objections or representations have been made (sections 6, 9 and 10).

FC 3.1 Licensing Act 2003 – Framework

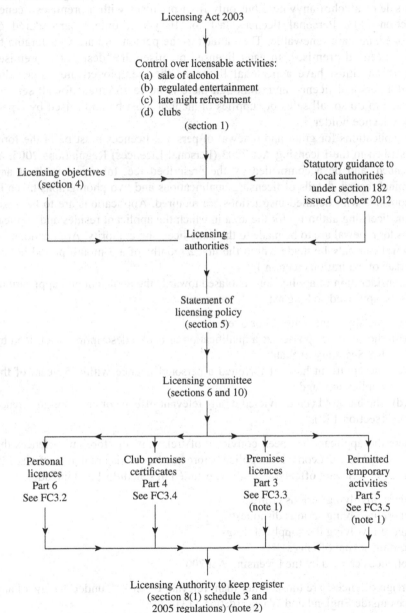

Licensing Act 2003

Control over licensable activities:
(a) sale of alcohol
(b) regulated entertainment
(c) late night refreshment
(d) clubs
(section 1)

Licensing objectives
(section 4)

Statutory guidance to
local authorities
under section 182
issued October 2012

Licensing
authorities

Statement of
licensing policy
(section 5)

Licensing committee
(sections 6 and 10)

Personal
licences
Part 6
See FC3.2

Club premises
certificates
Part 4
See FC3.4

Premises
licences
Part 3
See FC3.3
(note 1)

Permitted
temporary
activities
Part 5
See FC3.5
(note 1)

Licensing Authority to keep register
(section 8(1) schedule 3 and
2005 regulations) (note 2)

Notes
1. Magistrates' court and police have powers of closure – see text (sections 160 and 161).
2. The Licensing Act 2003 (Licensing Authority's register) (other information) Regulations 2005.

PERSONAL LICENCES

Personal licences (FC 3.2) are licences permitting an individual to sell alcohol by retail or to supply alcohol by or on behalf of a club. They do not authorise the sale of alcohol anywhere but only from premises with a premises licence (section 111). Personal licences last for 10 years, unless surrendered or revoked, and are renewable. They attach to the person and are transferable to other licensed premises. In each licensed premises the 'designated premises supervisor' must have a personal licence. Others employed there may also hold a personal licence and, although there is no requirement for all serving alcohol to do so, all sales or supplies of alcohol must be authorised by a personal licence holder.

Applications for grant and renewal of personal licences must be in the form prescribed in the Licensing Act 2003 (Personal Licences) Regulations 2005, as amended, and be accompanied by the prescribed fee. Information about any criminal record, details of licensing qualifications and two photographs taken in accordance with detailed instructions are required. Applications are to be made to the licensing authority for the area in which the applicant resides and applications for renewal are to be made to the same licensing authority. Applications for renewal can only be made within the first 2 months of a 3-month period before the date of expiration (section 117).

Consideration of applications is biased towards the applicant and applications must be approved so long as:

(a) the applicant is age 18 or over;
(b) the applicant possesses a qualification or is of a description prescribed by the Secretary of State;
(c) the applicant has not forfeited a personal licence within 5 years of the application; and
(d) she has not been convicted of any relevant offence or any foreign offence (section 120).

Where the applicant has been convicted of 'relevant' or 'foreign' offences the Licensing Act must consult the police before any licence is granted (sections 120 and 121). 'Relevant offences' are those detailed in schedule 4 and include:

• those involving serious crime
• those involving serious dishonesty
• those involving the supply of drugs
• certain sexual offences
• offences created by the Licensing Act 2003.

'Foreign offences' are those, other than 'relevant offences', under the law of any place outside England and Wales (section 113).

The police may issue a notice of objection if it is considered that the grant of the licence would undermine the prevention of crime and disorder. In the absence of such objection the licence must be granted. If the licensing authority receives a police objection notice it must arrange a hearing and must then give reasons for its decision on the application (section 120).

FC 3.2 Personal licences – Sections 111 to 135 Licensing Act 2003

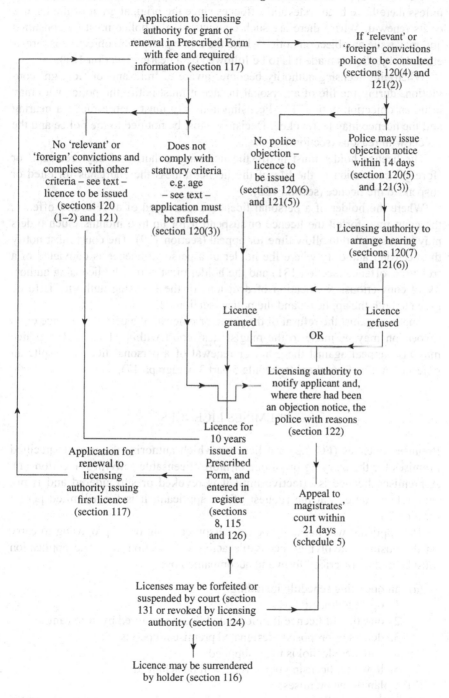

Application to licensing authority for grant or renewal in Prescribed Form with fee and required information (section 117)

If 'relevant' or 'foreign' convictions police to be consulted (sections 120(4) and 121(2))

No 'relevant' or 'foreign' convictions and complies with other criteria – see text – licence to be issued (sections 120 (1–2) and 121)

Does not comply with statutory criteria e.g. age – see text – application must be refused (section 120(3))

No police objection – licence to be issued (sections 120(6) and 121(5))

Police may issue objection notice within 14 days (section 120(5) and 121(3))

Licensing authority to arrange hearing (sections 120(7) and 121(6))

Licence granted

Licence refused

OR

Licensing authority to notify applicant and, where there had been an objection notice, the police with reasons (section 122)

Application for renewal to licensing authority issuing first licence (section 117)

Licence for 10 years issued in Prescribed Form, and entered in register (sections 8, 115 and 126)

Appeal to magistrates' court within 21 days (schedule 5)

Licenses may be forfeited or suspended by court (section 131 or revoked by licensing authority (section 124)

Licence may be surrendered by holder (section 116)

Where the application is one for renewal, the application must be approved unless there have been 'relevant' offences since the original grant of the licence or its renewal. Where there are such convictions, the police must be consulted and may issue an objection notice where the crime prevention objective is threatened. If objection is made it is to be followed by a hearing (section 121).

Where the licensing authority becomes aware of 'relevant' or 'foreign' convictions during the life of a personal licence it must notify the police who may issue an objection notice. The licensing authority must then arrange a hearing and the licence may be revoked. Decisions must be notified to the police and the holder with reasons (section 124).

The licence holder must notify the licensing authority of any 'relevant' or 'foreign' conviction in the life of the licence unless the court has forfeited or suspended the licence (section 132).

Where the holder of a personal licence is convicted of a 'relevant' offence, the court may forfeit the licence or suspend it for up to 6 months. Such orders may be suspended to allow time for appeal (section 129). The court must notify the licensing authority where the holder of a personal licence is convicted of a 'relevant' offence (section 131) and the holder must notify the licensing authority of convictions. Notification of decisions of the licensing authority is to be given to both the applicant and the police (section 122).

Appeals against the refusal of the grant, or renewal of a personal licence or its revocation may be made to the magistrates' court within 21 days. The police may also appeal against the grant or renewal of a personal licence despite an objection notice from them (schedule 5 part 3 paragraph 17).

PREMISES LICENCES

Premises licences (FC 3.3) are licences which authorise the use of specified premises for the carrying on of one or more 'licensable activities' (section 11). A premises licence is effective until it is revoked or surrendered and is not time related unless, at the request of the applicant, it is for a limited period (section 26).

The applicant will normally be the person carrying on or proposing to carry on the business involving 'licensable activities' (section 16). The application must be in the prescribed form and accompanied by:

(a) an operating schedule including:
    1. opening hours;
    2. duration of licence if time related licence required by applicant;
    3. details of proposed 'designated premises supervisor';
    4. whether alcohol is to be supplied;
    5. how the 'licensing objectives' will be achieved.
(b) a plan of the premises;
(c) in the case of premises supplying alcohol, the consent of the 'designated premises supervisor' named in the application (section 17); and
(d) the prescribed fee.

FC 3.3 Premises licences – Sections 11 to 59 Licensing Act 2003

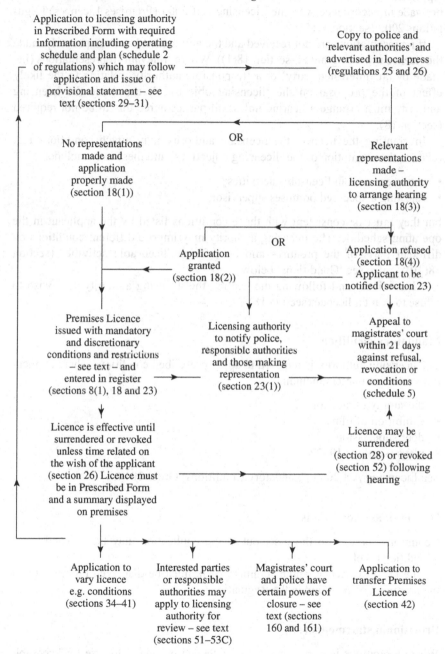

Application to licensing authority in Prescribed Form with required information including operating schedule and plan (schedule 2 of regulations) which may follow application and issue of provisional statement – see text (sections 29–31)

Copy to police and 'relevant authorities' and advertised in local press (regulations 25 and 26)

OR

No representations made and application properly made (section 18(1))

Relevant representations made – licensing authority to arrange hearing (section 18(3))

OR

Application granted (section 18(2))

Application refused (section 18(4)) Applicant to be notified (section 23)

Premises Licence issued with mandatory and discretionary conditions and restrictions – see text – and entered in register (sections 8(1), 18 and 23)

Licensing authority to notify police, responsible authorities and those making representation (section 23(1))

Appeal to magistrates' court within 21 days against refusal, revocation or conditions (schedule 5)

Licence is effective until surrendered or revoked unless time related on the wish of the applicant (section 26) Licence must be in Prescribed Form and a summary displayed on premises

Licence may be surrendered (section 28) or revoked (section 52) following hearing

Application to vary licence e.g. conditions (sections 34–41)

Interested parties or responsible authorities may apply to licensing authority for review – see text (sections 51–53C)

Magistrates' court and police have certain powers of closure – see text (sections 160 and 161)

Application to transfer Premises Licence (section 42)

Note

1. Regulation numbers are of the Licensing Act 2003 (Premises Licences and Club Premises Certificates) Regulations 2005 as amended.

Applications must be advertised and arrangements made for representations to be made in accordance with the Licensing Act 2003 (Premises Licences) Regulations 2005 (section 17(5)).

If representations are not received and the application has been properly made the licence must be granted (section 18(1)). Where 'relevant' representations (i.e. made by an 'interested party' or a 'responsible authority' relating to the likely effect of the proposal on the 'licensing objectives') are made the licensing authority must arrange a hearing unless all parties agree this to be not required (section 18).

In granting the licence, the licensing authority may attach conditions that relate to the promotion of the 'licensing objectives' and these may include:

* ruling out certain licensable activities;
* rejecting a specified premises supervisor,

but they must be consistent with those conditions listed by the applicant in the operating schedule. The licensing authority may impose different conditions on different parts of the premises and on different licensable activities (section 18(10)) – also see 'Conditions' below.

Alternatively and following the hearing the licensing authority may wish to refuse to grant a licence (section 18).

Mandatory conditions

The licensing authority is required to apply prescribed conditions where applications relate to matters including:

* the supply of alcohol
* exhibition of films
* door supervision
* plays (sections 19–22).

See Licensing Act 2003 (Mandatory Conditions) Order 2014.

Discretionary conditions

Conditions relevant to those set out in the application may be attached – see 'Conditions' below.

The form of licence, and a summary of it, must be displayed prominently at the premises, as specified in the Regulations.

Provisional statements (PS)

Where a premises is to be constructed, extended or altered for use for licensable activities, application may be made to the licensing authority for a provisional statement. This should be in the prescribed form and accompanied by the fee. The application must include details of the work to be carried out (section 29). This allows the person to establish if the works will be acceptable to the licensing authority when completed and lead to the issue of a premises licence or variation.

Applications for provisional statements must be advertised in the prescribed manner to give responsible authorities, interested parties and local residents the chance to make representations (section 30).

Where no representations are made, the licensing authority is to issue a provisional statement indicating that no representations have been made (section 31(1–2). Where there are 'relevant representations' (see above) the licensing authority must arrange a hearing unless this is agreed as being unnecessary by all parties (section 31(3), (6) and (7)). Following the hearing the licensing authority must notify the decision to the applicant, police and anyone who made representations (section 31).

Variations

There is provision for premises licences to be varied on application made in the prescribed way. This may include variation in any way, e.g. alteration of conditions or of the licensable activities, but not:

(i)   to change the designated premises supervisor for which there is a separate procedure in section 37;
(ii)   to extend the date of expiration of the premises licence where there is a time limit; or
(iii) to vary substantially the premises.

Applications are subject to fee and advertising requirements (section 34).

If there are no representations the licensing authority must make the variation subject to the mandatory conditions. Where relevant representations are made a hearing must be held unless dispensed with by agreement of all parties. Notification of the decision is to be made to the applicant, police and all those making representations (section 36).

Transfers

Application may be made to the licensing authority for the transfer of a premises licence using the prescribed form and accompanied by the fee. The application must be copied to the police who may object within 14 days if they believe that the transfer would undermine the crime protection objective (section 42).

The licensing authority must not consider the application unless:

(a) the transfer is to have immediate effect;
(b) the existing holder has consented; or
(c) the applicant has taken all reasonable steps to gain that consent and is in a position to use the premises immediately (section 44).

Where the police have objected a hearing must be held (section 44(5)) and the applicant and police notified of the decision with the reasons for it (section 45).

Reviews

Interested parties or a responsible authority (which includes a local authority as environmental health authority) may apply to the licensing authority at any time for a review, e.g. residents concerned about public nuisance or police consider that measures to prevent crime or disorder are not adequate. Hearings must then be held and the licensing authority must, if it considers necessary for the promotion of licensing objectives:

- modify the conditions;
- exclude any licensable activity;
- remove the designated supervisor;
- suspend the licence for no more than 3 months;
- revoke the licence.

The licensing authority decision following reviews does not have effect until after the period for appeal has expired or the appeal has been heard (sections 51–53C).

CLUB PREMISES CERTIFICATES

Club premises certificates (CPCs) (FC 3.4) give authorisation for the 'qualifying clubs' (defined in sections 61 to 62) to use club premises for certain 'qualifying club activities' which are:

(a) the supply of alcohol to a member of the club and to their guests for consumption on the premises (supply off the premises is only allowed in certain circumstances – section 73); and

(b) the provision of regulated entertainment for the members and their guests.

Such premises will include party political clubs, British Legion, Working Men's Clubs and sports clubs.

The holding of a CPC exempts the club from the requirements for any member or employee to be the holder of a personal licence and have a designated premises supervisor in order to supply or sell alcohol. However, there are special provisions concerning conditions to be attached regarding the sale of alcohol including its management by a committee (section 64).

Clubs with a CPC are not subject to immediate closing orders by the police and are not affected by magistrates' orders for an area – see below in relation to premises licences and temporary event notices.

Applications for a CPC are to be made in the form prescribed in Schedule 9 of the Licensing Act 2003 (Premises Licences and Club Premises Certificates) Regulations 2005 and must be accompanied by the necessary supporting information including an operating schedule, a plan of the premises and a copy of the club rules. The operating schedule will set out the activities to be undertaken, the hours of opening and the arrangements proposed to achieve the licensing objectives. The application has to be advertised (see applications below) and is subject to representations and hearings as for premises licences above (section 71). In the absence of relevant representations the application must be granted (section 72). Conditions relevant to those set out in the application may be attached – see 'Conditions' below – and the CPC will also be subject to mandatory conditions dealing with the exhibition of films (sections 73 to 75). CPCs may be subject to variation as for premises licences.

FC 3.4 Club premises certificates (CPC) – Sections 60 to 97 Licensing Act 2003

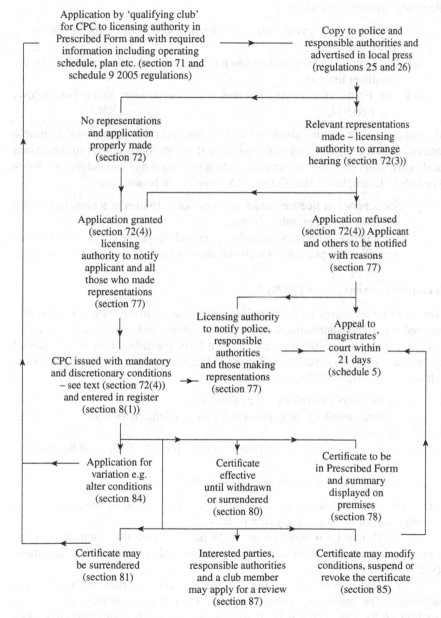

1. The regulations in this procedure are the Licensing Act 2003 (Premises Licences and Club Premises Certificates) Regulations 2005.

PERMITTED TEMPORARY ACTIVITIES

A permitted temporary activity (PTA) (FC 3.5) is one that goes ahead in accordance with a temporary event notice submitted by the 'premises user' to the licensing authority and where:

   (i)   the temporary event notice has been acknowledged by the licensing authority;

  (ii) the temporary event notice has not been subsequently withdrawn by the premises user; and

 (iii) the licensing authority has not issued a counter notice (see below) (section 98).

A temporary event notice allows premises users over 18 to carry out licensable activities on a temporary basis for no more than 96 hours subject to conditions and restrictions specified in the Act, including a maximum attendance of 500 – see below. Examples of situations where these might be used are:

   (i)   for a personal licence holder to carry out activities at a premises which does not have a premises licence;

  (ii) for a person who does not hold a personal licence to carry out licensable activities at a premises which may or may not be licensed (section 100).

Temporary event notice (TEN)

This must be submitted by the premises user to the licensing authority in the prescribed form and containing specified information, and copied to the chief of police at least 10 days before the event in a form prescribed by the Secretary of State in the Licensing Act 2003 (Permitted Temporary Activities) (Notices) Regulations 2005. This will include details of:

  (a) the licensable activities to be carried out;

  (b) the total length of the event which must not exceed 96 hours;

  (c) the times;

  (d) the maximum number of people to be admitted which must be less than 500;

  (e) whether alcohol is to be served (section 100).

The licensing authority must acknowledge the TEN before the time of the event by returning the duplicate copy of it (section 102).

The TEN may be withdrawn up to 24 hours before the event and, in this case, the intended duration does not count towards maximum periods above (section 103).

A PTA event does not require the specific authorisation of the licensing authority. The temporary event notice is a notification process and not an application for approval. It is for the licensing authority to check that the various limitations as imposed by the Act on frequency of use, etc. are not to be breached and to deal with any police objections on the grounds of a threat to crime prevention objective. The licensing authority may only intervene if a counter notice is necessary – see below.

FC 3.5 Permitted temporary activities (PTA) – Sections 98 to 110 Licensing Act 2003

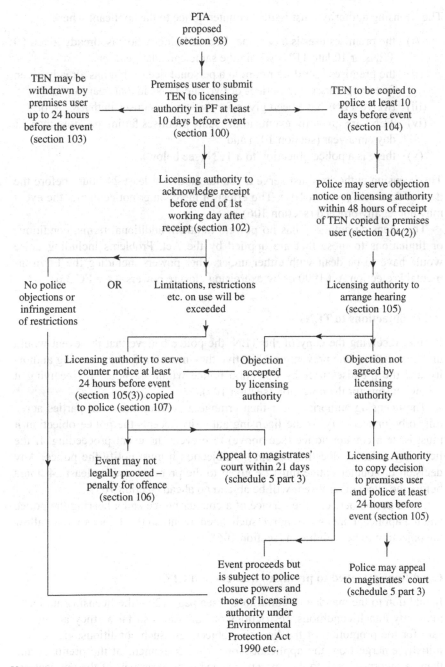

Notes
1. TEN = Temporary Event Notice.
2. PF = Prescribed Form.

Counter notice (CN)

The licensing authority must issue a counter notice to the applicant where:

(i) the premises user is a personal licence holder who has already given 50 TENs or 10 late TENs within the same calendar year, or

(ii) the premises user does not hold a personal licence but has already given five temporary event notices or two late TENs in that year, or

(iii) twelve TENs have been given for the same premises in that year, or

(iv) TENs are given for events at the same premises falling on more than 21 days in a year (section 107) and

(v) there is a police objection to a TEN (see below).

The licensing authority must serve a counter notice at least 24 hours before the date of the event (section 104). The effect of the counter notice is that the event may not legally proceed (section 106).

The licensing authority has no powers to attach additional terms, conditions or limitations to those that are applied by the Act. Problems including noise would have to be dealt with either under other powers including the Environmental Protection Act 1990 or by exercising closure powers – see FC 3.6.

Police objections to TENs

If, after receiving the copy of the TEN, the police believe that the event would undermine the crime prevention objective they must notify the licensing authority and the premises user by objection notice within 48 hours of receiving it giving reasons for the objection (section 104).

The licensing authority must then arrange a hearing unless all parties agree this to be unnecessary. If the licensing authority accepts the police objection it must issue a counter notice (see above) to prevent the event proceeding. If the licensing authority does not accept the objection it must notify the police. Any decision or counter notice must be given to the premises user at least 24 hours before the event otherwise it will be able to go ahead (section 105).

In the period between the service of a counter notice and a hearing the police and the applicant are able to agree such amendments to the TEN as would allow the objection to be withdrawn (section 106).

Conditions attached to premises licences and CPCs

In addition to the mandatory conditions – see page 120 – the licensing authority may only attach conditions, restrictions or limitations so far as they are necessary for the promotion of the licensing objectives. Such conditions, etc. should initially emerge from the applicant's own risk assessment of the premises and then be incorporated into the operating schedule to form part of the application. Early consultation with responsible authorities and residents is desirable and the applicant's proposed conditions, etc. will also form part of the statutory consultation that forms part of the application process and in any subsequent hearings. The licensing authority will therefore be unable to include any discretionary conditions unless the need for them has arisen through the making of relevant representations and then only in relation to the licensing objective under which the representation was made (sections 18, 71 and 72).

There is no provision for the attachment of conditions, etc. to permitted temporary activities.

CLOSURE OF PREMISES ASSOCIATED WITH
NUISANCE OR DISORDER, ETC.

References

Anti-social Behaviour, Crime and Policing Act 2014.
Reform of anti-social behaviour powers: Statutory guidance for frontline professionals – July 2014.

Extent

These provisions apply to England and Wales.

Scope

These provisions apply to the closure of premises which are a nuisance to members of the public, or where there is disorder associated with the use of those premises.

Closure notices

A police inspector, or the local authority, can issue a closure notice if satisfied that the use of particular premises has resulted, or is likely to result, in nuisance to members of the public, or disorder near those premises associated with the use of those premises, and that the notice is necessary to prevent the nuisance or disorder from continuing, recurring or occurring (section 76(1)).

A closure notice is a notice prohibiting access to the premises for a period specified in the notice for up to 48 hours (section 76(2) and 77) and can prohibit access by persons except those specified, at all times except those specified, or in all circumstances except those specified (section 76(3)). It cannot prohibit access by people living on the premises, or the owner of the premises (section 76(4)).

A closure notice must:

(a) identify the premises;
(b) explain the effect of the notice;
(c) state that failure to comply with the notice is an offence;
(d) state that an application will be made for a closure order;
(e) specify when and where the application will be heard;
(f) explain the effect of a closure order; and
(g) give information about the names of, and means of contacting, persons and organisations in the area that provide advice about housing and legal matters (section 76(5)).

FC 3.6 Closure of premises associated with nuisance or disorder, etc. – Sections 76 to 92 Anti-social Behaviour, Crime and Policing Act 2014

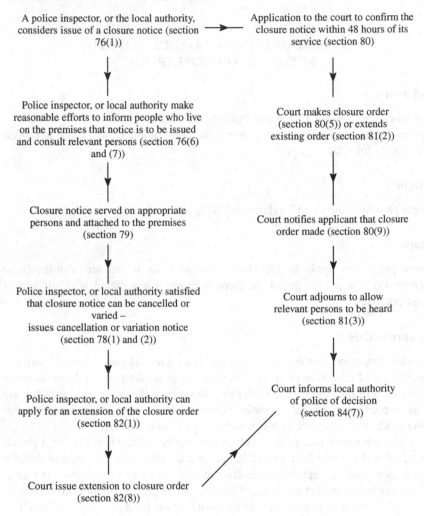

A police inspector, or the local authority, considers issue of a closure notice (section 76(1))

Application to the court to confirm the closure notice within 48 hours of its service (section 80)

Police inspector, or local authority make reasonable efforts to inform people who live on the premises that notice is to be issued and consult relevant persons (section 76(6) and (7))

Court makes closure order (section 80(5)) or extends existing order (section 81(2))

Closure notice served on appropriate persons and attached to the premises (section 79)

Court notifies applicant that closure order made (section 80(9))

Police inspector, or local authority satisfied that closure notice can be cancelled or varied – issues cancellation or variation notice (section 78(1) and (2))

Court adjourns to allow relevant persons to be heard (section 81(3))

Police inspector, or local authority can apply for an extension of the closure order (section 82(1))

Court informs local authority of police of decision (section 84(7))

Court issue extension to closure order (section 82(8))

Notes
1. Appeals are to the crown court within 21 days (section 84).
2. A person who without reasonable excuse remains on or enters premises in contravention of a closure notice commits an offence (section 86(1)). Obstruction of a person acting under these powers commits an offence (section 86(3)). Offences are liable on summary conviction to imprisonment for a period not exceeding 3 months (or 51 weeks), or to a fine, or to both (section 86(4) and (5)).
3. The Police or a local authority can apply for reimbursement of expenditure for the purpose of clearing, securing or maintaining premises (section 88(1) and (2)).

A closure notice may be issued only if reasonable efforts have been made to inform people who live on the premises and any person who has control of or responsibility for the premises or who has an interest in them that the notice is going to be issued (section 76(6)).

Before issuing a closure notice the police officer or local authority must ensure that any body or individual the officer or authority thinks appropriate has been consulted (section 76(7)).

Cancellation or variation of closure notices

Where the relevant officer or authority is no longer satisfied the closure notice is appropriate (section 78(1)) they must issue a cancellation notice cancelling the closure notice (section 78(2)) or a variation notice varying the closure notice so that it does not apply to the part of the premises (section 78(3)).

A cancellation notice or a variation notice (or an extension of the notice) must be signed by the person who signed the closure notice (or, if that person is not available, by another person who could have signed it (section 78(4)).

Service of notices

A closure notice, an extension notice, a cancellation notice or a variation notice must be served by a constable, in the case of a notice issued by a police officer or a representative of the authority that issued the notice, in the case of a notice issued by a local authority (section 79(1)).

The constable or local authority representative must if possible:

(a) fix a copy of the notice to at least one prominent place on the premises;
(b) fix a copy of the notice to each normal means of access to the premises;
(c) fix a copy of the notice to any outbuildings that appear to the constable or representative to be used with or as part of the premises;
(d) give a copy of the notice to at least one person who appears to the constable or representative to have control of or responsibility for the premises; and
(e) give a copy of the notice to the people who live on the premises and to any person who does not live there but was informed (section 79(2)).

If there are persons occupying another part of the building or other structure in which the premises are situated whose access to that part will be impeded if a closure order is made, the constable or representative must also if possible serve the notice on those persons (section 79(3)).

The constable or local authority representative may enter any premises, using reasonable force if necessary, to serve notices (section 79(4)).

Power of court to make closure orders

Whenever a closure notice is issued an application must be made to a magistrates' court for a closure order (section 80(1)).

An application for a closure order must be made:

(a) by a constable, if the closure notice was issued by a police officer;
(b) by the authority that issued the closure notice, if the notice was issued by a local authority (section 80(2)).

The application must be heard by the magistrates' court not later than 48 hours after service of the closure notice (section 80(3)).

The court may make a closure order if it is satisfied:

(a) that a person has engaged, or is likely to engage, in disorderly, offensive or criminal behaviour on the premises, or
(b) that the use of the premises has resulted, or is likely to result, in serious nuisance to members of the public, or
(c) that there has been, or is likely to be, disorder near those premises associated with the use of those premises,

and that the order is necessary to prevent the behaviour, nuisance or disorder from continuing, recurring or occurring (section 80(5)).

The period of closure cannot exceed 3 months (section 80(6)).

The court must notify the relevant licensing authority if it makes a closure order in relation to premises in respect of which a premises licence is in force (section 80(9)).

Temporary orders

If the court does not make a closure order it can order that the closure notice continues in force for a specified further period of not more than 48 hours (section 81(2)).

The court can adjourn the hearing of the application for a period up to 14 days to enable:

(a) the occupier of the premises, or
(b) the person with control of or responsibility for the premises, or
(c) any other person with an interest in the premises,

to show why a closure order should not be made (section 81(3)) but it may order that the closure notice continues in force until the end of the period of the adjournment (section 81(4)).

Extension of closure orders

At any time before the expiry of a closure order, an application can be made to a justice of the peace, by complaint, for an extension of the period for which the order is in force (section 82 (1)).

A police officer or local authority may make an application if satisfied on reasonable grounds that it is necessary for the period of the order to be extended to prevent the occurrence, recurrence or continuance of:

(a) disorderly, offensive or criminal behaviour on the premises;

(b) serious nuisance to members of the public resulting from the use of the premises; or

(c) disorder near the premises associated with the use of the premises,

and also satisfied that the local authority or chief officer of police (as appropriate) has been consulted about the intention to make the application (section 82 (3) and (4)).

The justice of the peace can issue a summons directed to:

(a) any person on whom the closure notice was served, or

(b) any other person who appears to the justice to have an interest in the premises but was not served the closure notice,

requiring the person to appear before the magistrates' court to respond to the application (section 82(5)).

If the magistrates' court is satisfied, it may make an order extending the period of the closure order by a period not exceeding 3 months (section 82(7)), but the period of a closure order cannot be extended so that the order lasts for more than 6 months (section 82(8)).

Discharge of closure orders

At any time before the expiry of a closure order, an application may be made to a justice of the peace, by complaint, for the order to be discharged (section 83(1)).

Where a person other than a constable or local authority (as appropriate), makes an application for the discharge of an order that was made on the application of a constable or local authority, the justice can issue a summons for the constable or local authority requiring him or her to appear before the magistrates' court to respond to the application (section 83(3) and (5)).

Appeals

An appeal to the crown court within 21 days against a decision to make or extend a closure order can be made by:

(a) a person on whom the closure notice was served;

(b) anyone else who has an interest in the premises but on whom the closure notice was not served (section 84(1)).

A constable or local authority can appeal against:

(a) a decision not to make a closure order;

(b) a decision not to extend a closure order;

(c) a decision not to order the continuation in force of a closure notice issued by them (section 84(2) and (3)).

The Crown Court can make whatever order it thinks appropriate (section 84(6)).

The Crown Court must notify the relevant licensing authority if it makes a closure order in relation to premises in respect of which a premises licence is in force (section 84(7)).

Enforcement of closure orders

An authorised person can enter premises, and use reasonable force, in respect of which a closure order is in force or do anything necessary to secure the premises against entry (section 85(1) and (3)). This includes entry to carry out essential maintenance or repairs to the premises (section 85(5)).

A person seeking to enter premises must, if required to do so by or on behalf of the owner, occupier or other person in charge of the premises, produce evidence of his or her identity and authority before entering the premises (section 85(4)).

Offences

A person who without reasonable excuse remains on or enters premises in contravention of a closure notice commits an offence (section 86(1)).

A person who without reasonable excuse obstructs a person acting under these powers commits an offence (section 86(3)). Persons committing these offences are liable on summary conviction to imprisonment for a period not exceeding 3 months (or 51 weeks), or to a fine, or to both (section 86(4) and (5)).

Access to other premises

Where access to premises is prohibited or restricted by an order and those premises are part of a building or structure, and another part is not subject to the prohibition or restriction, an occupier or owner of that other part can apply to the appropriate court (crown court or magistrates' court) for an order (section 87(1) and (2)).

Notice of an application must be given to:

(a) whatever constable the court thinks appropriate;
(b) the local authority;
(c) a person on whom the closure notice was served;
(d) anyone else who has an interest in the premises but on whom the closure notice was not served (section 87(3)).

The court may make whatever order it thinks appropriate in relation to access to any part of the building or structure (section 87(4)).

Reimbursement of costs

The Police or a local authority that incurs expenditure for the purpose of clearing, securing or maintaining premises in respect of which a closure order is in force may apply to the court (section 88(1)) for whatever order the court thinks appropriate for the reimbursement (in full or in part) by the owner or occupier of the closed premises (section 88(2)).

This application cannot be heard after 3 months starting with the day on which the closure order ceases to have effect (section 88(3)).

An application must also be served on the local policing body for the area in which the premises are situated, if the application is made by a local authority or the local authority, if the application is made by a local policing body (section 88(5)).

Compensation

Section 90 provides for compensation for a person who claims to have incurred financial loss in consequence of a closure notice.

Definition

authorised person means

(a) in relation to a closure order made on the application of a constable, means a constable or a person authorised by the chief officer of police for the area in which the premises are situated, and

(b) in relation to a closure order made on the application of a local authority, means a person authorised by that authority (section 85(2)).

HYPNOTISM LICENCE

Reference

Hypnotism Act 1952.

Extent

These provisions apply to England, Scotland and Wales.

Control of demonstrations of hypnotism at places licensed for public entertainment

Where an authority have power to grant licences for the regulation of theatres or other places of public amusement or public entertainment with conditions includes the power to attach conditions regulating or prohibiting the giving of an exhibition, demonstration or performance of hypnotism on any person at the place to which the licence relates (section 1(1)).

Control of demonstrations of hypnotism at other places

No person can give an exhibition, demonstration or performance of hypnotism on any living person at or in connection with an entertainment to which the public are admitted, whether on payment or otherwise, at any place, unless:

(a) the controlling authority have authorised that exhibition, demonstration or performance; or

(b) the place is in Scotland and a licence mentioned in section 1 is in force in relation to it (section 2(1)).

A licence is not needed for an exhibition, demonstration or performance of hypnotism that takes place in the course of a performance of a play given at premises in Scotland where a licence has been issued (section 2(1A)).

Conditions can be applied to the authorisation (section 2(2)).

A person giving an exhibition, demonstration or performance of hypnotism without an authorisation or licence is liable on summary conviction to a fine not exceeding level 3 on the standard scale (section 2(3)).

A fee is payable to the controlling authority for an authorisation for a hypnotism licence (section 2A).

A person who gives an exhibition, demonstration or performance of hypnotism on a person under 18 at or in connection with an entertainment to which the public are admitted, whether on payment or not, is, unless he had reasonable cause to believe that that person was over 18, liable on summary conviction to a fine not exceeding level 3 on the standard scale (section 3).

These provisions do not apply to the use of hypnotism for scientific or research purposes or for the treatment of mental or physical disease (section 5).

Definitions

hypnotism includes hypnotism, mesmerism and any similar act or process which produces or is intended to produce in any person any form of induced sleep or trance in which the susceptibility of the mind of that person to suggestion or direction is increased or intended to be increased but does not include hypnotism, mesmerism or any such similar act or process which is self-induced (section 6).
controlling authority means:

(a) in relation to a place in England and Wales, the licensing authority in whose area the place, or the greater or greatest part of it, is situated, and
(b) in relation to a place in Scotland, the authority having power to grant licences of the kind mentioned in section 1 in that area.

licensing authority has the meaning given by the Licensing Act 2003 (section 2(4)).

CHILDREN'S PERFORMANCE LICENCE

References

Children And Young Persons Act 1933, sections 23 and 24.
Children And Young Persons Act 1963, section 37.
The Children (Performances) Regulations 1968 (as amended).
Children (Performances and Activities) (England) Regulations 2014.

Scope

This procedure applies to the licensing of children's performances.

Extent

This procedure applies in England, Scotland and Wales. There are similar provisions in Northern Ireland.

Restriction on children under 16 taking part in public performances, etc. (FC 3.7)

A child cannot take part in a performance or otherwise take part in a sport, or work as a model, where payment in made, other than expenses, is made to him or to another person, unless a licence is granted by the local authority where he resides or, if he does not reside in Great Britain, by the local authority in whose area the applicant or one of the applicants for the licence resides or has his place of business (section 37(1) Children And Young Persons Act 1963 (CYPA 1963)).

A licence is not required if no payment is made to him or to another person, and in the 6 months preceding the performance he has not taken part in other performances on more than 3 days; or the performance is arranged by a school.

This also may include a condition requiring sums earned by the child to be paid into the county court (or, in Scotland, consigned in the sheriff court) or dealt with in a manner approved by the local authority (section 37(6) CYPA 1963).

The power to grant licences includes applying restrictions and conditions in regulations and a local authority cannot grant a licence for a child unless they are satisfied that he is fit to give the performance, that proper provision has been made to secure his health and kind treatment and that his education will not suffer. However, if they are satisfied, they must not refuse to grant the licence (section 37(4) CYPA 1963).

A licence must specify the times, if any, during which the child may be absent from school for the purposes authorised by the licence; and a child who is absent during any specified times will be deemed to be absent with leave granted by a person authorised by the managers, governors or proprietor of the school or, in Scotland, with reasonable excuse (section 37(7) CYPA 1963).

A licence may be varied on the application of the person holding it by the local authority granting it or by any local authority in whose area any activity takes place (section 39(1) CYPA 1963).

The local authority granting the licence and any local authority in whose area any activity takes place, may vary or revoke the licence if any condition is not observed or they are not satisfied that the child is fit to give the performance, that proper provision has been made to secure his health and kind treatment and that his education will not suffer. However, they must, before doing so, give to the holder of the licence such notice (if any) of their intention as may be practicable in the circumstances (section 39(2) CYPA 1963).

Where a local authority grant licence authorising a child to do something in the area of another local authority they must send the particulars the Secretary of State may prescribe to them; and where a local authority vary or revoke a licence which was granted by, or relates to an activity in the area of, another local authority, they must inform that other authority (section 39(3) CYPA 1963).

A local authority proposing to vary or revoke a licence granted by another local authority must, if practicable, consult that other authority (section 39(4) CYPA 1963).

The holder of a licence must keep records as the Secretary of State may by regulations prescribe and shall on request produce them to an officer of the authority who granted the licence, at any time not later than 6 months after the occasion or last occasion (section 39(5) CYPA 1963).

Where a local authority refuse an application for a licence or revoke or vary a licence they must state their grounds for doing so in writing to the applicant or, as the case may be, the holder of the licence. This does not apply to an application to vary the licence by the holder.

The applicant or holder may appeal to a magistrates' court or, in Scotland, the sheriff, against the refusal, revocation or variation, and against any condition on the licence but not a condition which the local authority are required to impose (section 39(6) CYPA 1963).

Offences

If any person:

(a) causes or procures any child or, being his parent or guardian, allows him, to do anything without a licence; or

(b) fails to observe any licence condition; or

(c) knowingly or recklessly makes any false statement in or in connection with an application for a licence,

he is liable on summary conviction to a fine not exceeding level 3 on the standard scale or imprisonment for a term not exceeding 3 months or both (section 40(1) CYPA 1963).

If any person fails to keep or produce any record which he is required to keep or produce he is liable on summary conviction to a fine not exceeding level 3 on the standard scale or imprisonment for a term not exceeding 3 months or both (section 40(2) CYPA 1963).

The court may revoke the licence if the holder or one of the holders of a licence is convicted of an offence (section 40(3) CYPA 1963).

It is a defence to prove that the accused believed that no payment for taking part in the performance, other than expenses, is made to him or to another person, and in the time restrictions did not apply; or the performance is arranged by a school and that he had reasonable grounds for that belief (section 40(4) CYPA 1963).

Restrictions on training for performances of a dangerous nature

No child under the age of 12 years can be trained to take part in performances of a dangerous nature, and a licence is required for children over 12 and every person who causes or procures a person, or being his parent or guardian allows him, to be trained to take part in performances of a dangerous nature, is liable on summary conviction to a fine not exceeding level 3 on the standard scale (section 24(1) C&YP Act 1933).

FC 3.7 Children's performance licence – Children And Young Persons Act 1963, Section 37. Children And Young Persons Act 1933. The Children (Performances) Regulations 1968 (as amended)

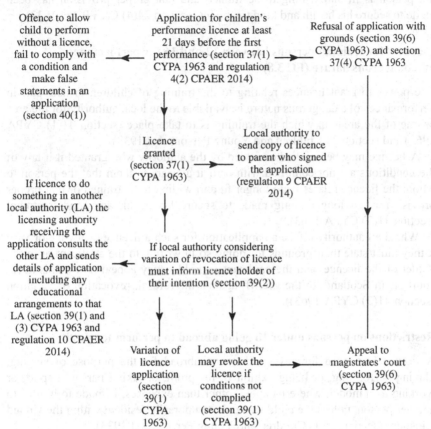

Offence to allow child to perform without a licence, fail to comply with a condition and make false statements in an application (section 40(1))

Application for children's performance licence at least 21 days before the first performance (section 37(1) CYPA 1963 and regulation 4(2) CPAER 2014)

Refusal of application with grounds (section 39(6) CYPA 1963) and section 37(4) CYPA 1963

Licence granted (section 37(1) CYPA 1963)

Local authority to send copy of licence to parent who signed the application (regulation 9 CPAER 2014)

If licence to do something in another local authority (LA) the licensing authority receiving the application consults the other LA and sends details of application including any educational arrangements to that LA (section 39(1) and (3) CYPA 1963 and regulation 10 CPAER 2014)

If local authority considering variation of revocation of licence must inform licence holder of their intention (section 39(2))

Variation of licence application (section 39(1) CYPA 1963)

Local authority may revoke the licence if conditions not complied (section 39(1) CYPA 1963)

Appeal to magistrates' court (section 39(6) CYPA 1963)

Notes
1. CYPA 1963 – Children and Young Persons Act 1963.
2. CPAER 2014 – Children (Performances and Activities) (England) Regulations 2014.
3. Where a child is taking part in an activity, performance, or rehearsal or whilst the child is living elsewhere than the place the child would otherwise live during the period to which the licence applies the licensing authority must approve a chaperone, the place of accommodation and places of rehearsal and performance (regulations 15 to 17 CPAER 2014).

A local authority may grant a licence for a child over 12 years to be trained to take part in performances of a dangerous nature (section 24(4) C&YP Act 1933).

A licence must specify the place or places at which the person is to be trained and must include conditions that are in the opinion of the authority, necessary for his protection, but a licence cannot be refused if the authority is satisfied that the person is fit and willing to be trained and that proper provision has been made to secure his health and kind treatment (section 24(4) C&YP Act 1933).

Licences for training persons between 12 and 16 for performances of a dangerous nature (FC 3.8)

The power to grant licences relating to the training of children to take part in performances of a dangerous nature is available to the local authority for the area or one of the areas in which the training is to take place (section 41(1) CYPA 1963) and section 24 Children and Young Persons Act 1933).

A licence may be revoked or varied by the authority who granted it if any of the conditions are not complied with or if it appears to them that the person to whom the licence relates is no longer fit and willing to be trained or that proper provision is no longer being made to secure his health and kind treatment (section 41(2) CYPA 1963).

Where an authority refuse an application for such a licence or revoke or vary it they must state their grounds for doing so in writing to the applicant or to the holder of the licence, and the applicant or holder may appeal to a magistrates' court or, in Scotland, to the sheriff, against the refusal, revocation or variation (section 41(3) CYPA 1963)).

Restrictions on persons under 18 going abroad to perform for profit

A licence is needed for any child, to go abroad for the purpose of singing, playing performing, or being exhibited, for profit, or taking part in a sport, or working as a model, where payment, other than expenses, is made to him or to another person, unless the child was only temporarily resident within the United Kingdom (section 25(1) Children and Young Persons Act 1933).

A justice of the peace can grant a licence on the prescribed form, and subject to restrictions and conditions the justice of the peace thinks fit, for any child who has attained the age of 14 years to go abroad for any purpose above. The justice of the peace must be satisfied:

(a) that the application for the licence is made by or with the consent of his parent or guardian;
(b) that he is going abroad to fulfil a particular engagement;
(c) that he is fit for the purpose, and that proper provision has been made to secure his health, kind treatment, and adequate supervision while abroad, and his return from abroad at the expiration or revocation of the licence;
(d) that a copy of the contract of employment or other document showing the terms and conditions of employment drawn up in a language understood by him has been provided for him (section 25(2) Children and Young Persons Act 1933).

FC 3.8 Children's performance licence – dangerous activities – Children And Young Persons Act 1968

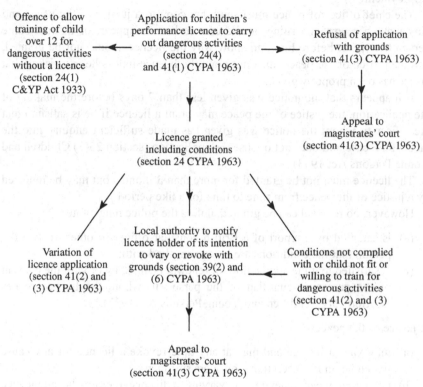

Offence to allow training of child over 12 for dangerous activities without a licence (section 24(1) C&YP Act 1933)

Application for children's performance licence to carry out dangerous activities (section 24(4) and 41(1) CYPA 1963)

Refusal of application with grounds (section 41(3) CYPA 1963)

Appeal to magistrates' court (section 41(3) CYPA 1963)

Licence granted including conditions (section 24 CYPA 1963)

Variation of licence application (section 41(2) and (3) CYPA 1963)

Local authority to notify licence holder of its intention to vary or revoke with grounds (section 39(2) and (6) CYPA 1963)

Conditions not complied with or child not fit or willing to train for dangerous activities (section 41(2) and (3) CYPA 1963)

Appeal to magistrates' court (section 41(3) CYPA 1963)

Note
CYPA 1963 – Children and Young Persons Act 1963.

The person applying for a licence must, at least 7 days before making the application, give the chief officer of police for the district in which the child lives, notice of the intended application together with a copy of the contract of employment or other document showing the terms and conditions of employment.

The chief officer of police must send that copy to a justice of the peace and may make a report in writing on the case to him or appear, or instruct some person to appear, before him and show why the licence should not be granted, and the justice of the peace must not grant the licence unless he is satisfied that notice has been properly given.

If it appears that the notice was given less than 7 days before the making of the application, the justice of the peace may grant a licence if he is satisfied that the officer to whom the notice was given has made sufficient enquiry into the facts of the case and does not oppose the application (section 25(3) Children and Young Persons Act 1933).

The licence must not be granted for more than 3 months but may be renewed by a justice of the peace from time to time for a like period.

However, no renewal can be granted, unless the police magistrate:

(a) is satisfied by a report of a British consular officer or other trustworthy person that the conditions are being complied with;
(b) is satisfied that the application for renewal is made by or with the consent of the parent or guardian of the person to whom the licence relates (section 25(4) Children and Young Persons Act 1933).

A justice of the peace:

(a) may vary a licence and may at any time revoke a licence for any cause which he, in his discretion, considers sufficient;
(b) need not, when renewing or varying a licence require the attendance before him of the person to whom the licence relates (section 25(5) Children and Young Persons Act 1933).

The justice of the peace to whom application is made for the grant, renewal or variation of a licence must, unless he is satisfied that in the circumstances it is unnecessary, require the applicant to give security as he may think fit. This can be either by entering into a recognisance with or without sureties or otherwise to comply with the restrictions and conditions in the licence or in the varies licence. The recognisance (see definitions below) may be enforced by the court (section 25(6) Children and Young Persons Act 1933).

If the justice of the peace is satisfied that in exceptional circumstances it is not in the interests of the person to whom the licence relates to require him to return from abroad at the expiration of the licence, then, the justice of the peace may release all persons concerned from any obligation to return that person from abroad (section 25(7) Children and Young Persons Act 1933).

Where a licence is granted, renewed or varied, the justice of the peace must send the prescribed particulars to the Secretary of State for transmission to the proper consular officer, and every consular officer must register the particulars and perform other duties the Secretary of State may direct (section 25(8) Children and Young Persons Act 1933).

Where a licence is granted, renewed or varied, the particulars which the justice of the peace must send to the Secretary of State for transmission to the proper consular officer are:

(a) the name and address of the child;
(b) the date, place of birth and nationality of the child;
(c) the name and address of the applicant for the licence;
(d) the name and address of the parent of the child;
(e) details of the engagement, including where and for how long the child is to participate;
(f) a copy of the contract of employment or other document showing the terms and conditions on which the child is engaged; and
(g) a copy of the licence (regulation 31 CPAER 2014).

Form of licence

A licence granted must include:

(a) the name of the child;
(b) the name of the applicant;
(c) details of the engagement that the child is going abroad to fulfil;
(d) the date on which the licence is granted and upon which it will expire;
(e) details of any security given by the applicant in accordance with section 25(6) CYPA 1933;
(f) details of any conditions considered necessary for the grant of the licence; and
(g) the signature of the person granting the licence (regulation 30 CPAER 2014).

Licences for children and young persons performing abroad

A licence may be granted for a child under the age of 14 if:

(a) the engagement which he is to fulfil is for acting and the application for the licence is accompanied by a declaration that the part he is to act cannot be taken except by a person of about his age; or
(b) the engagement is for dancing in a ballet which does not form part of an entertainment of which anything other than ballet or opera also forms part and the application for the licence is accompanied by a declaration that the part he is to dance cannot be taken except by a child of about his age; or
(c) the engagement is for taking part in a performance the nature of which is wholly or mainly musical or which consists only of opera and ballet and the nature of his part in the performance is wholly or mainly musical (section 42 (2) CYPA 1963).

FC 3.9 Children's performance licence – performing abroad – Children And Young Persons Act 1933

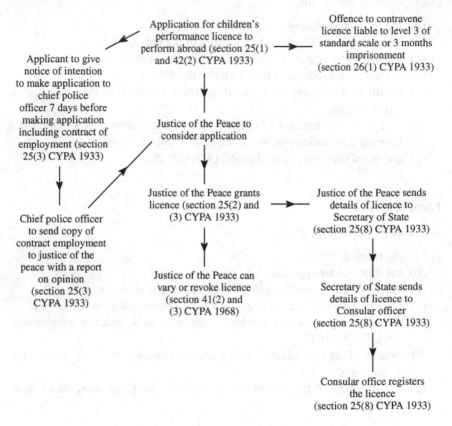

Note
CYPA 1933 – Children and Young Persons Act 1933.

Offences and penalties

It is an offence for any person to contravene a licence and is liable, on summary conviction, to a fine not exceeding level 3 on the standard scale, and/or, to imprisonment for any term not exceeding 3 months. If that person procured the child to go abroad by means of any false pretence or false representation, he is liable to imprisonment for a term not exceeding 2 years (section 26 (1) CYPA 1933).

Where it is proved that he caused, procured, or allowed a child to go abroad and that:

(a) that child has while abroad been singing, playing, performing, or being exhibited, for profit, or

(b) that child has while abroad taken part in a sport, or worked as a model, and payment in respect of his doing so, other than for expenses, was made to him or to another person,

the defendant is presumed to have caused, procured or allowed him to go abroad for that purpose.

However, where the contrary is proved, the court may order the defendant to take such steps as the court directs to secure the return of the child to the United Kingdom, or to enter into a recognisance as the court may direct to secure his health, kind treatment and adequate supervision while abroad, and his return to the United Kingdom at the expiration of a period the court may think fit (section 26(2)).

Proceedings for an offence can be taken within 3 months from the first discovery of the offence or, if after that period the person against whom it is proposed to institute the proceedings is outside the United Kingdom, within 6 months after his return to the United Kingdom (section 26(3)).

Children (Performances and Activities) (England) Regulations 2014

Application for licence

An application for a licence must:

(a) be made in writing by the person responsible for the organisation of, or the engaging of the child in, the activity; or the person responsible for the production of the performance in which the child is to take part;

(b) contain the information specified in Parts 1 and 2 of Schedule 2 of the regulations;

(c) be signed by the applicant and a parent of the child; and

(d) be accompanied by a copy of the child's birth certificate, two identical photographs of the child taken during the 6 months preceding the date of the application and a copy of the contract, draft contract or other documents containing particulars of the agreement regulating the child's appearance in the performances or regulating the activity for which the licence is requested (regulation 4(1)).

The licensing authority may refuse to grant a licence if the application is not received at least 21 days before the day on which the first performance or activity, for which the licence is requested, is to take place (regulation 4(2)).

Licence conditions

The licensing authority must impose conditions which it considers necessary in order to ensure that:

(a) the child is fit to take part in the performance or activity;
(b) proper provision is made to secure the child's health and kind treatment; and
(c) proper provision is made to ensure that the child's education will not suffer (regulation 5).

Where the applicant requests a licence for a child to take part in a particular activity, performance or rehearsal, but is unable to specify the dates upon which the child will take part in that activity, performance or rehearsal at the time of the application, if the licensing authority decides to grant the licence, it must impose a condition that the child may only take part in that particular activity, performance or rehearsal for a specified number of days within a 6-month period (regulation 6).

Where the licensing authority considers that the information provided by the applicant is insufficient to enable it to decide whether to issue a licence or subject to conditions, the licensing authority must request additional information or documentation to enable it to make such a decision (regulation 7(1)).

In particular, the licensing authority may:

(a) request that a child be medically examined;
(b) request a report from the head teacher or principal of the school that the child attends;
(c) interview any proposed private teacher;
(d) interview the applicant, the child, the child's parents, or the proposed chaperone, as appropriate (regulation 7(2)).

Form of licence

A licence must include:

(a) the name of the child;
(b) the name of the parents of the child;
(c) the name of the applicant;
(d) the names, times, nature and location of the activity or performance (and location of any rehearsal if different) for which the licence has been granted;
(e) the dates of the activity, performance or rehearsal, or instead of the dates, the number of days on which the child will participate in the activity, performance or rehearsal and the period, not exceeding 6 months, in which the activities, performances or rehearsals may take place;

(f) any conditions, which the licensing authority considers necessary for the grant of the licence; and

(g) a statement that the licence is subject to the restrictions and conditions in these Regulations (regulation 8(1)).

A photograph of the child must be attached to the licence (regulation 8(2)).

Particulars that a licensing authority must provide in respect of a licence

The licensing authority must send a copy of the licence to the parent who signed the application form (regulation 9).

Where a performance or activity is to take place in the area of a host authority other than the licensing authority, the licensing authority must send to that host authority a copy of the application form, licence, any additional information or documentation obtained by it and, where the licensing authority approves any arrangements for the child's education, details of the days during the period of the licence on which the child subject to the licence would ordinarily be required to attend school if that child were attending a school maintained by the licensing authority (regulation 10).

Records to be kept by the licence holder under section 39(5) of the 1963 Act

For 6 months from the date of the last performance or activity to which the licence relates, the licence holder must retain the records specified in Part 1 of Schedule 3, where the licence is granted in respect of a performance; or Part 2 of Schedule 3, where the licence is granted in respect of an activity (regulation 11).

Production of licence

The licence holder must, on request, produce the licence at all reasonable hours during the period beginning with the first and ending with the last performance or activity to which the licence relates, at the place of performance (or any place of rehearsal), or the place where the activity to which the licence relates takes place, to an authorised officer of the host authority or a constable (regulation 12).

Education

The licensing authority must not grant a licence unless it:

(a) is satisfied that the child's education will not suffer by reason of taking part in the performances or activities;

(b) has approved the arrangements (if any) for the education of the child during the period to which the licence applies; and

(c) has approved the place where the child is to receive education, subject to such conditions as it considers necessary to ensure that the place is suitable for the child's education (regulation 13(1)).

The licence holder must ensure that any arrangements approved by the licensing authority for the child's education are carried out (regulation 13(2)).

The licensing authority must not approve any arrangements for the education of a child by a private teacher unless it is satisfied that:

(a) the proposed course of study for the child is satisfactory and will be properly taught by the private teacher;

(b) the private teacher is a suitable person to teach the child in question and will teach no more than six children (including the child in question) at any time, or twelve children if all the children being taught have reached a similar standard in the subject being taught to the child in question; and

(c) the child will, during the period to which the licence applies, receive education for periods, which when aggregated, total not less than 3 hours on each day on which the child would be required to attend school if the child were attending a school maintained by the licensing authority (regulation 13(3)).

These requirements are deemed to have been met if the licensing authority is satisfied that the child will receive education:

(a) for not less than 6 hours a week;

(b) during each complete period of 4 weeks, or if there is a period of less than 4 weeks, during that period, for periods not less than the aggregate periods of education required by paragraph (3)(e) in respect of the period;

(c) on days on which the child would be required to attend school if the child were a pupil attending a school maintained by the licensing authority; and

(d) for not more than 5 hours on any such day (regulation 13(4)).

Any period of education does not include:

(a) any period which takes place other than during the hours when a child is permitted to be present at a place of performance or rehearsal; and

(b) any period of less than 30 minutes (regulation 13(5)).

Earnings

The licensing authority may include a condition in the licence that any or all of the sums earned by the child for taking part in the performance or activity be dealt with in a particular manner by the licence holder (regulation 14).

Chaperones

A licensing authority must approve a person to be a chaperone to:

(a) have care and control of the child; and

(b) safeguard, support and promote the well-being of the child,

whilst the child is taking part in an activity, performance, or rehearsal or whilst the child is living elsewhere than the place the child would otherwise live during the period to which the licence applies (regulation 15(1)). This does not apply if a child is being cared for by a parent or teacher who would ordinarily provide the child's education (regulation 15(2)).

The maximum number of children a chaperone may take care of at any one time is twelve; or where the person approved to act as a chaperone is the private teacher of the child in question, three (regulation 15(3)).

The licensing authority must not approve a person as a chaperone unless it is satisfied that the person:

(a) is suitable and competent to exercise proper care and control of a child of the age and sex of the child in question; and

(b) will not be prevented from carrying out duties towards the child by duties towards other children (regulation 15(4)).

Where a child suffers any injury or illness while under the care of the chaperone, the licence holder must ensure that the parent of the child named in the application form and the licensing and host authorities are notified immediately of such injury or illness (regulation 15(5)).

Accommodation

Where a child is required to live somewhere other than where that child would usually live during the period to which the licence applies, the licensing authority must approve that place as being suitable for that child (regulation 16(1)).

The licensing authority's approval may be subject to any of the following conditions:

(a) that transport will be provided for the child between the place of performance, rehearsal or activity, and the accommodation;

(b) that suitable arrangements are made for meals for the child; and

(c) any other condition conducive to the welfare of the child in connection with that accommodation (regulation 16(2)).

Place of performance and place of rehearsal

The licensing authority must approve any place where the child will perform, rehearse or take part in any activity (regulation 17(1)).

The licensing authority must not approve the place of performance, rehearsal or activity unless it is satisfied that, having regard to the age of the child and the nature, time and duration of the performance, rehearsal or activity:

(a) suitable arrangements have been made for:
   (i) the provision of meals for the child;
   (ii) the child to dress for the performance, rehearsal or activity (arrangements for a child who has attained the age of 5 years to dress for a performance, rehearsal or activity are not suitable unless such a child can dress only with children of the same sex as the child in question (regulation 17(4))); and
   (iii) the child's rest and recreation, when not taking part in a performance, rehearsal or activity;

(b) the place has suitable and sufficient toilets and washing facilities; and

(c) the child will be adequately protected against inclement weather (regulation 17(2)).

The licensing authority may give its approval subject to such conditions as it considers necessary (regulation 17(3)).

Travel arrangements

The licence holder must ensure that suitable arrangements are made to get the child home or to any other destination after the last performance or rehearsal, or the conclusion of any activity on any day (regulation 18).

The following requirements apply to all licensed performances and to all performances, which are exempted from the requirement to obtain a licence (regulation 19).

A child taking part in a performance must not be employed in any other employment on the day of that performance or the following day (regulation 20).

Regulation 21 sets out the earliest and latest times a child may be at a place of performance or rehearsal (regulation 21(1)). This regulation does not apply where the place of performance or rehearsal is the place where the child ordinarily lives or receives education (regulation 21(2)).

Regulation 22 sets out the maximum number of hours a child may be at a place of performance or rehearsal, may perform or rehearse in one day and may perform or rehearse continuously.

When calculating the number of hours on any day during which a child is present at a place of performance or rehearsal, any periods of education required to comply with arrangements approved must be taken into account, even if that education is provided elsewhere than at the place of performance or rehearsal (regulation 22(2)).

Regulation 23 gives break times for children under the age of five or over five present at the place of performance or rehearsal.

Regulations 25 to 29 specify the restrictions on the number of days and frequency of performance including the discretion of a chaperone.

Definitions

performance means:

(a) any performance in connection with which a charge is made (whether for admission or otherwise);

(b) any performance in premises may be used for the supply of alcohol or are licensed premises;

(c) any broadcast performance;

(d) any performance recorded with a view to its use in a broadcast or such a service or in a film intended for public exhibition;

(e) if the child takes the place of a performer in any rehearsal or in any preparation for the recording of the performance (section 37(2)).

recognisance – a bond by which a person undertakes before a court or magistrate to observe some condition, especially to appear when summoned.

FIREWORKS AND EXPLOSIVES LICENCES

References

Fireworks Regulations 2004.
Explosives Regulations 2014.

Extent

These Regulations extend to England, Wales and Scotland.

Fireworks Regulations 2004 – summary

Those intending to supply fireworks to the public outside the traditional selling periods (i.e. all year round) are required to hold a licence to supply fireworks, either from the local authority, Fire Service or the Health and Safety Executive (HSE). All three bodies have the discretion to refuse such an application on the grounds of either a potential increase in anti-social behaviour or injuries.

The traditional periods where selling without a licence is permitted are:

* 5 November (from 15 October to 10 November);
* New Year (from 26 December to 31 December);
* Chinese New Year (on the first day of the Chinese New Year and the 3 days immediately preceding it);
* Diwali (on the day of Diwali and the 3 days immediately preceding it).

It is an offence under the Health and Safety at Work, etc. Act 1974 to sell fireworks by retail without a licence or to store fireworks unsafely. The penalty is a fine of up to £20,000 and/or 12 months' imprisonment.

Licensing of fireworks suppliers (FC 3.10)

No person can supply or expose for supply any adult firework, without:

  (a) a licence granted for each premises under his control at which the fireworks are supplied or exposed for supply; or
  (b) a licence granted to him, if the fireworks which he supplies or exposes for supply are kept at premises which are not under his control (regulation 9(1)).

However, fireworks can be sold by unlicensed traders for Chinese New Year and the preceding 3 days, Diwali and the preceding 3 days, Bonfire Night celebrations (15 October to 10 November) and for New Year celebrations (26 to 31 December) (regulation 9(2)).

The following people do not need a licence:

* a person who is employed in, or whose trade or business is the supply of fireworks or assemblies, in accordance with the Pyrotechnic Articles (Safety) Regulations 2010;

- a person who is employed by, or in business as, a professional organiser or operator of firework displays; or
- a person who is employed in, or whose trade or business is, the transport of fireworks (regulation 9(2A)).

An application for a licence must be made to the local licensing authority in whose area:

(a) the premises concerned are located in the case of a licence; or
(b) the principal business premises of the applicant are (regulation 9(3)).

A local licensing authority must not grant a licence unless it is satisfied:

(a) that the premises which are the subject of the application, are licensed or registered under the Explosives Act 1875; or
(b) in the case of an application that the fireworks which will be supplied or exposed for supply by the applicant, will be kept at premises which are licensed or registered under that Act (regulation 9(4)).

A local licensing authority may refuse to grant a licence, or may revoke a licence which it has granted, if the applicant has committed:

(a) an offence under section 11 of the Fireworks 2003 Act;
(b) an offence under section 12 of the Consumer Protection Act 1987 arising from a contravention of the Pyrotechnic Article (Safety) Regulations 2010;
(c) an offence under sections 4, 5 or 32 of the Explosives Act 1875; or
(d) an offence in relation to the use, storage or keeping of fireworks under the Health and Safety at Work, etc. Act 1974 (regulation 9(5)).

Where a local licensing authority refuses to grant a licence, or revokes a licence, it must notify the applicant of its decision (regulation 9(6)).

A local licensing authority must charge a fee of £500 a year in connection with the grant of a licence (regulation 9(7)).

A person may appeal to the court against a decision of a local licensing authority to refuse to grant him a licence, or to revoke a licence, and any such appeal must be made within 28 days of the decision in question being notified to that person (regulation 9(8)).

Information about fireworks

The licensee must maintain for a period of 3 years, beginning with the date on which he supplies fireworks, a record of the following information:

(a) the name and address of the person who supplied the firework to him;
(b) the name and address of the person to whom he is supplying the firework;
(c) the date when the firework was supplied to him;
(d) the date when he supplied the firework to another person; and
(e) the total amount of explosives contained in the firework supplied (regulation 10(3)).

FC 3.10 Fireworks licence – Fireworks Regulations 2004

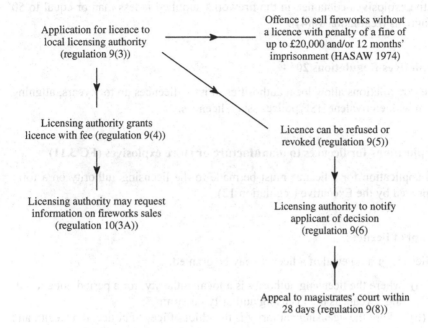

A person who supplies adult fireworks must, if requested by a local licensing authority within the 3-year period, provide any of the information specified in the request (regulation 10(3A)).

These requirements do not apply if, in a single transaction, the total amount of the explosives contained in the fireworks supplied is less than or equal to 50 kilograms (regulation 10(4)).

Explosives Regulations 2014

These regulations allow local authorities to issue licences up to 5 years, aligning them with equivalent HSE/police-issued licences.

Applications for licences to manufacture or store explosives (FC 3.11)

An application for a licence must be made to the licensing authority on a form approved by the Executive (regulation 12).

Grant of licences

A licence or a renewal of a licence may be granted:

(a) where the licensing authority is a local authority, for a period not exceeding 5 years as the licensing authority determines;

(b) where the licensing authority is the chief officer of police, the Health and Safety Executive or the Office of Nuclear Regulation (ONR), for a period not exceeding 5 years as that licensing authority determines. However, where the applicant for the licence or renewal of a licence has been granted an explosives certificate, the licence or renewal may only be granted for any period not exceeding the due expiry date of that explosives certificate; or

(c) for any period or without a time limit in a case:

  (i) where local authority consent is required; or

  (ii) under existing licence, police operations and training, no significant additional health and safety issues arise; or

  (iii) where the application is for a licence, or a renewal of a licence, relating only to the manufacture or storage of ammonium nitrate blasting intermediate (regulation 13(1)).

The licensing authority must grant a licence or renewal of a licence unless any of the grounds for refusing apply – see below regulation 20 (regulation 13(2)).

Where the Executive or the ONR is the licensing authority in respect of an application for a licence the procedure in regulation 14 applies for obtaining the assent of the local authority, or each local authority where the proposed site which is the subject of the application for a licence is situated partly within the area of one local authority and partly within the area of another, and the Executive or the ONR, must refuse to grant the licence unless the local authority, or each local authority, agrees (regulation 13(3)).

This does not apply to an application:

(a) for a licence to store no more than 2000 kilograms; and the applicant has not notified the relevant licensing authority that the required separation distances could not be complied with;

(b) for a licence relating to the manufacture of explosives by means of on-site mixing;

(c) for a licence relating to the manufacture or storage of ammonium nitrate blasting intermediate;

(d) for a licence relating to the manufacture or storage of explosives by a person who wishes to carry on such manufacture or storage within a part of a site where another person already holds a licence for the manufacture or storage of explosives; and either:

  (i) the application relates to manufacturing or storage activities which would be permitted at that part of the site under the existing licence; or

  (ii) in the opinion of the relevant licensing authority or a local authority whose assent would otherwise be required, no significant new health and safety issues are raised by the application;

(e) for a licence relating to the manufacture of explosives by a police force maintained for their operational purposes or the training of members of that police force in relation to those purposes;

(f) for a licence for the manufacture or storage of explosives at a site which, immediately before any grant of that application, is one which the disapplication applies to and, in the opinion of the relevant licensing authority, no significant new health and safety issues are raised by the application; or

(g) for a licence to follow, without a gap in time, a licence above and in the opinion of the relevant licensing authority, no significant new health and safety issues are raised by the application (regulation 13(4)).

Every licence must include conditions which specify:

(a) the site and, within it, the places where the explosives may be stored, or, in the case of a licence to manufacture explosives, where they may be manufactured;

(b) the hazard type, if any, the description and maximum amount of explosives which may be:

  (i) stored or otherwise present; or

  (ii) in the case of licence to manufacture explosives, manufactured,

at any one time at any place so specified (regulation 13(5)).

In addition, a licence which is granted by the relevant licensing authority in cases where the assent of the local authority was required above (paragraph 3) or not required (paragraph 4):

(a) must be granted subject to such conditions as the relevant licensing authority considers appropriate which relate to separation distances;

(b) may be granted subject to such conditions as the relevant licensing authority considers appropriate which relate to:
  (i) the construction, siting or orientation of any building (including any protective works around the building) where the activity will be carried on;
  (ii) the activities which may be undertaken in specified buildings, rooms within those buildings, other structures or other places within the site; and
  (iii) the manufacture and storage of the ingredients of explosives or articles or substances which are liable to ignite spontaneously or are flammable or otherwise dangerous in ways which could initiate or aggravate a fire or explosion;
(c) may, where both the manufacture and storage of explosives at the same site was applied for, cover both that manufacture and storage (regulation 13(6)).

Where a licensing authority grants a licence which relates to the storage of pyrotechnic articles at any site where those articles are to be offered for sale, the licensing authority may attach conditions to the licence it considers appropriate relating to:

(a) the storage and display of those articles in areas where they can be purchased;
(b) the prevention of risk of fire arising in respect of those articles; and
(c) the safe use of fire escapes in that area (regulation 13(7)).

Every person who is granted a licence to manufacture or store explosives must ensure that the relevant licensing authority and the local planning authority in whose area the manufacture or storage takes place is, within 28 days of the licence being granted or varied in a way which affects the separation distances required to be maintained, given a plan of the site and its immediate surrounding area showing the separation distances required to be maintained pursuant to the licence or varied licence (regulation 13(8)).

A licence granted must be in a form approved by the Executive (regulation 13(9)).

Local authority assent procedure in relation to licence applications

The relevant licensing authority (the Executive or the ONR – regulation 13(10)) must issue the applicant with a draft licence containing the conditions which that licensing authority proposes to attach to the licence (regulation 14(1)).

The applicant must as soon as reasonably practicable send a copy of the application and draft licence to the local authority in whose area the manufacture or storage is proposed to take place (regulation 14(2)).

Within 28 days of sending to the local authority this information, the applicant must:

(a) publish in a newspaper circulating in the locality where the manufacture or storage of explosives is proposed to take place a notice:

FC 3.11 Explosives licence – local authority assent procedure – Regulation 14 Explosives Regulations 2014

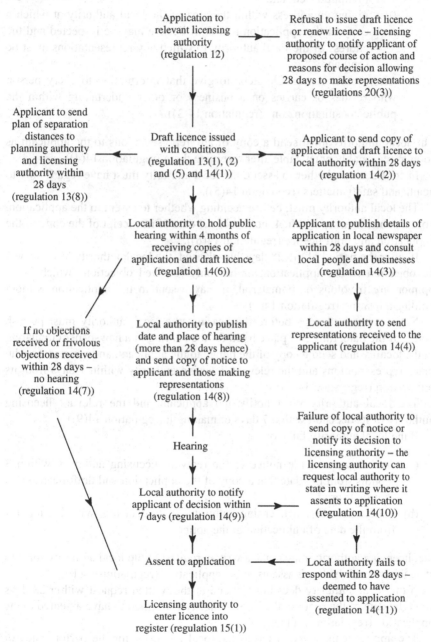

Application to relevant licensing authority (regulation 12)

Refusal to issue draft licence or renew licence – licensing authority to notify applicant of proposed course of action and reasons for decision allowing 28 days to make representations (regulations 20(3))

Applicant to send plan of separation distances to planning authority and licensing authority within 28 days (regulation 13(8))

Draft licence issued with conditions (regulation 13(1), (2) and (5) and 14(1))

Applicant to send copy of application and draft licence to local authority within 28 days (regulation 14(2))

Local authority to hold public hearing within 4 months of receiving copies of application and draft licence (regulation 14(6))

Applicant to publish details of application in local newspaper within 28 days and consult local people and businesses (regulation 14(3))

If no objections received or frivolous objections received within 28 days – no hearing (regulation 14(7))

Local authority to publish date and place of hearing (more than 28 days hence) and send copy of notice to applicant and those making representations (regulation 14(8))

Local authority to send representations received to the applicant (regulation 14(4))

Hearing

Failure of local authority to send copy of notice or notify its decision to licensing authority – the licensing authority can request local authority to state in writing where it assents to application (regulation 14(10))

Local authority to notify applicant of decision within 7 days (regulation 14(9))

Assent to application

Local authority fails to respond within 28 days – deemed to have assented to application (regulation 14(11))

Licensing authority to enter licence into register (regulation 15(1))

    (i) giving details of the application;

    (ii) inviting representations on matters affecting the health and safety of persons other than the applicant's employees to be made in writing to the local authority within 28 days of the date that the notice is first published; and

    (iii) giving an address within the area of the local authority at which a copy of the application and draft licence may be inspected and the address of the local authority to which any representations must be sent; and

  (b) taking other reasonable steps to give that information to every person who resides or carries on a business or other undertaking within the public consultation zone (regulation 14(3)).

The local authority must send a copy of any representations to the applicant as soon as reasonably practicable after receiving them (regulation 14(4)).

In considering whether to assent, the local authority must have regard only to health and safety matters (regulation 14(5)).

The local authority must, before deciding whether to assent to the application, hold a public hearing within 4 months of the date of its receipt of the copy of the application and draft licence (regulation 14(6)).

If, after the period of the 28 days has elapsed, the local authority has received no objection to the application, or has only received objections which in its opinion are frivolous or immaterial, it may assent to the application without holding a hearing (regulation 14(7)).

Not less than 28 days before the hearing, the local authority must publish notice of the date, time and place fixed for the hearing in a newspaper circulating in the locality and send a copy of the notice to the applicant, and any person who made representations and the relevant licensing authority within 7 days from its publication (regulation 14(8)).

The local authority must notify the applicant and the relevant licensing authority of its decision within 7 days of making it (regulation 14(9)).

If the local authority fails to:

  (a) send a copy of the notice to the relevant licensing authority within 3 months from the date that a copy of the application and draft licence was sent to it; or

  (b) notify the relevant licensing authority of its decision, within 2 months from the date of publication of the notice,

that licensing authority may make a written request to the local authority for it to state in writing whether it assents to the application (regulation 14(10)).

If the local authority does not respond to the written request within 28 days from the date of the request, the local authority is deemed to have assented to the application (regulation 14(11)).

The applicant must pay a fee to the local authority for the performance of their functions, not exceeding the sum of the costs reasonably incurred in performing those functions (regulation 14(12)).

Registers and retention of documents

The licensing authority must:

(a) maintain a register in accordance with Schedule 4;
(b) keep a copy of any licence granted by it (together with a copy of the application for the licence) for as long as the licence remains valid; and
(c) if a local authority or the ONS, send the Executive on request a copy of any part of the register or other document within such time as the Executive may direct (regulation 15(1)).

Variation of licences (FC 3.12)

The licensing authority which grants a licence may vary it:

(a) where there has been a change in circumstances where the separation distances can no longer be maintained and a consequent reduction in the maximum amount of explosive that may be stored is required;
(b) where the Executive or the ONR is the licensing authority in cases where the assent of the local authority was required, or in cases where that assent was not required above, before the grant of the licence and where there has been a material change in circumstances so that a variation is necessary to ensure safety; or
(c) in relation to any of the matters it relates to, by agreement with the licensee (regulation 16(1)).

A licence may be varied on the grounds without the agreement of the licensee (regulation 16(2)).

Where the Executive or the ONR is the licensing authority in cases where the assent of the local authority is needed or not as the case may be the assent procedure above applies (regulation 16(3)).

A proposed variation is one which:

(a) relates to changes in the permitted quantities or types of explosive as a result of which the licensee could be required to maintain a separation distance greater than the separation distance required before the variation and, in the opinion of the Executive or the ONR, or the local authority concerned, significant new health and safety issues are raised by that proposed variation;
(b) would increase the period of the licence by more than 12 months; or
(c) would remove the period of the licence so that it would be unlimited as to time,

and the Executive or the ONR must refuse to grant a varied licence unless the local authority, or each local authority, has assented (regulation 16(4)).

Where a licensing authority proposes to vary a licence without the agreement of the licensee it must, before taking any such action, notify the licensee of its proposed course of action and allow the licensee the opportunity of making representations to the licensing authority about it, within a period of 28 days from the date of the notification (regulation 16(5)).

Representations may be made in writing, or both in writing and orally (regulation 16(6)).

Where the licensing authority decides to vary a licence without the agreement of the licensee it must provide the licensee with written reasons for its decision (regulation 16(7)).

Where the licensing authority varies a licence without the agreement of the licensee, that variation takes effect from a date to be determined by the licensing authority which must be a date after the 28 day period above (regulation 16(8)).

Transfer of licences

A licence may on application be transferred in writing by the licensing authority which issued the licence to any other person who wishes to manufacture or store explosives in place of the licensee (regulation 17(1)).

A licensing authority must grant an application for a transfer of a licence unless it is of the opinion that the applicant is not a fit person:

(a) to store explosives, where the application is to transfer a licence to store explosives; or

(b) to manufacture explosives, where the application is to transfer a licence to manufacture explosives (regulation 17(2)).

Where a licensing authority is of this opinion, it must refuse the application to transfer the licence (regulation 17(3)).

Death, bankruptcy or incapacity of a licensee

If a licensee dies or becomes incapacitated, a person manufacturing or storing explosives in accordance with the conditions of the licence is to be treated the licensee until either:

(a) the expiration of a period of 60 days starting with the date of such death or incapacity;

(b) the grant or refusal of a new licence; or

(c) the transfer of, or a refusal to transfer, a licence,

whichever is the earlier (regulation 18(1)).

If a licensee becomes bankrupt or, in the case of a company, goes into liquidation, administration or receivership or has a receiving order made against it, any liquidator, administrator, receiver or trustee in bankruptcy is to be treated as being the licensee (regulation 18(2)).

Refusal of a licence and draft licence and refusal of a renewal or transfer of a licence

The licensing authority must refuse an application for a licence or issue the draft licence where the licensing authority is of the opinion that:

(a) the proposed site or, within it, any place where the manufacture or storage of explosives is proposed to take place is unsuitable for that manufacture or storage; or

(b) the applicant is not a fit person to store or manufacture explosives, in the case of the application for a relevant licence (regulation 20(1) and (2)).

Where a licensing authority proposes to refuse an application for a licence; a renewal of a licence; a variation of a licence; or a transfer of a licence it must, before taking any such action, notify the applicant of its proposed course of action and allow that applicant the opportunity of making representations to the licensing authority about it, within a period of 28 days from the date of the notification and it must provide the applicant with written reasons for its decision (regulation 20(3) and (5)).

Representations may be made in writing, or both in writing and orally (regulation 20(4)).

A refusal by the licensing authority to issue the draft licence is to be treated as a refusal of an application for a licence (regulation 20(6)).

Revocation of a licence

The licensing authority which grants a licence may revoke that licence:

(a) where there has been a change in circumstances such that the site or, within it, any place in which explosives are manufactured or stored which the licence relates to is no longer suitable for that manufacture or storage;

(b) where it appears to the licensing authority on information obtained by it after the grant of the licence that the licensee is not a fit person to store or manufacture explosives, or

(c) by agreement with the licensee (regulation 23(1)).

A person whose licence is revoked must ensure that:

(a) all explosives are removed from a site as soon as is practicable after revocation of a licence;

(b) those explosives are deposited at a site which is the subject of a licence which permits any storage resulting from that depositing, or suitable arrangements to dispose of the explosives; and

(c) the licence is returned to the licensing authority within 28 days of the date that the revocation takes effect (regulation 23(2)).

Where a licensing authority proposes to revoke a licence, it must, before taking any such action, notify the licensee of its proposed course of action and allow that person the opportunity of making representations to the licensing authority about it, within a period of 28 days from the date of the notification (regulation 23(3)).

Representations may be made in writing, or both in writing and orally (regulation 23(4)).

Where the licensing authority decides to revoke a licence, it must provide in writing to the licensee the reasons for its decision (regulation 23(5)).

Where the licensing authority revokes a licence, that revocation takes effect from a date to be determined by the licensing authority which must be a date after the 28 day period (regulation 23(6)).

FC 3.12 Explosives licence – variation Explosives Regulations 2014

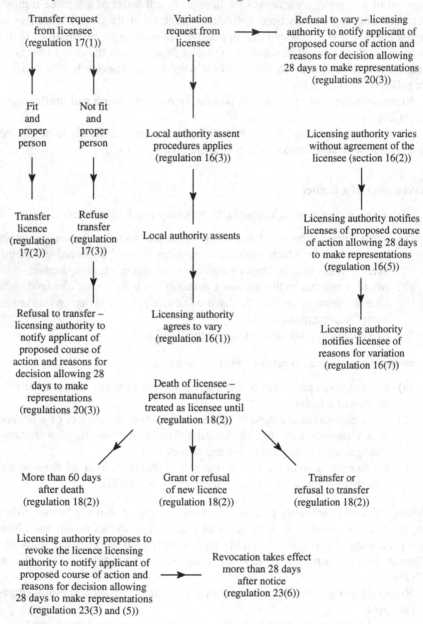

Transfer request
from licensee
(regulation 17(1))

Variation
request from
licensee

Refusal to vary – licensing
authority to notify applicant of
proposed course of action and
reasons for decision allowing
28 days to make representations
(regulations 20(3))

Fit
and
proper
person

Not fit
and
proper
person

Local authority assent
procedures applies
(regulation 16(3))

Licensing authority varies
without agreement of the
licensee (section 16(2))

Transfer
licence
(regulation
17(2))

Refuse
transfer
(regulation
17(3))

Local authority assents

Licensing authority notifies
licenses of proposed course
of action allowing 28 days
to make representations
(regulation 16(5))

Refusal to transfer –
licensing authority to
notify applicant of
proposed course of
action and reasons for
decision allowing 28
days to make
representations
(regulations 20(3))

Licensing authority
agrees to vary
(regulation 16(1))

Licensing authority
notifies licensee of
reasons for variation
(regulation 16(7))

Death of licensee –
person manufacturing
treated as licensee until
(regulation 18(2))

More than 60 days
after death
(regulation 18(2))

Grant or refusal
of new licence
(regulation 18(2))

Transfer or
refusal to transfer
(regulation 18(2))

Licensing authority proposes to
revoke the licence licensing
authority to notify applicant of
proposed course of action and
reasons for decision allowing
28 days to make representations
(regulation 23(3) and (5))

Revocation takes effect
more than 28 days
after notice
(regulation 23(6))

Licences for sites which cease to be ONR regulated sites

Where a site with a licence granted by the ONR ceases to be an ONR regulated site and the licence remained in force immediately before the relevant date, it is to be treated as a licence granted by the relevant licensing authority (regulation 24(1) and (2)).

Licences for sites which become ONR regulated sites

Where a site with a licence becomes an ONR regulated site and the licence remained in force immediately before the date on which the site became an ONR regulated site, a licence granted by a licensing authority (other than the ONR) is to be treated as a licence granted by the ONR (regulation 25).

Definitions

licensing authority means a local authority or chief officer of police for applications for the storage within a site of no more than 2000 kilograms of explosives.

The Office for Nuclear Regulation is a licensing authority for applications to manufacture or store explosives on an 'ONR regulated site'.

The Health and Safety Executive is the licensing authority where explosives are to be stored on the surface of a mine or within a harbour. It is also the licensing authority in relation to the manufacture of explosives and the manufacture and storage of ANBI and in other cases.

The licensing authority is determined in accordance with Schedule 1 of the regulations (regulation 24(3)).

GAMBLING

Responsible authorities must be consulted on Gambling Act and Licensing Act applications. The responsible authority for the prevention of risk of pollution of the environment or harm to human health is a statutory consultee for these purposes and in local authorities this is placed in the environmental health service.

Licensing Committee

Each licensing authority must establish a licensing committee consisting of between 10 and 15 members of the authority (section 6 Licensing Act 2003). This committee will deal with all licences and permits including those required by the Gambling Act 2005.

The **licensing objectives** for gambling are:

(a) preventing gambling from being a source of crime or disorder, being associated with crime or disorder or being used to support crime;
(b) ensuring that gambling is conducted in a fair and open way; and
(c) protecting children and other vulnerable persons from being harmed or exploited by gambling.

The Gambling Commission issues guidance to local authorities on operating the Act, and in particular, the principles to be applied by local authorities in exercising their functions under the Act – Guidance to licensing authorities (4th Edition) (with sections 20 and 27 updated and published February 2013).

The functions of licensing authorities with respect to gambling are:

- to license premises for gambling activities (casinos, adult gaming centres, betting shops, tracks, bingo premises, family entertainment centres);
- to consider notices given for the temporary use of premises for gambling;
- to grant permits for gaming and gambling machines in clubs and miners' welfare institutes;
- to regulate gaming and gaming machines in alcohol-licensed premises;
- to grant permits to family entertainment centres (FECs) for the use of certain lower stake gaming machines;
- to grant permits for prize gaming;
- to consider occasional use notices for betting at tracks; and
- to register small societies' lotteries.

Definitions

Responsible authorities

(a) a licensing authority in England and Wales in whose area the premises are wholly or partly situated;

(b) the Commission;

(c) the chief officer of police or chief constable for a police area in which the premises are wholly or partly situated;

(d) the fire and rescue authority for an area in which the premises are wholly or partly situated;

(e) the local planning authority, for an area in which the premises are wholly or partly situated;

(f) an authority which has functions by virtue of an enactment in respect of minimising or preventing the risk of pollution of the environment or of harm to human health in an area in which the premises are wholly or partly situated;

(g) a body which is designated in writing, by the licensing authority for an area in which the premises are wholly or partly situated, as competent to advise the authority about the protection of children from harm;

(h) Her Majesty's Commissioners of Customs and Excise; and

(i) any other person prescribed for the purposes of this section by regulations made by the Secretary of State (section 157).

premises licence – a licence which states that it authorises premises to be used for:

(a) the operation of a casino (a 'casino premises licence');

(b) the provision of facilities for the playing of bingo (a 'bingo premises licence');

(c) making Category B gaming machines available for use (an 'adult gaming centre premises licence');

(d) making Category C gaming machines available for use (a 'family entertainment centre premises licence'); or

(e) the provision of facilities for betting, whether by making or accepting bets, by acting as a betting intermediary or by providing other facilities for the making or accepting of bets (a 'betting premises licence') (section 150).

interested party – a person is an interested party in relation to a premises licence or in relation to an application for or in respect of a premises licence if, in the opinion of the licensing authority which issues the licence or to which the application is made, the person:

(a) lives sufficiently close to the premises to be likely to be affected by the authorised activities;

(b) has business interests that might be affected by the authorised activities; or

(c) represents persons who satisfy paragraph (a) or (b) (section 158).

provisional statement provides a level of assurance about the outcome of a subsequent premises licence application and provides the holder with some protection against representations when they make an application for a premises licence.

gambling means gaming, betting and participating in a lottery (section 3). Facilities for gambling means:

(a) inviting others to gamble in accordance with arrangements made by him;

(b) providing, operating or administering arrangements for gambling by others; or

(c) participating in the operation or administration of gambling by others,

but does not include providing an article other than a gaming machine to a person who intends to use it, or may use it, in the course of any of the above activities or making facilities for remote communication available for use by persons carrying on any of those activities, or persons gambling in response to those activities with certain conditions (section 5).

gaming means playing a game of chance for a prize.

game of chance includes a game that involves both an element of chance and an element of skill, a game that involves an element of chance that can be eliminated by superlative skill, and a game that is presented as involving an element of chance, but does not include a sport.

prize in relation to gaming (except in the context of a gaming machine) means money or money's worth, and includes both a prize provided by a person organising gaming and winnings of money staked (section 6).

casino is an arrangement whereby people are given an opportunity to participate in one or more casino games.

casino game means a game of chance which is not equal chance gaming (section 7).

equal chance gaming is equal chance gaming if it does not involve playing or staking against a bank, and the chances are equally favourable to all participants (section 8).

betting means making or accepting a bet on:

(a) the outcome of a race, competition or other event or process;
(b) the likelihood of anything occurring or not occurring; or
(c) whether anything is or is not true (section 9).

betting intermediary means a person who provides a service designed to facilitate the making or acceptance of bets between others. Acting as a betting intermediary is providing facilities for betting (section 13).

simple lottery is where persons are required to pay in order to participate in an arrangement, and in the course of the arrangement one or more prizes are allocated to one or more members of a class, and the prizes are allocated by a process which relies wholly on chance (section 14).

complex lottery is where persons are required to pay in order to participate in the arrangement and in the course of the arrangement one or more prizes are allocated to one or more members of a class and the prizes are allocated by a series of processes, and the first of those processes relies wholly on chance.

prize in relation to lotteries includes any money, articles or services – whether or not described as a prize, and whether or not consisting wholly or partly of money paid, or articles or services provided, by the members of the class among whom the prize is allocated (section 14).

All sections referred to above are of the Gambling Act 2005.

PREMISES LICENCES (GAMBLING)

References

Gambling Act 2005.
Gambling Act 2005 (Premises Licences and Provisional Statements) Regulations 2007.
Gambling Act 2005 (Mandatory and Default Conditions) (England and Wales) Regulations 2007.
Gambling Act 2005 (Operating Licence Conditions) Regulations 2007.
Gambling Act 2005 (Premises Licences) (Review) Regulations 2007.
Gambling (Premises Licence Fees) (England and Wales) Regulations 2007.
Gambling (Operating Licence and Single-Machine Permit Fees) Regulations 2006.

Extent

These provisions apply to England, Wales and Scotland.

Scope

Most functions under the Act are usually delegated to a Licensing Committee in England and Wales. Licensing authorities can determine an application for a

FC 3.13 Premises licences – Gambling Act 2005 and Gambling Act 2005 (Premises Licences and Provisional Statements) Regulations 2007

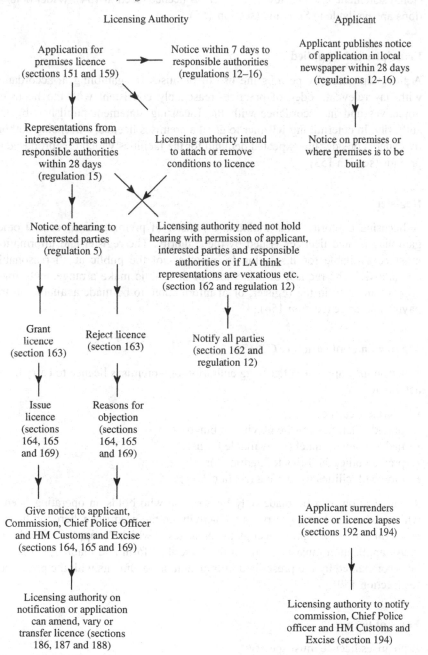

Licensing Authority   Applicant

Application for premises licence (sections 151 and 159) → Notice within 7 days to responsible authorities (regulations 12–16)

Applicant publishes notice of application in local newspaper within 28 days (regulations 12–16)

Representations from interested parties and responsible authorities within 28 days (regulation 15)

Licensing authority intend to attach or remove conditions to licence

Notice on premises or where premises is to be built

Notice of hearing to interested parties (regulation 5)

Licensing authority need not hold hearing with permission of applicant, interested parties and responsible authorities or if LA think representations are vexatious etc. (section 162 and regulation 12)

Grant licence (section 163)

Reject licence (section 163)

Notify all parties (section 162 and regulation 12)

Issue licence (sections 164, 165 and 169)

Reasons for objection (sections 164, 165 and 169)

Give notice to applicant, Commission, Chief Police Officer and HM Customs and Excise (sections 164, 165 and 169)

Applicant surrenders licence or licence lapses (sections 192 and 194)

Licensing authority on notification or application can amend, vary or transfer licence (sections 186, 187 and 188)

Licensing authority to notify commission, Chief Police officer and HM Customs and Excise (section 194)

Note
Sections refer to the Gambling Act 2005 and regulations to Gambling Act 2005 (Premises Licences and Provisional Statements) Regulations 2007 or Gambling Act 2005 (Proceedings of Licensing Committees and Sub-committees) (Premises Licences and Provisional Statements) (England and Wales) Regulations 2007.

premises licence, for the variation or transfer of a premises licence, for a provisional statement and a review of a premises licence (section 154). Wider delegations are available in Scotland (section 155).

Principles to be applied

A licensing authority permits the use of premises for gambling in accordance with any relevant codes of practice, reasonably consistent with the licensing objectives and in accordance with the Licensing statement published by the authority. In determining whether to grant a premises licence, a licensing authority cannot consider the expected demand for the facilities which it is proposed to provide (section 153).

Register

A licensing authority must maintain a register of premises licences and other gambling related licences issued by the authority. The register and information must be available for inspection by members of the public at all reasonable times, and at the request of a member of the public make arrangements for a copy of an entry in the register, or of information, to be made available on the payment of a fee (section 156).

Making an application (FC 3.13)

A person may apply to a licensing authority for a premises licence to carry on an activity to:

- operate a casino;
- provide facilities for the playing of bingo;
- make a gaming machine available for use;
- provide other facilities for gaming; or
- provide facilities for betting (section 37(1)).

An application may be made only by a person who holds an operating licence which authorises him to carry on the activity on the premises where the licence is sought and has a right to occupy the premises to which the application relates.

An application must be made in the prescribed form and manner, contain or be accompanied by the prescribed information or documents with the prescribed fee (section 159).

Form of licence

A premises licence must specify:

(a) the name of the person to whom it is issued;
(b) a home or business address of that person;
(c) the premises to which it relates;
(d) the activities for which it authorises the premises to be used;

(e) any condition attached by the licensing authority; and

(f) any exclusion of a default condition effected by the licensing authority.

It must also include a plan of the premises, and the date the licence will expire (section 151).

For the detail of the form and content of a premises licence see the Gambling Act 2005 (Premises Licences and Provisional Statements) Regulations 2007 as amended.

Note that a premises licence cannot refer to more than one activity except in the case of a track but special arrangements must be made for this (section 152).

Notice of application

Applications for a premises licence, variation, transfer, reinstatement of a premises licence and for a provisional statement must give notice of the application to each of the responsible authorities. The notice must be given within 7 days of the date on which the application is made.

The applicant must publish notice of his application in a local newspaper and display the notice on the exterior of the premises to which the application refers for 28 consecutive days starting on the day on which the application is made to the licensing authority.

If the premises has not yet been built the applicant must display the notice at a place which is as near as reasonably practicable to the premises or proposed premises where it can conveniently be read by members of the public.

Where a person fails to give proper notice of his application or publish the notice of application within the relevant period in the correct form or manner, the responsible authority concerned must make any representations to the licensing authority within 28 days of the day on which it receives the notice.

The licensing authority cannot grant the application until the correct notice has been given by the applicant.

An application or a notice must be made or given in writing which includes facsimile transmission or electronic mail with the prescribed fee (regulations 12 to 16 Gambling Act 2005 (Premises Licences and Provisional Statements) Regulations 2007).

Representations

An interested party or responsible authority may make representations in writing to the licensing authority concerning an application (section 161), within 28 days beginning on the date on which the application was made to the licensing authority (regulation 15 Gambling Act 2005 (Premises Licences and Provisional Statements) Regulations 2007).

Hearings

In this section, regulation references arc to the Gambling Act 2005 (Proceedings of Licensing Committees and Sub-committees) (Premises Licences and Provisional Statements) (England and Wales) Regulations 2007.

Requirement for hearing

A licensing authority must hold a hearing to consider an application for a premises licence if:

(a) an interested party or responsible authority has made representations about the application;

(b) the authority propose to attach a condition to the licence; or

(c) the authority propose to exclude a condition that would otherwise be attached to the licence.

However, a licensing authority may determine an application for a premises licence without a hearing:

(a) with the consent of the applicant, and any interested party or responsible authority who has made representations about the application;

(b) if the authority think that the representations are vexatious, frivolous, or will certainly not influence the authority's determination of the application.

If a licensing authority propose to determine an application without a hearing they shall as soon as is reasonably practicable notify any person who made representations (section 162).

The hearing must be held subject to the Gambling Act 2005 (Proceedings of Licensing Committees and Sub-committees) (Premises Licences and Provisional Statements) (England and Wales) Regulations 2007.

The hearing is to be held as soon as is reasonably practicable after the expiry of any period for representations (regulation 4).

Notice of hearing

The committee must give notice of any hearing specifying the date on which, the place at which and the time when the hearing is to take place. The notice must also state that the relevant committee will make available the relevant documents to any person who has made representations and the licensee (for transfers) so that it is received no later than 10 working days before the hearing is to be held (regulation 5).

Information and documents to accompany the notice of hearing

The notice of hearing must be accompanied by information in writing explaining the following:

(a) the consequences where a party informs the relevant committee that he does not wish to attend or be represented at the hearing, or fails to inform the relevant committee whether he wishes to attend or be represented at the hearing;

(b) the requirements imposed on the relevant committee in conducting a hearing;

(c) the consequences where a party has indicated that he wishes to attend or be represented at the hearing, but fails to attend or be represented at the hearing;

(d) the procedure to be followed at the hearing;

(e) the time limit and method, if any, by which a party should inform the relevant committee that he wishes to attend or address the hearing;

(f) the time limit and method, if any, by which a party should inform the relevant committee that he wishes to be assisted or represented by another person;

(g) the time limit and method, if any, by which a party should inform the licensing authority that he will want to call a witness to give evidence at the hearing, and the matters in relation to which he wishes that witness to give evidence;

(h) the time limit and method, if any, by which a party should inform the relevant committee that he wishes to withdraw any representations;

(i) the time limit and method, if any, by which a party should inform the relevant committee that he is willing to consent to the application being determined without a hearing;

(j) the matters, if any, on which the relevant committee considers at the time that it will want clarification at the hearing from a party.

The documents must be sent to the applicant, the licensee (for an application or a review) and a person who has made representations (where requested) (regulation 6).

Power to postpone

The committee may postpone a hearing to a specified date, or arrange for a hearing to be held on a date specified by the committee if it considers it necessary to enable it to consider any information or documents provided by any party in response to a notice, or at the hearing, or having regard to the ability of any party, person representing a party or witness to attend the hearing.

Where the committee has adjourned a hearing to a specified date or specified an additional date it must notify the parties of the new date, time and place for the hearing (regulation 7).

Proceedings of a relevant committee in conducting a hearing

The hearing must take place in public; however, the committee may direct that all or part of a hearing must be in private if it is satisfied that it is necessary in all the circumstances of the case, having regard to any unfairness to a party that is likely to result from a hearing in public and the need to protect as far as possible, the commercial or other legitimate interests of a party (regulation 8).

The committee must allow a party to attend a hearing and be assisted or represented by any person whether or not that person is legally qualified.

At the beginning of the hearing the committee must explain the procedure that it proposes to follow in conducting the hearing.

In conducting a hearing the committee must ensure that each party is given the opportunity to address the relevant committee or call witnesses to give evidence on any matter relevant to the application or review, or any representations made on the application or review.

The committee must also allow any party to question any other party or person representing a party and take into consideration documentary or other information in support of the application or representations produced by a party before the hearing; or at the hearing, with the consent of all the other parties attending the hearing (regulation 4).

Failure of parties to attend the hearing

The committee can proceed with a hearing in the absence of a party or a party's representative, if the party has:

(a) informed the committee that he does not intend to attend or be represented at the hearing;

(b) failed to inform the committee whether he intends to attend or be represented at the hearing; or

(c) left the hearing in circumstances enabling the committee reasonably to conclude that he does not intend to participate further.

If a party has indicated that he does intend to attend or be represented at the hearing, but fails to attend or be represented, the relevant committee may adjourn the hearing to a specified date if it considers it to be in the public interest, or proceed with the hearing in the party's absence.

Where the hearing proceeds in the absence of a party, the committee must consider the application or representations made by that party.

Where the committee adjourns the hearing to a specified date it must notify the parties of the date, time and place to which the hearing has been adjourned (regulation 10).

Exclusion of disruptive persons

The committee can require any person attending the hearing who in their opinion is behaving in a disruptive manner, to leave the hearing and can refuse to permit him to return, or permit him to remain or return only on such conditions as the committee specify.

Where a person is required to leave the hearing, the committee must permit him to submit in writing, before the end of the hearing, any information which he would have been entitled to give orally had he not been required to leave and take into account that information in reaching a determination of the application or review (regulation 11).

Procedure where a hearing is not to take place

Where the parties have notified the committee that they consent to the application or review being determined without a hearing the committee must notify all the parties that the hearing has been dispensed with, and determine the application or review (regulation 12).

Determination of an application or a review

Following a hearing the committee must determine the application or review within 5 working days after the hearing (regulation 13).

Power to extend time

The committee may extend the 5 day time limit for a specified period where it considers an extension to be in the public interest. Where the committee has extended the time limit it must give a notice of the extension to the parties stating the period of the extension and the reasons for it (regulation 14).

Record of proceedings

The committee must ensure that a record of the hearing is taken in a permanent and intelligible form and that the record is kept for 6 years from the date that the application or review is finally determined (regulation 15).

Irregularities and clerical mistakes

The committee may disregard any irregularity resulting from a failure to comply with a provision of the Regulations where that irregularity comes to its attention prior to it making a determination of the application or review. If the committee considers that any person may have been prejudiced by any such irregularity, it must take such steps that it considers necessary to remedy the consequences of the irregularity, before reaching its determination (regulation 16).

The committee may correct clerical mistakes in any document recording a determination of the committee, or errors arising in such a document from an accidental slip or omission (regulation 17).

Determination of application

On considering an application for a premises licence a licensing authority must grant it or reject it (section 163).

Where a licence is granted or rejected they must give notice of the grant to the applicant, the Commission, any person who made representations about the application, the chief officer of police and Her Majesty's Commissioners of Customs and Excise.

The licensing authority must issue a premises licence to the applicant with a summary of the terms and conditions of the licence in the prescribed form, and include reasons for conditions and if representations were made about the application the authority's response to the representations. Where the application is rejected, the reasons for objection must be given to the applicant, etc. (sections 164, 165, 169).

Conditions

Conditions cannot be imposed requiring membership in clubs and other bodies nor can conditions be included imposing limits on stakes, fees, winnings or prizes (sections 170 and 171). Conditions on licences are included in the Gambling Act 2005 (Mandatory and Default Conditions) (England and Wales) Regulations 2007.

Door supervision

Where a condition for door supervision is attached to a premises licence, if the person carrying out the guarding is required to hold a licence under the Private Security Industry Act 2001 authorising the guarding it becomes a condition of the premises licence (section 178).

Annual fee

Premises licences are subject to an annual fee (section 184).

Availability of licence

The licence must be kept on the premises it refers to and made available on request to a constable, an enforcement officer or an authorised local authority officer. It is an offence if he fails without reasonable excuse for failure to comply with these requirements and liable on summary conviction to a fine not exceeding level 2 on the standard scale (section 185).

Change of circumstance

Where the holder of a premises licence ceases to reside or attend at the address specified in the licence he must notify the licensing authority, and inform it of a home or business address at which he resides or attends accompanied by a fee.

Where notification is accompanied by the licence, the licensing authority must make the alteration to the information contained in the licence, and return the licence to the licensee.

It is an offence to fail to comply with these requirements and liable on summary conviction to a fine not exceeding level 2 on the standard scale (section 186).

Variation

The holder of a premises licence may apply to the licensing authority to vary the licence by:

(a) adding, amending or removing an authorised activity;
(b) amending another detail of the licence;
(c) adding, amending, excluding or removing a condition attached to the licence.

The application for variation must be accompanied by a statement of the variation sought, the licence to be varied, or both a statement explaining why it is not reasonably practicable to produce the licence, and an application for the issue of a copy of the licence.

In granting an application for variation a licensing authority must specify a time when the variation shall begin to have effect, and may make transitional provision (section 187).

Transfer

A person may apply to a licensing authority for a premises licence to be transferred to him. An application for transfer must specify the time when the transfer is to take effect, and be accompanied by a written statement by the licensee consenting to the transfer.

A licensing authority shall grant an application for transfer unless they think it would be wrong to do so having regard to any representations.

On the grant of an application for the transfer of a premises licence, the licensing authority must alter the licence so that the applicant for the transfer becomes the licensee, specify in the licence the time when the transfer takes effect and make any other alteration of the licence as appears to them to be required (section 188).

An application for transfer must be accompanied by the licence, or both a statement explaining why it is not reasonably practicable to produce the licence, and an application by the licensee for the issue of a copy of the licence.

If an application for transfer states that the applicant has failed to contact the licensee having taken all reasonable steps to do so, the licensing authority must disregard the requirement for a written statement by the licensee consenting to the transfer and take all reasonable steps to notify the licensee, or notify the applicant of their determination and the reasons for it.

At the request of the applicant the licence shall have effect as if the applicant for transfer were the licensee during the period beginning with the receipt of the application for transfer by the licensing authority, and ending with the determination of the application by the licensing authority (section 189).

Copy of licence

Where a premises licence issued, or a summary given, is lost, stolen or damaged, the licensee may apply to the licensing authority for a copy accompanied by the prescribed fee.

If satisfied that the licence or summary to which the application relates has been lost, stolen or damaged, and where the licence or summary has been lost or stolen, that the loss or theft has been reported to the police, after granting the application, the authority must issue a copy of the licence or summary to the applicant certified by the authority as a true copy, and the form in which the licence had effect before the loss, theft or damage (section 190).

Surrender

A premises licence shall cease to have effect if the licensee notifies the licensing authority of his intention to surrender the licence, and gives the licensing authority either the licence, or a written statement explaining why it is not reasonably practicable to produce the licence.

The licensing authority must notify the Commission, and the chief officer of police and Her Majesty's Commissioners of Customs and Excise (section 192).

Revocation for failure to pay fee

Where the holder of a premises licence fails to pay the annual fee the licensing authority must revoke the licence but the licensing authority may ignore this if they think that a failure to pay is attributable to administrative error (section 193).

Lapse

A premises licence will lapse if the licensee dies, the licensee becomes, in the opinion of the licensing authority, incapable of carrying on the licensed activities by reason of mental or physical incapacity, the licensee becomes bankrupt or the licensee's estate sequestrated and in the case of a company or other legal body it ceases to exist, or goes into liquidation.

Where the premises licence has lapsed, they must notify the Commission, the chief officer of police and Her Majesty's Commissioners of Customs and Excise (section 194).

Reinstatement

For 6 months after the lapse a person may apply to the licensing authority for the licence to be reinstated with the applicant as the licensee.

The licensing authority must grant an application for reinstatement unless they think it would be wrong to do so having regard to any representations.

The licensing authority must alter the licence so that the applicant for reinstatement becomes the licensee, specify in the licence that the reinstatement takes effect at the time when the application is granted, and make any other alteration of the licence as appears to them to be required (section 195).

The application for reinstatement of a premises licence must be accompanied by the licence, or both a statement explaining why it is not reasonably practicable to produce the licence, and an application for the issue of a copy of the licence.

The licence remains valid while the application is being considered by the licensing authority (section 196).

Review

In this section, reference to sections are in the Gambling Act 2005 and regulations to the Gambling Act 2005 (Premises Licences) (Review) Regulations 2007.

These Regulations make provision about the form of, and procedure for, applications for a review of a premises licence.

Application for review

A responsible authority or interested party may apply to the licensing authority for a review by the authority of a premises licence.

The application must be made in the prescribed form and manner, specify the grounds on which the review is sought, and contain or be accompanied by the prescribed information or documents (section 197).

Giving of notice of an application

A person making an application must give notice (7 days) of the application to the licence holder and the responsible authorities. The application must be in the specified form including the period of 28 days when representations about the application may be made to the licensing authority by the person who holds the premises licence, a responsible authority, or an interested party in relation to the premises (regulation 4). An application, and associated notices must be made or given in writing which includes facsimile transmission or electronic mail (regulation 12).

Publication of notice of an application and notice of review by the licensing authority

The authority must publish notice of the application

(a) either:
    (i) in a local newspaper or, if there is none, a local newsletter, circular or similar document, circulating within the licensing authority's area on at least one occasion within 10 working days after the application is made to the authority; or
    (ii) on the licensing authority's internet website for at least 28 consecutive days; and
(b) by displaying the notice at a place, for 28 consecutive days:
    (i) which is as near as reasonably practicable to the premises; and
    (ii) where it can conveniently be read by members of the public (regulation 5 and 8).

Rejection of application

A licensing authority may reject an application for the review of a premises licence if they think that the grounds on which the review is sought:

(a) do not raise an issue relevant to the principles to be applied;
(b) are frivolous or vexatious;
(c) will certainly not cause the authority to wish to:
    • revoke the licence;
    • suspend the licence for a specified period not exceeding 3 months;

- exclude a condition attached to the licence or remove or amend an exclusion;
- add, remove or amend a condition;

(e) are substantially the same as the grounds specified in an earlier application; or

(f) are substantially the same as representations made in relation to the application for the premises licence.

A licensing authority shall consider the length of time that has elapsed since the making of the earlier application or since the making of the representations.

If a licensing authority consider that paragraphs (a) to (f) apply to some but not all of the grounds on which a review is sought, they may reject the application (section 198).

If the licensing authority do not reject the application they must grant it (section 199).

Initiation of review by licensing authority

A licensing authority may review the use of a premises licences of a particular class and, in particular, arrangements made by licensees to ensure compliance with the conditions attached.

A licensing authority may review any matter connected with the use of premises in reliance on a premises licence if the authority have reason to suspect that the premises may have been used in purported reliance on a licence but not in accordance with a condition of the licence, or for any reason (which may relate to the receipt of a complaint about the use of the premises) think that a review would be appropriate.

Before reviewing a premises licence the licensing authority must give notice of their intention to hold the review to the licensee, and publish notice of their intention to hold the review (section 200).

Review (FC 3.14)

As soon as the period for representations has expired, the licensing authority must review the premises licence.

The purpose of the review will be to consider whether to:

(a) revoke the licence;

(b) suspend the licence for a specified period not exceeding 3 months;

(c) exclude a condition attached to the licence or remove or amend an exclusion;

(d) add, remove or amend a condition;

specifying the time at which the action will take effect.

The licensing authority must hold a hearing about the review unless the applicant for the review, and each person who has made representations about the review have consented to the conduct of the review without a hearing, or the licensing authority think that each representation made about the review is frivolous, vexatious, or will certainly not influence the review.

The licensing authority must consider any representations and any grounds specified in the application for the review (sections 201 and 202).

As soon as possible after completion of a review, the licensing authority must give notice of their decision on the review to the licensee, the applicant for the review, the Commission, any person who made representations, the chief officer of police and Her Majesty's Commissioners of Customs and Excise.

The notice must be in the prescribed form, and give the authority's reasons for their decision (section 203).

Failure to give proper notice of an application

Where a person fails to give proper notice or publish a proper notice of an application within the requirements of these Regulations as to the form and manner in which it is to be given, the person holding the licence may make representations about the application within 28 days of receiving the notice. The licensing authority cannot grant the application until notice has been given by the applicant in accordance with the regulations and the 28 day period has elapsed (regulation 6 and 7).

There are similar provisions for failure to properly publish notice of intention to hold a review (regulation 10).

Provisional statements

Application

A person may make an application for a provisional statement about premises:

(a) that he expects to be constructed;
(b) that he expects to be altered; or
(c) that he expects to acquire a right to occupy.

An application for a provisional statement must include plans and other information about the construction, alteration or acquisition (section 204).

Where a licensing authority issues a provisional statement or an application is made for a premises licence in respect of the premises, the licensing authority must disregard any representations made in relation to the application for the premises licence unless they think that the representations address matters that could not have been addressed in representations in relation to the application for the provisional statement, or reflect a change in the applicant's circumstances.

The licensing authority may refuse the application, or grant it on terms or conditions not included in the provisional statement but only on consideration of the above matters (section 205).

Rights of appeal

Where a licensing authority reject an application the applicant may appeal.

FC 3.14 Review of a premises licence – Gambling Act 2005 and Gambling Act 2005 (Premises Licences) (Review) Regulations 2007.

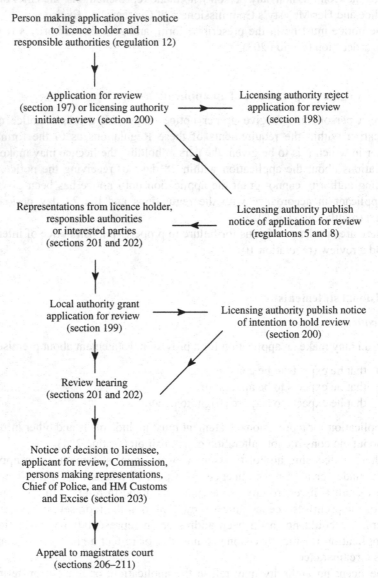

Person making application gives notice
to licence holder and
responsible authorities (regulation 12)

Application for review
(section 197) or licensing authority
initiate review (section 200)

Licensing authority reject
application for review
(section 198)

Representations from licence holder,
responsible authorities
or interested parties
(sections 201 and 202)

Licensing authority publish
notice of application for review
(regulations 5 and 8)

Local authority grant
application for review
(section 199)

Licensing authority publish notice
of intention to hold review
(section 200)

Review hearing
(sections 201 and 202)

Notice of decision to licensee,
applicant for review, Commission,
persons making representations,
Chief of Police, and HM Customs
and Excise (section 203)

Appeal to magistrates court
(sections 206–211)

Note
References to sections are Gambling Act 2005 and regulations are to the Gambling Act 2005 (Premises Licences) (Review) Regulations 2007.

Where a licensing authority grant an application a person who made represen-tations in relation to the application, and the applicant may appeal.

Where a licensing authority take action as a result of a review of a premises licence, or determine to take no action as a result of a review the licensee, a person who made representations in relation to the review, the person who applied for the review and the Commission may appeal.

Where a licensing authority take action or make a determination regarding transfer the licensee or the applicant for transfer may appeal (section 206).

Appeals are heard in the magistrates' court. The magistrates' court may dismiss the appeal, substitute for the decision appealed against any decision that the licensing authority could have made, send the case to the licensing authority to decide in accordance with a direction of the court or make an order about costs (section 207).

Any decision of the licensing authority will have no effect while an appeal could be brought, or has been brought and has not yet been either finally deter-mined or abandoned. However, the licensing authority making a determination may direct that the action can be taken in which case the magistrates' court or sheriff determining an appeal may make any order that it or he thinks appropri-ate (section 208).

A party to an appeal to the magistrates' court or sheriff may appeal on a point of law to the High Court or to the Court of Session (section 209).

Types of licence

Gaming machines

An adult gaming centre premises licence authorises the holder to make available for use on the premises up to 20 per cent Category B machines which are avail-able for use on the premises and to make any number of Category C or D gaming machines available for use on the premises.

Family entertainment centre

A family entertainment centre premises licence authorises the holder to make any number of Category C and D gaming machines available for use on the premises.

Bingo licences

A bingo premises licence authorises the holder to make available for use on the premises a number of Category B gaming machines not exceeding 20 per cent of the total number of gaming machines which are available for use on the premises and any number of Category C and D gaming machines available for use on the premises.

Betting licences

A betting premises licence authorises the holder to make up to four gaming machines, each of which must be of Category B, C or D, available for use (section 172).

Virtual gaming

A casino premises licence, and a betting premises licence authorises the holder to make facilities available for betting on the outcome of a virtual game, race, competition or other event or process (section 173).

Casino licences

A casino premises licence for a **regional casino** using at least 40 gaming tables authorises the holder to make gaming machines available for use on the premises provided that each gaming machine is of Category A, B, C or D, and the number of gaming machines is not more than 25 times the number of gaming tables used in the casino, and no more than 1250.

A casino premises licence for a **large casino** using at least one gaming table, or for a regional casino using fewer than 40 gaming tables, authorises the holder to make gaming machines available for use on the premises provided that each gaming machine is of Category B, C or D, and the number of gaming machines is not more than 5 times the number of gaming tables used in the casino, and no more than 150.

A casino premises licence for a **small casino** using at least one gaming table authorises the holder to make gaming machines available for use on the premises provided that each gaming machine is of Category B, C or D, and the number of gaming machines is not more than twice the number of gaming tables used in the casino, and no more than 80 (Gambling Act 2005 (Gaming Tables in Casinos) (Definitions) Regulations 2009).

A casino premises licence may be issued only in respect of a regional casino, a large casino or a small casino.

A casino premises licence authorises the holder to use the premises to make available any number of games of chance other than casino games and to use the premises for the provision of facilities for bingo, betting or both (bingo is not authorised in a small casino) (section 174).

CASINO LICENCE APPLICATIONS

References

Gambling Act 2005 (Gaming Tables in Casinos) (Definitions) Regulations 2009.
Gambling (Inviting Competing Applications for Large and Small Casino Premises Licences) Regulations 2008.
Gambling (Geographical Distribution of Large and Small Casino Premises Licences) Order 2008.
Categories of Casino Regulations 2008.

Extent

These provisions apply to England, Wales and Scotland.

Applications for a casino licence (Gambling Act 2005 schedule 9)

The Gambling (Inviting Competing Applications for Large and Small Casino Premises Licences) Regulations 2008 and the Gambling (Geographical Distribution of Large and Small Casino Premises Licences) Order 2008 together provide the method of application and geographical spread and number of casinos throughout the country.

Where a licensing authority are able to grant one or more casino licences, but not all, of the competing applications the licensing authority must first consider whether they would grant a premises licence. They must disregard whether any of the other competing applications is more deserving of being granted. Each competing applicant is an interested party in relation to each of the other competing applications, and the application is subject to the requirements of a premises licence (paragraph 3 and 4).

If a licensing authority determine that they would grant a number of competing applications greater than the number in the casino limits in the Gambling Act 2005 section 175 they can do so.

The licensing authority:

(a) must determine which of the competing applications would, in the authority's opinion, be likely if granted to result in the greatest benefit to the authority's area;
(b) may enter into a written agreement with an applicant, whether as to the provision of services in respect of the authority's area or otherwise;
(c) may determine to attach conditions to any licence issued so as to give effect to an agreement entered into under paragraph (b); and
(d) may have regard to the effect of an agreement entered into under paragraph (b) in making the determination specified in paragraph (a).

Having determined to grant one or more applications the authority must grant that application and reject the other competing applications (paragraph 5).

The applicant can appeal against the decision of the licensing authority (paragraph 8).

This procedure also applies to provisional statements (paragraph 9 and 10).

Resolution not to issue casino licences

A licensing authority may resolve not to issue casino premises licences.

A resolution must apply to the issue of casino premises licences generally, must specify the date on which it takes effect and last for 3 years. The Authority may revoke the resolution at any time. The resolution must be published by being included in a statement (section 166).

TEMPORARY USE OF PREMISES FOR GAMBLING

References

Gambling Act 2005.
Gambling Act 2005 (Temporary Use Notices) Regulations 2007.

Extent

These provisions apply to England and Wales. Similar provisions apply in Scotland.

Scope

Premises can be used temporarily for gambling if a temporary use notice is in effect and the activity is carried on in accordance with the terms of the notice.

Nature and form of notice

A temporary use notice is a notice given by the holder of an operating licence, and stating his intention to carry on one or more specified prescribed activities. A notice given by a person may specify an activity only if the person's operating licence authorises him to carry on the activity (section 215).

A temporary use notice must:

- be in the prescribed form;
- specify the activity to be carried on in reliance on the notice;
- specify the premises on which the activity is to be carried out on;
- specify the period of time during which the notice is to have effect;
- specify times of day during that period at which the activity is to be carried on;
- specify periods during the previous 12 months during which a temporary use notice has had effect in respect of the premises or any part of the premises;
- specify date on which the notice is given;
- contain any other prescribed information (section 216).

A temporary use notice will have effect during the period specified in the notice (section 217).

A set of premises may not be the subject of temporary use notification for more than 21 days in a period of 12 months. More than one temporary use notice can be in place if the aggregate of the periods for which the notices have effect does not exceed 21 days.

If a temporary use notice is given to a licensing authority and the period would be exceeded, the licensing authority must give a counter notice stating the temporary use notice is not valid.

Provisions exist for use of the premises for a period which would not exceed the time limits. This would be included in the counter notice.

Where there is a choice as to which part of the specified period to exclude, the licensing authority must consult the person who gave the temporary use notice before giving a counter notice (section 218).

FC 3.15 Temporary use of premises for gambling – Gambling Act 2005 and Gambling Act 2005 (Temporary Use Notices) Regulations 2007

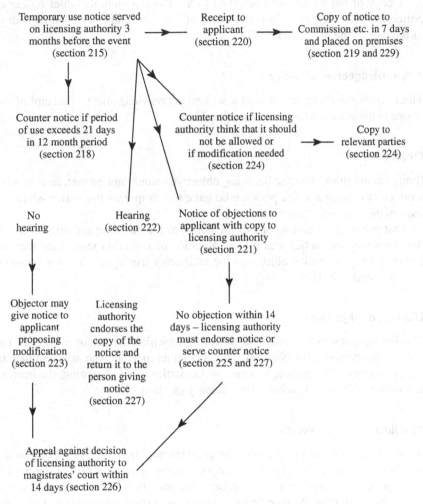

Temporary use notice served on licensing authority 3 months before the event (section 215) → Receipt to applicant (section 220) → Copy of notice to Commission etc. in 7 days and placed on premises (section 219 and 229)

Counter notice if period of use exceeds 21 days in 12 month period (section 218)

Counter notice if licensing authority think that it should not be allowed or if modification needed (section 224) → Copy to relevant parties (section 224)

No hearing

Hearing (section 222)

Notice of objections to applicant with copy to licensing authority (section 221)

Objector may give notice to applicant proposing modification (section 223)

Licensing authority endorses the copy of the notice and return it to the person giving notice (section 227)

No objection within 14 days – licensing authority must endorse notice or serve counter notice (section 225 and 227)

Appeal against decision of licensing authority to magistrates' court within 14 days (section 226)

Notes
1. Reference to sections are to the Gambling Act 2005 and to regulations are to the Gambling Act 2005 (Temporary Use Notices) Regulations 2007.
2. Temporary use notice proceedings must be dealt with within six weeks of receiving the notice (section 228).
3. A licensing authority must maintain a register of temporary use notices (section 234).

Giving notice (FC 3.15)

A temporary use notice must be given to the licensing authority. It must be given 3 months before the use is proposed.

The notice must be accompanied by a copy of the notice and the relevant fee.

A copy of the notice must be given to the Commission, the chief officer of police and Her Majesty's Commissioners of Customs and Excise within 7 days of the date of the notice (section 219).

Acknowledgement of notice

The licensing authority must send a written acknowledgement of receipt of the notice to the person who gave it (section 220).

Objections

If any person thinks that the licensing objectives would not be met, he may give a notice of objection to the person who gave the temporary use notice within 14 days of the temporary use notice.

That person must give a copy of the notice to the licensing authority to which the temporary use notice was given. A notice of objection must state that the person giving the notice objects to the temporary use notice, and the person's reasons (section 221).

Hearing of objections

The licensing authority must hold a hearing at which the person who gave the notice, the person who objected, and any other person who was entitled to receive a copy of the notice, however if the parties agree in writing, the hearing need not be heard if a hearing is unnecessary (section 222).

Modification by agreement

If the hearing has not taken place, the objector may by notice in writing to the person who gave the temporary use notice propose a modification of the notice. If the person who gave the notice accepts the modification he must give a new notice, incorporating the modification and the objection will be treated as withdrawn (section 223).

Counter notice

If the licensing authority think that the temporary use notice should not be allowed or should have effect only with modification, the authority may give a counter notice to the person who gave the notice.

A counter notice may provide for the temporary use notice:

(a) not to have effect;
(b) to have effect only for a specified activity;

(c) to have effect only for an activity carried on during a specified period of time or at specified times of day;

(d) to have effect with a specified condition.

A counter notice must be in the prescribed form, contain the prescribed information, be given as soon as is reasonably practicable and state the licensing authority's reasons for giving it.

A copy must be given to the relevant parties (section 224).

Dismissal of objection

Where a counter notice has not been served, the licensing authority must give notice of their determination to the person who gave the temporary use notice, and each person who received a copy of the notice (section 225).

Appeal

The person who gave the temporary use notice, and a person who was entitled to receive a copy of the temporary use notice may appeal against the decision of the licensing authority in the magistrates' court within 14 days of the appellant receiving notice of the decision. The magistrates' court may dismiss the appeal, direct the licensing authority to take action of a specified kind, return the case to the licensing authority, action the direction of the court or make an order about costs.

A party to an appeal may bring a further appeal to the High Court on a point of law (section 226).

Endorsement of notice

If no objection is made within 14 day period, the licensing authority must endorse the copy submitted and return the endorsed copy to the person giving the notice. If a notice of objection is pending when the 14 day period expires, the licensing authority must either give a counter notice or endorse the copy as above (section 227).

Consideration by licensing authority: timing

Temporary use notice proceedings must be dealt with within 6 weeks of receiving the notice (proceedings are considering whether to give a notice of objection, holding a hearing or agreeing to dispense with a hearing and giving a counter notice) (section 228).

Availability of notice

A person who gives a temporary use notice must arrange for a copy of the notice to be displayed prominently on the premises at any time when an activity is being carried on and arrange for the notice endorsed by the licensing authority to be produced on request to a constable, an officer of customs and excise, an enforcement officer or an authorised local authority officer.

It is an offence to fail to comply liable on summary conviction to a fine not exceeding level 2 on the standard scale (section 229).

Withdrawal of notice

A temporary use notice can be withdrawn by person who makes notification and the notice will have no effect and no further proceedings will take place (section 230).

Register

A licensing authority must maintain a register of temporary use notices given to them, make the register and information available for inspection by members of the public at all reasonable times, and make arrangements for the provision of a copy of an entry in the register, or of information, to a member of the public on request on payment of a fee (section 234).

The Gambling Act 2005 (Temporary Use Notices) Regulations 2007 makes provision for the form and content of notices, fees to be charged and the way in which notices are given to licensing authorities.

GAMING AND GAMING MACHINES IN ALCOHOL-LICENSED PREMISES

References

Gambling Act 2005 schedules 12 and 13.

Gambling Act 2005 (Licensed Premises Gaming Machine Permits) (England and Wales) Regulations 2007 *(these Regulations make provision for licensed premises gaming machine permits including the form of a permit and various fees).*

Gambling Act 2005 (Club Gaming and Club Machine Permits) Regulations 2007 *(these Regulations make provision for club gaming and club machine permits, including the application form, the form of each permit, and various fees).*

Gambling Act 2005 (Exempt Gaming in Alcohol-Licensed Premises) Regulations 2007 *(these Regulations prescribe the maximum amounts that may be staked and won in equal chance games).*

Gambling Act 2005 (Gaming in Clubs) Regulations 2007 *(these Regulations prescribe the kinds of gaming for which a club may be established or conducted to provide facilities, in order to be a members' club or a commercial club).*

Gambling Act 2005 (Exempt Gaming in Clubs) Regulations 2007 *(these Regulations prescribe the maximum amounts that may be staked and won and the maximum participation fee that may be charged in games of equal chance).*

Gambling Act 2005 (Club Gaming Permits) (Authorised Gaming) Regulations 2007 *(these Regulations prescribe the games of chance that may be played under a club gaming permit and the maximum amount that may be charged by way of a participation fee in those games).*

Gambling Act 2005 (Gaming Machines) (Definitions) Regulations 2007 *(these Regulations assign meanings to the terms 'dual-use computer' and 'domestic computer').*

Categories of Gaming Machines Regulations 2007 *(these Regulations provide the necessary definitions for each category, A to D).*

Extent

These provisions apply to England, Wales and Scotland.

Club and gaming permits – schedule 12 Gambling Act 2005

A members' club or miners' welfare institute can apply to a licensing authority for a club gaming permit and a club machine permit. A commercial club can apply for a club machine permit (paragraph 1).

The application must:

(a) be made to a licensing authority in whose area the premises are wholly or partly situated;

(b) specify the premises for which the permit is sought;

(c) be made in the prescribed form and manner;

(d) contain or be accompanied by the prescribed information and documents; and

(e) be accompanied by the prescribed fee (£100 or £200) (paragraph 1).

The applicant must within the prescribed time send a copy of the application and of any accompanying documents to the Commission, and the chief officer of police. Failure to comply with these requirements will mean any permit granted has no effect (paragraphs 2 and 3).

A person who receives a copy of an application for a permit may object to the application within the prescribed period of time and in the prescribed manner (paragraph 4).

The licensing authority must consider the application and grant or refuse it (paragraph 5).

The form, manner and fees for permit applications is included in the Gambling Act 2005 (Club Gaming and Club Machine Permits) Regulations 2007.

Refusal

The licensing authority may refuse an application for a permit only on one or more of the following grounds:

(a) that the applicant is not a members' club, a miners' welfare institute, or a commercial club as the case may be;

(b) that the premises on which the applicant conducts its activities are used wholly or mainly by children, by young persons or by both;

(c) that an offence, or a breach of a condition of a permit, has been committed in the course of gaming activities carried on by the applicant;

(d) that a permit held by the applicant has been cancelled in the last 10 years; or

(e) that an objection to the application has been made (paragraph 6).

Before refusing an application for a permit a licensing authority must hold a hearing to consider the application and any objection made.

But a licensing authority may dispense with the requirement for a hearing with the consent of the applicant, and any person who has made an objection (paragraph 7).

Where a licensing authority grant an application for a permit they must give notice of their decision and, where an objection was made of the reasons for it to the applicant, to the Commission, and the chief officer of police and issue the permit to the applicant (paragraph 8).

Where a licensing authority reject an application for a permit they must give notice of their decision and the reasons for it to the applicant, the Commission and the chief officer of police (paragraph 9).

If an applicant for a permit is the holder of a club premises certificate he need not make application for a club gaming permit and a club machine permit but the licensing authority must give notice of their decision as above (this does not apply to objectors).

The authority to whom an application is made shall grant it unless they think:

(a) that the applicant is established or conducted wholly or mainly for the purposes of the provision of facilities for gaming, other than gaming of a prescribed kind, and also provides facilities for gaming of another kind;

(b) that a club gaming permit or club machine permit issued to the applicant has been cancelled in the last 10 years.

This paragraph does not apply to Scotland (paragraph 10).

A permit must be in the prescribed form and must specify:

(a) the name of the club or institute in respect of which it is issued;
(b) the premises to which it relates;
(c) whether it is a club gaming permit or a club machine permit;
(d) the date on which it takes effect; and
(e) other information as may be prescribed.

The permit is valid for 10 years (paragraph 17).

If the application for the permit was made with reference to a club premises certificate, the permit must also identify the certificate (paragraph 11).

The permit holder must keep the permit on the premises (paragraph 12).

It is an offence for the occupier of premises to which a permit relates to fail to produce the permit on request for inspection by a constable, or an enforcement officer and is liable on summary conviction to a fine not exceeding level 2 on the standard scale (paragraph 13).

An annual fee is payable to the licensing authority (paragraph 14).

The permit holder must apply for changes to the permit together with the prescribed fee, and either the permit, or a statement explaining why it is possible to produce the permit. The licensing authority must issue a varied copy of the permit. But if the authority think that they would refuse an application for the permit were it made anew, they can refuse the application for variation, and cancel the permit (paragraph 15).

Where a permit is lost, stolen or damaged, the holder may apply to the issuing licensing authority for a copy with the prescribed fee. The permit must be issued if the licensing authority is satisfied that the permit has been lost, stolen or damaged, and where lost or stolen, that the loss or theft has been reported to the police (paragraph 16).

The permit lapses if the holder ceases to be a member of the relevant club (paragraph 18).

A permit ceases when surrendered by notice given by the holder to the licensing authority (paragraph 19).

The licensing authority must notify the Commission and the chief officer of police where it believes that the permit has lapsed or it receives notice of surrender (paragraph 20).

Cancelling a permit

The licensing authority can cancel it if it thinks that the premises on which the holder of the permit conducts its activities are used wholly or mainly by children, by young persons or by both, or that an offence, or a breach of a condition of a permit, has been committed in the course of gaming activities carried on by the permit holder.

Before cancelling a permit the licensing authority must:

(a) give the holder of the permit at least 21 days' notice of the authority's intention to consider cancelling the permit;
(b) consider any representations made by the holder;
(c) hold a hearing if the holder requests one; and
(d) comply with any prescribed requirements for the procedure to be followed in considering whether to cancel a permit.

If a licensing authority cancels a permit they must give notice of the cancellation and the reasons for it to the holder, the Commission and the chief officer of police. The cancellation of a permit does not take effect for 21 days or until any appeal brought has been determined (paragraph 21).

The licensing authority must cancel the permit if the holder fails to pay the annual fee, but not if they think that a failure to pay is an administrative error (paragraph 22).

Forfeiture

The court may order forfeiture of the permit if a holder or an officer of the permit holder is convicted of an offence. The court must notify the licensing authority of the order (paragraph 23).

Renewal

An application for the renewal of a permit can only be made between 3 months and 6 weeks before its expiry (paragraph 24).

Appeals

Where a licensing authority:

- rejects an application for the issue or renewal of a permit, the applicant may appeal;
- grants an application for the issue or renewal of a permit in relation to which an objection was made, the person who made the objection may appeal;
- cancels a permit, the holder may appeal;
- determines not to cancel a permit, any person who made representations to the authority in connection with their consideration whether to cancel the permit may appeal.

Appeals are to the magistrates' court within 21 days of the appellant receiving notice of the decision against which the appeal is brought.

The magistrates' court may:

(a) dismiss the appeal;
(b) substitute for the decision appealed against any decision that the licensing authority could have made;
(c) restore a permit;
(d) return the case to the licensing authority to decide in accordance with a direction of the court;
(e) make an order about costs (paragraph 25).

Register

The licensing authority must maintain a register of permits issued by the authority together with prescribed information and make the register and information available for inspection by members of the public at all reasonable times, and make arrangements for the provision of a copy of an entry in the register, or of information, to a member of the public on request who must pay the prescribed fee (paragraph 26).

**Licenced premises gaming machines permits – schedule 13
Gambling Act 2005**

A person who applies to a licensing authority for an on-premises alcohol licence or who holds an on-premises alcohol licence may apply to for a **licensed premises gaming machine permit** but not if a premises licence is in effect (paragraph 1).

An application for a permit must:

(a) be made in the form and manner required by the licensing authority;
(b) specify the premises;
(c) specify the number and category of gaming machines in respect of which the permit is sought;
(d) contain or be accompanied by such other information or documents as the licensing authority may direct; and
(e) be accompanied by the prescribed fee (paragraph 2).

On considering the licensing objectives regarding the application for a permit the licensing authority must:

(a) grant the application;
(b) refuse the application; or
(c) grant it in respect of:
  (i) a smaller number of machines than that specified in the application;
  (ii) a different category of machines from that specified in the application; or
  (iii) both (paragraph 4).

On granting an application, the licensing authority must issue a permit to the applicant.

If refused, the licensing authority must notify the applicant of the refusal, and the reasons for it (paragraph 5).

The licensing authority may grant an application only if the applicant holds an on-premises alcohol licence.

The licensing authority cannot refuse an application, or grant an application in respect of a different category or smaller number of gaming machines than that specified in the application, unless they have:

(a) notified the applicant of their intention to refuse the application, or grant the application in respect of:
  (i) a smaller number of machines than that specified in the application;
  (ii) a different category of machines from that specified in the application; or
  (iii) both, and
(b) given the applicant an opportunity to make oral or written representations or both (paragraph 6).

A permit must be in the prescribed form and must specify:

(a) the person to whom it is issued;
(b) the premises;
(c) the number and category of gaming machines which it authorises; and
(d) the date on which it takes effect (paragraph 7).

The holder of a permit must keep it on the premises (paragraph 8).

The permit holder must pay an annual fee (paragraph 9).

It is an offence to fail to produce the permit on request for inspection by a constable, an enforcement officer or an authorised local authority officer and liable on summary conviction to a fine not exceeding level 2 on the standard scale (paragraph 10).

Lost, stolen or damaged permits can be copied on payment of a relevant fee.

The licensing authority must consider an application and grant it if satisfied that the permit has been lost, stolen or damaged, and where the permit has been lost or stolen, that the loss or theft has been reported to the police (paragraph 11).

A permit shall cease to have effect:

(a) if an on-premises alcohol licence ceases to have effect with respect to the premises to which it relates, or the permit holder ceases to be the holder of an on-premises alcohol licence (paragraph 13);

(b) if the permit holder gives to the licensing authority notice of surrender, and either the permit, or a statement explaining why it is not reasonably practicable to produce the permit (paragraph 14).

Cancellation, variation and transfer

The licensing authority may vary the number or category (or both) of gaming machines authorised by the permit on application from the permit holder (paragraph 15).

The licensing authority may cancel the permit, or may vary the number or category (or both) of gaming machines authorised by it, if they think that:

(a) it would not be reasonably consistent with pursuit of the licensing objectives for the permit to continue to have effect;

(b) gaming has taken place on the premises in purported reliance on the permit but otherwise than in accordance with the permit or a condition of the permit;

(c) the premises are mainly used or to be used for making gaming machines available; or

(d) an offence under this Act has been committed on the premises.

Before cancelling or varying a permit a licensing authority must give the permit holder at least 21 days' notice of the authority's intention to consider cancelling or varying the permit, consider any representations made by the holder, and hold a hearing if the holder requests one.

If the licensing authority cancels or varies a permit they must give notice of the cancellation or variation and the reasons for it to the permit holder and the Commission.

The cancellation or variation of a permit shall not take effect for 21 days without an appeal being brought, or until any appeal brought has been determined (paragraph 16).

Failure to pay the annual fee will result in the licensing authority cancelling it, but not if they think that a failure to pay is attributable to administrative error (paragraph 17).

Where a permit holder, or the officer of a permit holder, is convicted of a relevant offence the court may order forfeiture of the permit. The court must notify the licensing authority of the forfeiture (paragraph 18).

A person may apply for the transfer of a permit to him and must supply the permit, or statement explaining why it is not reasonably practicable to produce the permit with the application (paragraph 19).

Appeals

The applicant for or holder of a permit may appeal if the licensing authority:

(a) rejects an application for a permit;

(b) grants an application for a permit in respect of a smaller number of machines than that specified in the application or a different category of machines from that specified in the application (or both); or

(c) gives a notice that the permit is to be cancelled.

The appeal must be made to the magistrates' court within 21 days of receiving notice of the decision against which the appeal is brought.

The magistrates' court may:

(a) dismiss the appeal;
(b) substitute for the decision appealed against any decision that the licensing authority could have made;
(c) restore a permit;
(d) return the case to the licensing authority to decide in accordance with a direction of the court;
(e) make an order about costs (paragraph 21).

The licensing authority must maintain a register of permits, make the register and information available for inspection by members of the public at all reasonable times, and make arrangements for the provision of a copy of an entry in the register, or of information, to a member of the public on request on payment of the relevant fee (paragraph 22).

Offences

Making machine available for use

It is an offence to make a gaming machine available for use by another without an operating licence except:

• in family entertainment centres;
• where no prize or limited prize is made available.

The requirements of club gaming permits, club machine permits, automatic entitlement to gaming machine permits and licensed premises gaming permits still apply (section 242).

Linked machines

It is an offence to make a gaming machine ('the first gaming machine') available for use by another, and the amount or value of a prize available through use of the first gaming machine is or may be wholly or partly determined by reference to use made of another gaming machine ('the linked gaming machine'). This does not apply in a licensed casino premises (section 245).

Penalty

Persons guilty of an offence are liable on summary conviction to imprisonment for up to 51 weeks (6 months in Scotland), a fine not exceeding level 5 on the standard scale, or both (section 246).

Family entertainment centre gaming machine permit

It is not an offence where a Category D gaming machine is available for use with a family entertainment centre gaming machine permit.

A family entertainment centre gaming machine permit is a permit issued by a licensing authority authorising a person to make Category D gaming machines available for use in a specified family entertainment centre (section 247 and schedule 10).

No prize or limited prize

It is not an offence for a gaming machine to be used by an individual, and the individual does not, by using the machine, acquire an opportunity to win a prize or a prize of a value in excess of the amount that he pays for or in connection with his use of the machine (section 248 and 249).

PERMITS FOR GAMING AND GAMBLING MACHINES IN CLUBS AND MINERS' WELFARE INSTITUTES

Exempt gaming

Gaming is exempt in these premises:

1. where the arrangements for the gaming satisfy the prescribed requirement amounts that may be staked, or the amount or value of a prize;
2. where no amount is deducted or levied from sums staked or won;
3. where any prescribed participation fee does not exceed such maximum or there is no fee;
4. where a game played on one set of premises is not linked with a game played on another set of premises;
5. where each person who participates is a member of the club or institute who applied for membership, was nominated for membership or became a member, at least 48 hours before he participates, or is a guest of a member of the club or institute who would be entitled to participate;
6. in alcohol licenced premises children and young persons are excluded from participation (section 269 and 279).

Club gaming permit

A club gaming permit is a permit issued by a licensing authority authorising the provision of facilities for gaming:

(a) on premises on which a members' club or a miners' welfare institute operates; and
(b) in the course of the activities of the club or institute.

A club gaming permit authorises:

(a) making up to three gaming machines available for use, each of which must be of Category B, C or D;

(b) the provision of facilities for gaming which are exempt but the stakes and prizes are within the prescribed amounts; and

(c) the provision of facilities for games of chance, of such class or description are in accordance with the conditions specified below.

Those conditions are:

(a) that prescribed participation fee is charged;

(b) that amount is deducted or levied from sums staked or won is in accordance with regulations;

(c) that the public and children are excluded from any area of the club's or institute's premises where gaming is taking place (section 271 and Gambling Act 2005 (Club Gaming and Club Machine Permits) Regulations 2007).

Club machine permit

A club machine permit is a permit issued by a licensing authority authorising up to three gaming machines, each of which must be of Category B, C or D, to be made available for use on premises on which a members' club, a commercial club or a miners' welfare institute operates, and in the course of the activities of the club or institute. Only members and their guest may use the machines. Children cannot use Category B or C gaming machine (section 273 and Gambling Act 2005 (Club Gaming and Club Machine Permits) Regulations 2007).

The provisions of schedule 12 Gambling Act 2005 on page 187 apply to these clubs.

Gaming machines: automatic entitlement

Clubs and Institutes can make one or two gaming machines, each of which is of Category C or D, available for use on premises provided that the person who holds the on-premises alcohol licence or the relevant Scottish licence sends the licensing authority written notice of his intention to make gaming machines available for use with the prescribed fee.

The club or institute must comply with the relevant code of practice about the location and operation of a gaming machine (section 282).

Licensed premises gaming machine permits

It is an offence to make a gaming machine of Category C or D available without obtaining a licensed premises gaming machine permit.

A licensed premises gaming machine permit is a permit issued by a licensing authority authorising a person to make gaming machines of Category C or D (or both) available for use on premises. See Gambling Act 2005 (Exempt Gaming in Alcohol-Licensed Premises) Regulations 2007.

The provisions of schedule 13 Gambling Act 2005 on page 190 apply to these clubs but do not apply in Scotland (section 282).

Removal of exemption

A licensing authority may make an order removing exemptions from specified premises.

A licensing authority may make an exemption order only if they think that:

(a) the application of the exemption is not reasonably consistent with pursuit of the licensing objectives;
(b) gaming has taken place on the premises in purported reliance on the exemption but in breach of a condition of that exemption;
(c) the premises are mainly used or to be used for gaming; or
(d) an offence under this Act has been committed on the premises.

Before making an order a licensing authority must:

(a) give the holder of the on-premises alcohol licence or of the relevant Scottish licence at least 21 days' notice of the authority's intention to consider making an order;
(b) consider any representations made by the licensee;
(c) hold a hearing if the licensee requests one; and
(d) comply with any prescribed requirements for the procedure to be followed in considering whether to make an order.

If a licensing authority makes an order, it must give the licensee a copy of the order, and written reasons for the decision to make the order.

A licensee may appeal against the making of an order to the magistrates' court within 21 days of receiving a copy of the order against which the appeal is brought.

The magistrates' court may dismiss the appeal, allow the appeal and quash the order made by the licensing authority and make an order about costs (section 284).

Definitions

gaming machine means a machine which is designed or adapted for use by individuals to gamble (section 235).

adult gaming centre means premises which have an adult gaming centre premises licence (section 237).

family entertainment centre means premises (other than an adult gaming centre) wholly or mainly used for making gaming machines available for use (section 238).

prize in relation to a gaming machine includes any money, article, right or service won, whether or not described as a prize, but does not include an opportunity to play the machine again (section 239).

members' club means a club:

(a) which is established and conducted wholly or mainly for purposes other than the provision of facilities for gaming;
(b) which is established and conducted for the benefit of its members;

(c) which is not established with the purpose of functioning only for a limited period of time; and

(d) which has at least 25 individual members (section 266).

commercial club means a club:

(a) which is established and conducted wholly or mainly for purposes other than the provision of facilities for gaming;

(b) which is not established with the purpose of functioning only for a limited period of time; and

(c) which has at least 25 individual members (section 267).

miners' welfare institute means an association:

• which is established and conducted for social and recreational purposes; and

• which affairs are managed by a group of individuals of whom at least two thirds are miners' representatives;

• which operates on premises the use of which is regulated in accordance with a charitable trust; and

• where the trust has received money from the Miners' Welfare Fund, the Coal Industry Social Welfare Organisation (section 268).

PERMITS TO FAMILY ENTERTAINMENT CENTRES (FECS) FOR THE USE OF CERTAIN LOWER STAKE GAMING MACHINES

References

Gambling Act 2005 schedule 10.

Gambling Act 2005 (Family Entertainment Centre Gaming Machine) (Permits) Regulations 2007.

Extent

These provisions apply to England, Wales and Scotland.

Application

Only adults who occupy or propose to occupy the premises may make an application for the use of the premises as an unlicensed family entertainment centre (paragraph 2).

An application for a permit must:

(a) be made in the form and manner required by the licensing authority;

(b) specify the premises for which the permit is sought;

(c) contain or be accompanied by such other information or documents required by the licensing authority; and

(d) include the prescribed fee (paragraph 5).

A licensing authority may prepare a statement of principles that they propose to apply in issuing permits. The statement may, in particular, specify matters that the licensing authority proposes to consider in determining the suitability of an applicant for a permit. The authority need not (but may) have regard to the licensing objectives, and have regard to any relevant guidance (paragraph 7).

On considering an application for a permit a licensing authority may grant or refuse it, but it cannot attach conditions.

The licensing authority must issue a permit to the applicant or, if refused, must notify the applicant of the refusal and the reasons for it (paragraph 8).

A licensing authority may grant an application for a permit only if they are satisfied that the applicant intends to use the premises as an unlicensed family entertainment centre, and have consulted the relevant chief officer of police. The permit has effect for 10 years and must be displayed on the premises (paragraphs 9,12 and 19).

A licensing authority cannot refuse an application unless they have notified the applicant of their intention to refuse the application and of their reasons, and given the applicant an opportunity to oral or written representations or both (paragraph 10).

A permit must be in the prescribed form and must specify:

(a) the person to whom it is issued;
(b) the premises; and
(c) the date on which it takes effect.

If the person to whom a permit is issued changes his name or wishes to be known by another name he can send the permit to the licensing authority with the prescribed fee, and a request that a new name be substituted for the old name. The licensing authority must comply with the request and return the permit to the holder (paragraph 11).

Lapsing

Permits lapse:

* if the holder ceases to occupy the premises (paragraph 13);
* if the licensing authority notify the holder that the premises are not being used as a family entertainment centre (paragraph 14);
* if the holder dies;
* the holder becomes, in the opinion of the licensing authority as notified to him, incapable of carrying on the activities authorised by the permit by reason of mental or physical incapacity;
* the holder becomes bankrupt.

The permit may still be valid for 6 months following the lapse and have effect to the personal representatives of the holder, the trustee of the bankrupt's estate or the liquidator of the company where relevant (paragraph 15).

Surrender

A permit ceases to have effect if the holder gives a notice of surrender to the licensing authority together with either the permit, or a statement explaining why it is not reasonably practicable to produce the permit (paragraph 16).

Forfeiture

Where the holder of a permit is convicted of a relevant offence the court by or before which he is convicted may order forfeiture of the permit. The court must notify the licensing authority of the forfeiture.

The terms on which forfeiture is ordered includes a requirement that the permit holder delivers the permit to the licensing authority or a statement explaining why it is not reasonably practicable to produce the permit (paragraph 17).

The holder of a permit may apply to the licensing authority for renewal of the permit between 6 and 2 months of the expiry date.

A licensing authority may refuse an application for renewal of a permit only on the grounds that an authorised local authority officer has been refused access to the premises without reasonable excuse, or that renewal would not be reasonably consistent with pursuit of the licensing objectives (paragraph 18).

It is an offence for an occupier of premises to fail to produce the permit on request for inspection by a constable, enforcement officer or authorised local authority officer liable on summary conviction to a fine not exceeding level 2 on the standard scale (paragraph 20).

Where a permit is lost, stolen or damaged, the holder may apply to the licensing authority for a copy with a prescribed fee (paragraph 21).

Appeal

The applicant for or holder of a permit may appeal to the magistrates' court within 21 days of notification, if the licensing authority reject an application for the issue or renewal of a permit or the permit lapses.

The magistrates' court may dismiss the appeal, substitute for any decision that the licensing authority could have made, restore a permit, return the case to the licensing authority to decide in accordance with a direction of the court and make an order about costs (paragraph 22).

Register

The licensing authority must maintain a register of permits issued by the authority, make the register and information available for inspection by members of the public at all reasonable times, and make arrangements for the provision of a copy of an entry in the register, or of information, to a member of the public on request on payment of the prescribed fee (paragraph 23).

PERMITS FOR PRIZE GAMING

References

Gambling Act 2005 schedule 14.

Gambling Act 2005 (Prize Gaming) (Permits) Regulations 2007 *(these Regulations make provisions about the fees relating to a prize gaming permit, and the form of the permit)*.

Gambling Act 2005 (Limits on Prize Gaming) Regulations 2009 *(these Regulations prescribed the limits on participation fees, and maximum prize amounts and values in respect of prize gaming)*.

Extent

These provisions apply in England, Wales and Scotland.

Prize gaming permits

Premises or facilities for prize gaming or equal chance gaming in an adult gaming centre, travelling fair or licensed family entertainment centre must have a prize gaming permit and comply with the conditions in that permit (section 289, 290 and schedule 14).

Such facilities are available in a bingo hall where it has a bingo premises licence (section 291).

Conditions for prize gaming

Participation fees must be within the prescribed limits. All the chances to participate in a particular game must be acquired or allocated on one day and in the place where the game is played.

The game must be played entirely on that day, and the result of the game must be made public in the place where the game is played, and as soon as is reasonably practicable after the game ends, and in any event on the day on which it is played.

A prize for which a game is played, or the aggregate of the prizes where all the prizes are money, must not exceed the prescribed amount, and in any other case, must not exceed the prescribed value. Participation in the game by a person does not entitle him or another person to participate in any other gambling (section 293).

Application

A person who applies for an on-premises alcohol licence or who holds one may apply for a licensed premises gaming machine permit (schedule 14 paragraph 1).

An application for a permit must be made in the form and manner that the licensing authority directs, specify the premises for which the permit is sought, the number and category of gaming machines for which the permit is sought, any other information or documents that the licensing authority directs, and the prescribed fee (schedule 14 paragraph 2).

The licensing authority must consider the application having regard to the licensing objectives, any relevant guidance issued by the Commission and such other matters as they think relevant.

After considering the application, the licensing authority must grant or refuse the application, or grant the permit for a smaller number of machines than that specified in the application, a different category of machines from that specified in the application, or both (schedule 14 paragraph 4).

The licensing authority cannot attach conditions to a permit. It must issue a permit to the applicant or where refused notify the applicant of the refusal, and the reasons for it (schedule 14 paragraph 5).

The licensing authority must notify the applicant of its intention to refuse the application or that it intends to grant a permit for a smaller number of machines or a different category of machines from that specified in the application, or both, and the applicant must be given an opportunity to make oral or written representations (schedule 14 paragraph 6).

Permit

A permit must be in the prescribed form and must specify:

(a) the person to whom it is issued;
(b) the premises;
(c) the number and category of gaming machines which it authorises; and
(d) the date on which it takes effect.

If the person to whom a permit is issued changes his name or wishes to be known by another name, he may send the permit to the licensing authority with the prescribed fee, and a request that a new name be substituted for the old name, and the licensing authority must issue the amended permit (schedule 14 paragraph 7).

The holder of a permit must keep it on the premises (schedule 14 paragraph 8).

Annual fees must be paid by the permit holder (schedule 14 paragraph 9).

It is an offence for the occupier of premises to fail to produce the permit on request for inspection by an enforcement officer, or an authorised local authority officer liable on summary conviction to a fine not exceeding level 2 on the standard scale (schedule 14 paragraph 10).

Where a permit is lost, stolen or damaged, the holder may apply to the licensing authority for a copy paying the prescribed fee (schedule 14 paragraph 11).

A permit ceases if:

- an on-premises alcohol licence ceases to have effect;
- the permit holder ceases to be the holder of an on-premises alcohol licence (schedule 14 paragraph 13); or
- the permit holder gives a notice of surrender to the licensing authority including either the permit, or a statement explaining why it is not reasonably practicable to produce the permit (schedule 14 paragraph 14).

The permit holder may apply to the licensing authority to vary the number or category (or both) of gaming machines authorised by the permit (schedule 14 paragraph 15).

Cancelling and variation

The licensing authority may cancel a permit, or may vary the number or category (or both) of gaming machines authorised by it, if they think that:

(a) it would not be reasonably consistent with the licensing objectives for the permit to continue to have effect;

(b) gaming has taken place on the premises in purported reliance on the permit but otherwise than in accordance with the permit or a condition of the permit;

(c) the premises are mainly used or to be used for making gaming machines available; or

(d) an offence under this Act has been committed on the premises.

Before cancelling or varying a permit the licensing authority must:

(a) give the permit holder at least 21 days' notice of the authority's intention to consider cancelling or varying the permit;

(b) consider any representations made by the holder;

(c) hold a hearing if the holder requests one; and

(d) comply with any prescribed requirements for the procedure to be followed in considering whether to cancel or vary a permit.

The licensing authority must give notice of the cancellation or variation and the reasons for it to the permit holder, and the Commission, subject to any appeal (schedule 14 paragraph 16).

If the permit holder fails to pay the annual fee, the licensing authority must cancel it unless failure to pay is attributable to administrative error (schedule 14 paragraph 17).

Forfeiture

Where a permit holder, or the officer of a permit holder, is convicted of a relevant offence the court by or before which he is convicted may order forfeiture of the permit. The court can apply terms on the permit. The court must notify the licensing authority of the court's action (schedule 14 paragraph 18).

Transfer

A person may apply for the transfer of a permit to him if he is applying for the transfer of an on-premises alcohol licence to him. He must supply the permit, or a statement explaining why it is not reasonably practicable to produce the permit with his application. The permit cannot be issued unless the alcohol licence is approved (schedule 14 paragraph 19).

Appeal

The applicant for or holder of a permit may appeal to the magistrates' court, if the licensing authority:

(a) rejects an application for a permit;

(b) grants an application for a permit in respect of a smaller number of machines than that specified in the application or a different category of machines from that specified in the application (or both); or

(c) gives a notice to cancel the permit

within 21 days of the action by the licensing authority.

The magistrates' court may:

(a) dismiss the appeal;
(b) substitute for the decision appealed against any decision that the licensing authority could have made;
(c) restore a permit;
(d) return the permit to the licensing authority to decide in accordance with a direction of the court; or
(e) make an order about costs (schedule 14 paragraph 21).

Register

The licensing authority must maintain a register of permits issued by the authority, make it and information available for inspection by members of the public at all reasonable times, and make arrangements for the provision of a copy of an entry in the register, or of information, to a member of the public on request with payment of the relevant fee (schedule 14 paragraph 22).

Definitions

prize gaming is where neither the nature nor the size of a prize played for is determined by reference to the number of persons playing, or the amount paid for or raised by the gaming (section 288).

REGISTERING SMALL SOCIETY LOTTERIES

References

Gambling Act 2005 schedule 11 part 5.
Gambling Act 2005 (Incidental Non-Commercial Lotteries) Regulations 2007.
Guidance to licensing authorities (4th Edition) *sets out the principles that licensing authorities should adopt in registering small society lotteries.*

Extent

These provisions apply to England, Wales and Scotland.

Scope

Society lotteries are lotteries promoted for the benefit of a non-commercial society. A society is non-commercial if it is established and conducted:

• for charitable purposes;
• for the purpose of enabling participation in, or of supporting, sport, athletics or a cultural activity;
• for any other non-commercial purpose other than that of private gain.

A small society lottery:

* does not have proceeds that exceed £20,000 for a single draw;
* does not have aggregated proceeds from lotteries in excess of £250,000 in any one year.

Small society lotteries do not require a licence but must be registered with the licensing authority in the area where the principal office of the society is located.

Licensing authorities should inform a society that they must apply for an operating licence if, in the course of running a small society lottery, proceeds exceed £20,000 for a single draw or aggregated proceeds from lotteries exceed £250,000 in any one year.

Society lotteries (schedule 11 part 4 Gambling Act 2005) (FC 3.16)

The arrangements for a small society lottery must ensure that at least 20 per cent of the proceeds of the lottery are applied to a purpose for which the society is conducted (paragraph 33).

The maximum prize in a small society lottery is £25,000 (paragraph 34).

The arrangements for a small society lottery may include a rollover (paragraph 35).

A lottery ticket in a small society lottery must identify the promoting society, state the price of the ticket, the name and an address of a member of the society who is designated as having responsibility within the society for the promotion of the lottery, or the external lottery manager, and either state the date of the draw or enable the date of the draw in the lottery to be determined (paragraph 36).

The price for each ticket must be the same, and must be paid to the promoter of the lottery before any person is given the ticket or any right in respect of membership of the class among whom prizes are to be allocated (paragraph 37).

The promoting society of a small society lottery must be registered with a local authority (paragraph 38 and part 5).

The society must send a statement to the local authority at least 3 months prior to the date of the lottery which includes:

(a) the arrangements for the lottery (including the dates on which tickets were available for sale or supply, the dates of any draw and the arrangements for prizes (including any rollover);
(b) the proceeds of the lottery;
(c) the amounts deducted by the promoters of the lottery in respect of the provision of prizes (including the provision of prizes in accordance with any rollover);
(d) the amounts deducted by the promoters of the lottery in respect of other costs incurred in organising the lottery;
(e) any amount applied to a purpose for which the promoting society is conducted; and
(f) whether any expenses in connection with the lottery were defrayed otherwise than by deduction from proceeds, and, if they were:
  (i) the amount of the expenses; and
  (ii) the sources from which they were defrayed.

FC 3.16 Registration of small society lotteries – Gambling Act 2005 Schedule 11 part 5

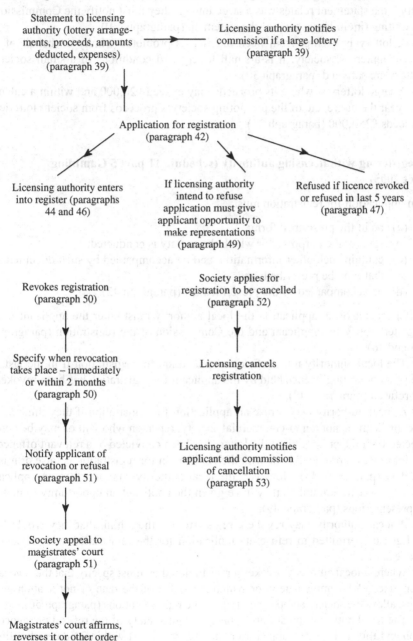

Statement to licensing authority (lottery arrangements, proceeds, amounts deducted, expenses) (paragraph 39)

Licensing authority notifies commission if a large lottery (paragraph 39)

Application for registration (paragraph 42)

Licensing authority enters into register (paragraphs 44 and 46)

If licensing authority intend to refuse application must give applicant opportunity to make representations (paragraph 49)

Refused if licence revoked or refused in last 5 years (paragraph 47)

Revokes registration (paragraph 50)

Society applies for registration to be cancelled (paragraph 52)

Specify when revocation takes place – immediately or within 2 months (paragraph 50)

Licensing cancels registration

Notify applicant of revocation or refusal (paragraph 51)

Licensing authority notifies applicant and commission of cancellation (paragraph 53)

Society appeal to magistrates' court (paragraph 51)

Magistrates' court affirms, reverses it or other order (paragraph 51)

The statement must be signed by two adult members of the society who are appointed for the purpose in writing by the society or, if it has one, its governing body, and accompanied by a copy of the appointment (paragraph 39).

If after receiving a statement the local authority thinks that the lottery to which the statement relates was a large lottery, they must notify the Commission in writing (including a copy of the statement (paragraph 40)).

A lottery is exempt from licensing if it is promoted wholly on behalf of a non-commercial society, it is a small lottery and conditions of a small society lottery are satisfied (paragraph 30).

A large lottery is where its proceeds may exceed £20,000 and within a calendar year the aggregate of the promoting society's proceeds from society lotteries exceeds £250,000 (paragraph 31).

Registering with licensing authority (schedule 11 part 5 Gambling Act 2005)

An application for registration must:

(a) be in the prescribed form;
(b) specify the purposes for which the society is conducted;
(c) contain such other information, and be accompanied by such documents, that may be prescribed; and
(d) be accompanied by the prescribed fee (paragraph 42).

After receipt of an application the local authority must enter the applicant in a register, notify the applicant and the Commission of the registration (paragraph 44 and 46).

The local authority must refuse an application for registration if in the last 5 years an operating licence held by the applicant for registration has been revoked or refused (paragraph 47).

A local authority may refuse an application for registration if they think that the applicant is not a non-commercial society, a person who will or may be connected with the promotion of the lottery has been convicted of a relevant offence, or information provided in or with the application for registration is false or misleading (paragraph 48). However, the local authority cannot refuse an application for registration unless they have given the applicant an opportunity to make representations (paragraph 49).

A local authority may revoke a registration if they think that they would be obliged or permitted to refuse an application for the registration were it being made anew.

Where a local authority revokes a registration they must specify that the revocation takes effect immediately, or within 2 months of the date of notification and must allow the society an opportunity to make representations (paragraph 50).

The local authority must notify the applicant society of the refusal or revocation and the society may appeal to a magistrates' court within 21 days of notification of the refusal or revocation of registration.

The magistrates' court may affirm the local authority's decision; reverse it; or make any other order (paragraph 51).

A registered society may apply in writing to the registering authority for the registration to be cancelled (paragraph 52). The local authority must then cancel the registration, notify the formerly registered society of the cancellation, and notify the Commission of the cancellation (paragraph 53).

An annual fee is payable to the registering local authority. If a registered society fails to pay the annual fee the local authority may cancel the society's registration and must notify the formerly registered society and the Commission (paragraph 54).

Statements must be retained for at least 18 months, made available for inspection by members of the public at all reasonable times, and make arrangements for the provision of a copy of it or part of it to any member of the public on request on payment of a fee (paragraph 55).

Promotion of or facilitating a lottery

It is an offence to promote or facilitate a lottery which is not exempt without an operating licence. It is not an offence if a person acts on behalf of the licence holder so long as he is not an external lottery manager (section 258 and 259).

Misusing profits of lottery

It is an offence if a person uses any part of the profits of a lottery for a purpose other than that stated on the lottery tickets, or in an advertisement for the lottery (section 260). This does not apply to:

(a) an incidental non-commercial lottery;
(b) a private society lottery; or
(c) a small society lottery.

It is an offence if a person uses any part of the profits of a lottery for a purpose other than one for which the lottery is permitted to be promoted (section 261 and schedule 11).

Breach of condition

A non-commercial society commits an offence if:

(a) a lottery, purporting to be an exempt lottery, is promoted on the society's behalf wholly or partly at a time when the society is not registered with a local authority;
(b) the society fails to comply with the registration records requirements; or
(c) the society provides false or misleading information (section 262).

Penalty

Persons guilty of offences are liable on summary conviction to imprisonment for a term not exceeding 51 weeks (6 months in Scotland), a fine not exceeding level 5 on the standard scale, or both (section 263).

Definitions

promoting a lottery – a person promotes a lottery if she:

(a) makes arrangements for the printing of lottery tickets;

(b) makes arrangements for the printing of promotional material;

(c) arranges for the distribution or publication of promotional material;

(d) possesses promotional material with a view to its distribution or publication;

(e) makes other arrangements to advertise a lottery;

(f) invites a person to participate in a lottery;

(g) sells or supplies a lottery ticket;

(h) offers to sell or supply a lottery ticket;

(i) possesses a lottery ticket with a view to its sale or supply;

(j) does or offers to do anything by virtue of which a person becomes a member of a class among whom prizes in a lottery are to be allocated; or

(k) uses premises for the purpose of allocating prizes or for any other purpose connected with the administration of a lottery (section 252).

lottery ticket means a document or article is a lottery ticket if it confers, or can be used to prove, membership of a class for the purpose of the allocation of prizes in a lottery (section 253).

proceeds and profits means the proceeds of a lottery is the aggregate of amounts paid for lottery tickets. The profits of a lottery are

(a) the proceeds of the lottery, minus

(b) amounts deducted by the promoters of the lottery for prizes, sums to be made available for allocation in another lottery in accordance with a rollover, or other costs reasonably incurred in organising the lottery (section 254).

draw includes any process by which a prize in the lottery is allocated (section 255).

rollover means an arrangement whereby the fact that a prize is not allocated or claimed in one lottery increases the value of the prizes available for allocation in another lottery (section 256).

external lottery manager is a person who acts as an external lottery manager if he makes arrangements for a lottery on behalf of a society or authority of which he is not a member, an officer, or an employee under a contract of employment (section 257).

CONTROL OF SCAFFOLDING AND HOARDINGS ON HIGHWAYS

Reference

Section 169 Highways Act 1980.

Extent

This procedure extends to England, Wales and Scotland.

Scope

This procedure applies to the licensing of obstructions of the highway.

FC 3.17 Control of scaffolding and hoardings on highways – Section 169 Highways Act 1980

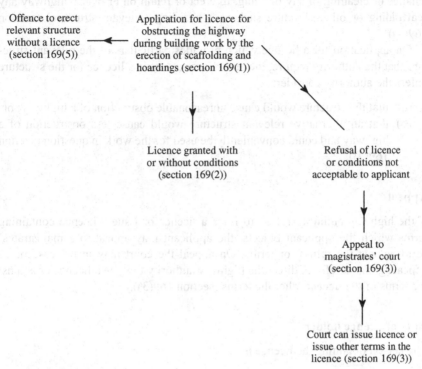

Offence to erect
relevant structure
without a licence
(section 169(5))

Application for licence for
obstructing the highway
during building work by the
erection of scaffolding and
hoardings (section 169(1))

Licence granted with
or without conditions
(section 169(2))

Refusal of licence
or conditions not
acceptable to applicant

Appeal to
magistrates' court
(section 169(3))

Court can issue licence or
issue other terms in the
licence (section 169(3))

Licences (FC 3.17)

A licence is needed from the highway authority for obstructing the highway in connection with any building or demolition work or the alteration, repair, maintenance or cleaning of any building, to erect or retain on or over a highway any scaffolding or other structure such as a hoarding (relevant structure) (section 169(1)).

On application for a licence which provides particulars of the relevant structure that the authority require, the authority must issue a licence for the structure unless the authority consider:

(a) that the structure would cause unreasonable obstruction of a highway; or
(b) that an alternative relevant structure would cause less obstruction of a highway and could conveniently be used for the work in question (section 169(2)).

Appeal

If the highway authority refuse to issue a licence or issue a licence containing terms which the applicant objects, the applicant may appeal to a magistrates' court against the refusal or terms. On appeal the court may in the case of an appeal against a refusal, direct the highway authority to issue a licence or against the terms of the licence, alter the terms (section 169(3)).

Duty of licence holder

The person obtaining the licence must:

(a) ensure that the structure is adequately lit at all times between half an hour after sunset and half an hour before sunrise;
(b) comply with any directions given to him in writing by the authority with respect to the erection and maintenance of traffic signs in connection with the structure; and
(c) do such things in connection with the structure as any statutory undertakers reasonably request him to do for the purpose of protecting or giving access to any apparatus belonging to or used or maintained by the undertakers (section 169(4)).

Offences

A person who does not obtain a licence for a relevant structure or fails without reasonable excuse to comply with the terms of a licence or fails to perform a duty imposed on him, is guilty of an offence and liable to a fine not exceeding level 5 on the standard scale (section 169(5)).

These provisions do not apply to a relevant structure if no part of it is less than 18 inches in a horizontal direction from a carriageway of the relevant highway and no part of it over a footway of the relevant highway is less than 8 feet in a vertical direction above the footway (section 169(6)).

Definition

relevant structure means any scaffolding or other structure in connection with any building or demolition work or the alteration, repair, maintenance or cleaning of any building, erected or retained on or over a highway which obstructs the highway.

CONTROL OF BUILDER'S SKIPS

References

Section 139 to 141 Highways Act 1980.

Extent

This procedure extends to England, Wales and Scotland.

Scope

This procedure applies to the licensing of builder's skips.

Permission (FC 3.18)

Permission of the highway authority is needed for a builder's skip to be deposited on a highway (section 139(1)). This does not apply to a skip on private land.

The permission applies to a skip on the highway specified in the permission, and the highway authority may issue the permission with no conditions or attach conditions in the permission including, in particular, conditions relating to:

(a) the siting of the skip;
(b) its dimensions;
(c) the manner in which it is to be coated with paint and other material for the purpose of making it immediately visible to oncoming traffic;
(d) the care and disposal of its contents;
(e) the manner in which it is to be lighted or guarded;
(f) its removal at the end of the period of permission (section 139(2)).

Duties of permission holder and offences

Where a builder's skip has been deposited on a highway with permission, the owner of the skip must ensure:

(a) that the skip is properly lighted during the hours of darkness and marked in accordance with any regulations;
(b) that the skip is clearly and indelibly marked with the owner's name and with his telephone number or address;
(c) that the skip is removed as soon as practicable after it has been filled;
(d) that each of the permission conditions are complied with and, if he fails to do so, he is guilty of an offence and liable to a fine not exceeding level 3 on the standard scale (section 139(4)).

FC 3.18 Builder's skip licence – Sections 139 to 141 Highways Act 1980

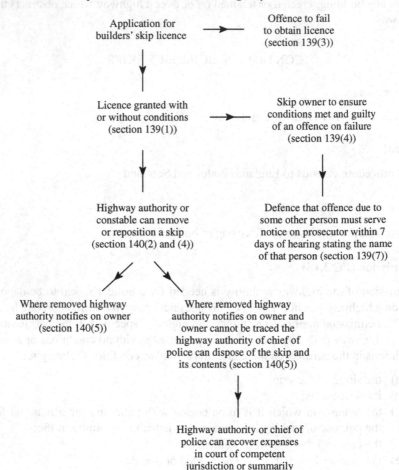

Application for
builders' skip licence

Offence to fail
to obtain licence
(section 139(3))

Licence granted with
or without conditions
(section 139(1))

Skip owner to ensure
conditions met and guilty
of an offence on failure
(section 139(4))

Highway authority or
constable can remove
or reposition a skip
(section 140(2) and (4))

Defence that offence due to
some other person must serve
notice on prosecutor within 7
days of hearing stating the name
of that person (section 139(7))

Where removed highway
authority notifies on owner
(section 140(5))

Where removed highway
authority notifies on owner and
owner cannot be traced the
highway authority of chief of
police can dispose of the skip and
its contents (section 140(5))

Highway authority or chief of
police can recover expenses
in court of competent
jurisdiction or summarily
as a civil debt (section 140(6))

Where an offence is due to the act or default of some other person, that other person is guilty of the offence, whether or not proceedings are taken against the permission holder (section 139(5)).

If a builder's skip is deposited on a highway without a permission, the owner of the skip is guilty of an offence and liable to a fine not exceeding level 3 on the standard scale (section 139(3)).

Defences

It is a defence, for the person charged to prove that the commission of the offence was due to the act or default of another person and that he took all reasonable precautions and exercised all due diligence to avoid the offence by himself or any person under his control (section 139(6)). To rely on this defence he must, within 7 clear days before the hearing, serve the prosecutor a notice in writing giving this information identifying or assisting in the identification of that other person (section 139(7)).

Where any person is charged with an offence under any other enactment for failing to secure that a builder's skip deposited on a highway was properly lighted during the hours of darkness, it is a defence for the person charged to prove that the offence was due to the act or default of another person and that he took all reasonable precautions and exercised all due diligence to avoid the offence by himself or any person under his control (section 139(8)).

Where a person is charged with obstructing, or interrupting any user of a highway by depositing a builder's skip on it, it is a defence for the person to prove that the skip was deposited on it in accordance with permission and either:

(a) that each of the permission requirements had been complied with; or
(b) that the commission of any offence was due to the act or default of another person and that he took all reasonable precautions and exercised all due diligence to avoid the offence by himself or any person under his control (section 139(9)).

Removal of builder's skips

The highway authority or a constable in uniform may require the owner of the skip to remove or reposition it or cause it to be removed or repositioned (section 140(2)).

The highway authority or a constable in uniform may themselves remove or reposition the skip or cause it to be removed or repositioned (section 140(4)).

Where a skip is removed, the highway authority or, the chief officer of police must, where practicable, notify the owner of its removal, but if the owner cannot be traced, or if after a reasonable period of time after being notified he has not recovered the skip, the highway authority or chief officer of police may dispose of the skip and its contents (section 140(5)).

Any expenses reasonably incurred by a highway authority or chief officer of police in the removal or repositioning of a skip or the disposal of a skip may be recovered from the owner of the skip in any court of competent jurisdiction or summarily as a civil debt (section 140(6)).

Any proceeds of the disposal of a skip must be used in the first place to meet the expenses reasonably incurred in the removal and disposal of the skip and any surplus must be given to the person entitled to it if he can be traced and if not may be retained by the highway authority or the chief officer of police, and any surplus retained by a chief officer of police must be paid into the police fund (section 140(7)).

Expenses include storing the skip until it is recovered by the owner or disposed of (section 140(8)).

The owner of a skip is not guilty of a section 139(4) offence if the failure resulted from the repositioning of the skip (section 140(9)).

Offences

A person required to remove or reposition, or cause to be removed or repositioned, a skip must comply with the requirement as soon as practicable, and if he fails to do so he is guilty of an offence and liable to a fine not exceeding level 3 on the standard scale (section 140(3)).

Definitions

builder's skip means a container designed to be carried on a road vehicle and to be placed on a highway or other land for the storage of builders' materials, or for the removal and disposal of builders' rubble, waste, household and other rubbish or earth.

owner means a builder's skip which is the subject of a hiring agreement, being an agreement for a hiring of not less than 1 month, or a hire purchase agreement, means the person in possession of the skip under that agreement (section 139(11)).

CONTROL OF CONSTRUCTION OF CELLARS, ETC. UNDER STREET

Reference

Section 179 and 180 Highways Act 1980.

Extent

These provisions apply to England and Wales.

Scope (FC 3.19)

Consent of the highway or local authority is required to construct works under any part of a street, and the authority may serve notice on a person who has constructed such works without consent requiring him to remove them, or to alter or deal with them in such a manner specified in the notice (section 179(1)).

A person aggrieved by the refusal of a consent, or by a requirement of a notice, may appeal to a magistrates' court (section 179(2)).

FC 3.19 Control of the construction of cellars, etc. under streets – Section 179 Highways Act 1980

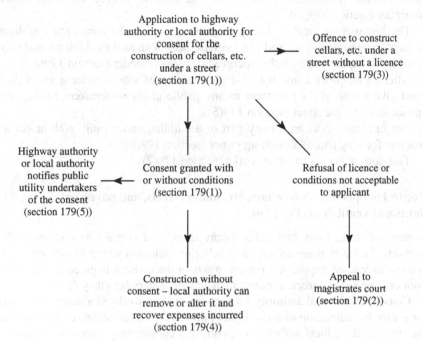

Application to highway authority or local authority for consent for the construction of cellars, etc. under a street (section 179(1))

Offence to construct cellars, etc. under a street without a licence (section 179(3))

Highway authority or local authority notifies public utility undertakers of the consent (section 179(5))

Consent granted with or without conditions (section 179(1))

Refusal of licence or conditions not acceptable to applicant

Construction without consent – local authority can remove or alter it and recover expenses incurred (section 179(4))

Appeal to magistrates court (section 179(2))

A person who constructs works without consent or contrary to a served notice is guilty of an offence and is liable to a fine not exceeding level 1 on the standard scale; and, subject to any order made on appeal, if he fails to comply with a requirement of a notice served on him he is guilty of a further offence and is liable to a fine not exceeding two pounds for each day during which the failure continues (section 179(3)).

The highway or local authority may also cause works constructed without consent to be removed, altered or otherwise dealt with as they think fit, and may recover the expenses reasonably incurred from the offender (section 179(4)).

After an authority consent to the construction of works under a street they must give notice of their consent to any public utility undertakers having any apparatus under the street (section 179(5)).

The relevant works are to any part of a building and a vault, arch or cellar, whether forming part of a building or not (section 179(6)).

This does not apply to street works (section 179(7)).

Control of openings into cellars, etc. under streets, and pavement lights and ventilators (FC 3.20)

Consent of the highway or local authority is required to make an opening in the footway of a street as an entrance to a cellar or vault, and where an authority give consent they must require the person to whom the consent is given to provide a door or covering constructed in such manner with materials as they direct.

Consent of the local authority for carrying out any works in a street to provide means for the admission of air or light to premises situated under, or abutting on, the street, and the local authority in giving any consent may impose any requirement about the construction of the works (section 180(2)).

A person aggrieved or a person who applies for consent by the refusal of a consent, or by a requirement, may appeal to a magistrates' court (section 180(3)).

It is an offence for any person to make an opening in the footway of a street or other works above without consent, or fail to comply with a requirement of the consent and is liable to a fine not exceeding level 1 on the standard scale (section 180(4)).

After an authority give consent they must give notice to any public utility undertakers having any apparatus under the street (section 180(5)).

Owner or occupier responsibilities

(a) Every vault, arch and cellar under a street;
(b) every opening in the surface of any street into any such vault, arch or cellar;
(c) every door or covering to any such opening;
(d) every cellar-head, grating, light and coal hole in the surface of a street; and
(e) all landings, flags or stones of the street by which any of the above are supported,

must be kept in good condition and repair by the owner or occupier of the vault, arch or cellar, or of the premises to which it belongs (section 180(6)).

FC 3.20 Control of openings into cellars, etc. under streets, and pavement lights and ventilators – Section 180 Highways Act 1980

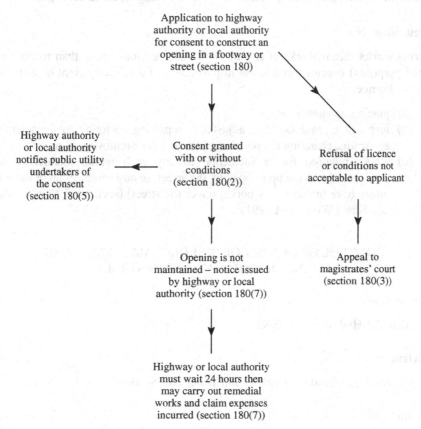

Application to highway authority or local authority for consent to construct an opening in a footway or street (section 180)

Consent granted with or without conditions (section 180(2))

Refusal of licence or conditions not acceptable to applicant

Highway authority or local authority notifies public utility undertakers of the consent (section 180(5))

Opening is not maintained – notice issued by highway or local authority (section 180(7))

Appeal to magistrates' court (section 180(3))

Highway or local authority must wait 24 hours then may carry out remedial works and claim expenses incurred (section 180(7))

If the owner or occupier does not maintain the building element above the appropriate authority may, after 24 hours from the service of a notice of their intention to do so on any person in default, cause such failure to be repaired or put into good condition, and may recover the expenses reasonably incurred by them from the owner or occupier of the premises to which it belongs (section 180(7)).

Definition

street works means works of any of the following kinds (other than works for road purposes) executed in a street in pursuance of a statutory right or a street works licence:

(a) placing apparatus; or
(b) inspecting, maintaining, adjusting, repairing, altering or renewing apparatus, changing the position of apparatus or removing it; or
(c) works required for or incidental to any such works (including, in particular, breaking up or opening the street, or any sewer, drain or tunnel under it, or tunnelling or boring under the street) (section 48 New Roads and Street Works Act 1991)

CONTROL OF DEPOSIT OF BUILDING MATERIALS AND MAKING OF EXCAVATIONS IN STREETS

References

Section 171 Highways Act 1980.

Extent

This procedure extends to England and Wales and Scotland.

Scope

This procedure applies to the licensing to prevent obstructions of the highway.

Consent (FC 3.21)

Consent of the highway authority is needed to temporarily deposit building materials, rubbish or other things or make a temporary excavation in a street that is a highway maintainable at the public expense (section 171(1)).

A highway authority may make the consent subject to such conditions as they think fit including in particular, conditions for preventing damage or ensuring access to apparatus of statutory undertakers (section 171(2)).

Appeal

A person aggrieved by the refusal of consent, and a person obtaining consent subject to conditions, may appeal to a magistrates' court against the refusal or the conditions (section 171(3)).

FC 3.21 Control of deposit of building materials and making of excavations in streets – Section 171 Highways Act 1980

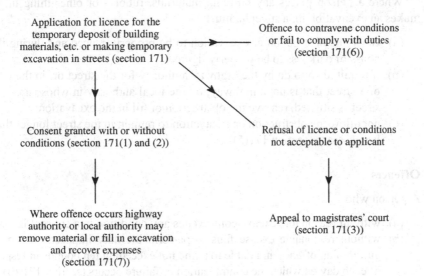

Application for licence for the temporary deposit of building materials, etc. or making temporary excavation in streets (section 171)

Offence to contravene conditions or fail to comply with duties (section 171(6))

Consent granted with or without conditions (section 171(1) and (2))

Refusal of licence or conditions not acceptable to applicant

Where offence occurs highway authority or local authority may remove material or fill in excavation and recover expenses (section 171(7))

Appeal to magistrates' court (section 171(3))

Duty of persons obtaining consent

The person obtaining consent must comply with any directions given to him in writing by the highway authority with respect to the erection and maintenance of traffic signs in connection with the deposit or excavation (section 171(4)).

Where a person places any building materials, rubbish or other thing in, or makes an excavation in, a street he must:

(a) cause the obstruction or excavation to be properly fenced and during the hours of darkness to be properly lighted;

(b) if required so to do by the highway authority for the street or, in the case of a street that is not a highway, by the local authority in whose area the street is situated, remove the obstruction or fill in the excavation;

(c) not allow the obstruction or excavation to remain in the street longer than is necessary (section 171(5)).

Offences

A person who:

(a) without reasonable excuse contravenes any condition in a consent is, or

(b) without reasonable excuse fails to perform the duty imposed on him is guilty of an offence and liable to a fine not exceeding ten pounds in respect of each day on which the contravention or failure occurs (section 171(6)).

Where an offence is committed in a street, the highway authority for the street or, in the case of a street that is not a highway, the local authority in whose area the street is situated, may remove the obstruction or, fill in the excavation and recover the expenses reasonably incurred by them in so doing from the person convicted of the offence (section 171(7)).

HIGHWAYS PROJECTION LICENCE

References

Section 177, 178 and 180 Highways Act 1980.

Extent

This act applies to England and Wales.

Scope

This procedure applies to all building projections over a highway.

Restriction on construction of buildings over highways

A licence with terms and conditions is required from the highway authority for the construction or alteration of a building over any part of a highway maintainable at the public expense.

Any person who contravenes the above is guilty of an offence and liable to a fine not exceeding level 5 on the standard scale; and if the offence is continued after conviction, he is guilty of a further offence and liable to a fine not exceeding fifty pounds for each day on which the offence continues (section 177(1)).

A licence may contain terms and conditions with respect to the construction (including the headway over the highway), maintenance, lighting and use of the building, as the highway authority think fit; and, any such term or condition is binding on the successor in title to every owner, and every lessee and occupier, of the building (section 177(2)).

A fee is payable for such a licence which is a reasonable sum in respect of legal or other expenses incurred in connection with the grant of the licence (from the applicant); and an annual charge of a reasonable amount for administering the licence (from the owner of the building (section 177(3)).

The licence must not cause any interference with the convenience of persons using the highway, or affect the rights of the owners of premises adjoining the highway, or the rights of statutory undertakers or the operator of an electronic communications code network or a driver information network (section 177(4)).

Where a licence makes provision for the execution of any works or the provision of any facilities which in the opinion of the highway authority require to be executed or provided by them in connection with the building or its construction or alteration, the authority may execute those works or, provide those facilities and may recover the expenses reasonably incurred by them in so doing from the licensee or from the owner of the building (section 177(5)).

Appeal

A person aggrieved by the refusal of a highway authority to grant a licence or by a term or condition of the licence may appeal to the Crown Court, except that no such appeal lies:

(a) if the land on which the highway in question is situated is owned by the highway authority, or

(b) against any term or condition which the highway authority declare to be necessary for the purpose of securing the safety of persons using the highway or of preventing interference with traffic (section 177(6)).

If a building has been constructed or altered without a licence or not in accordance with the terms and conditions of the licence, the highway authority may by notice served on the licensee or the owner of the building require him to demolish the building or make alterations within a time specified in the notice (section 177(7)).

Where there has been a failure to comply with any terms or conditions of a licence with respect to the maintenance or use of a building, the highway authority may by notice served on the licensee or the owner of the building require him to execute such works or take such steps necessary to secure compliance with those terms or conditions within a time specified in the notice (section 177(8)).

If a person served with a notice above fails to comply with the notice within the time specified, the highway authority may demolish the building or, execute

FC 3.22 Highway projection licence – Section 177, 178 and 180 Highways Act 1980

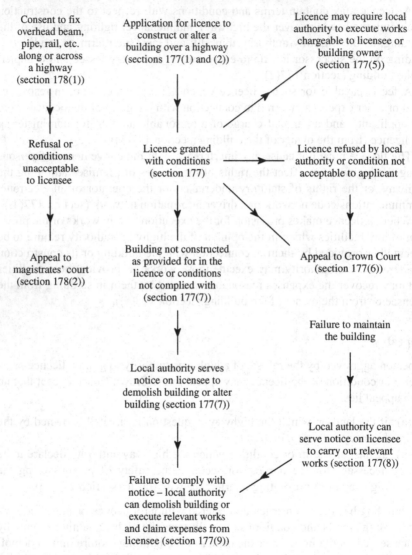

Notes
1. Offence to construct or alter a building over any part of a highway (section 177).
2. Offence for failure to obtain consent or does not comply with the terms or conditions of any consent (section 178(4)).

works or take steps necessary to comply with the notice and may recover the expenses reasonably incurred by them in so doing from that person (section 177(9)). Where highway authority demolish a building, they may dispose of the materials resulting from the demolition (section 177(10)).

Paragraph 23 of the electronic communications code which provides a procedure for certain cases where works involve the alteration of electronic communications apparatus applies, for the purposes of works authorised or required by a licence to be executed, to the licensee (section 177(12)).

Restriction on placing rails, beams, etc. over highways – consent

Consent of the highway authority is required to fix or place any overhead beam, rail, pipe, cable, wire or other similar apparatus over, along or across a highway and terms and conditions may be attached by the highway authority as they think fit (section 178(1)).

Appeal

A person aggrieved by the refusal of a consent, or by any terms or conditions attached to the consent, may appeal to a magistrates' court (section 178(2)). However, no appeal can be made against any term or condition attached by the Minister to a consent if he declares the term or condition to be necessary for the purpose of securing the safety of persons using the highway or of preventing interference with traffic on it (section 178(3)).

If a person does not obtain consent or does not comply with the terms or conditions of any consent, he is guilty of an offence and liable to a fine not exceeding level 1 on the standard scale; and if the offence is continued after conviction he is guilty of a further offence and liable to a fine not exceeding one pound for each day on which the offence continues (section 178(4)).

This does not apply to any works or apparatus belonging to any defined statutory undertakers (section 178(5)).

LICENCE TO PLANT TREES, SHRUBS, ETC. IN A HIGHWAY

References

Highways Act 1980.

Extent

These provisions apply to England and Wales.

Scope (FC 3.23)

The highway authority for a highway may grant a licence to permit the occupier or the owner of any premises adjoining the highway to plant and maintain, or to retain and maintain, trees, shrubs, plants or grass in such part of the highway specified in the licence (section 142(1)).

The highway authority may:

(a) grant a licence to the person who at the time of the grant is the occupier of premises and insert in the licence provisions prohibiting assignment of the licence and providing for its duration; or

(b) grant a licence to the owner of premises and his successors in title and insert in the licence provisions providing for the licence to be annexed to those premises and providing for its duration (section 142(2)).

A fee can be charged for the licence which is a reasonable sum for legal or other expenses incurred in connection with the grant of the licence and an annual charge of a reasonable amount for administering the licence (section 142(3)).

The licence must include a condition that within 1 month after any change in the ownership of the premises in question takes place, the licensee is to inform the highway authority of it (section 142(4)).

A highway authority can attach conditions to the licence they consider necessary to ensure the safety and convenience of passengers in the highway and to prevent traffic therein being delayed, to prevent any nuisance or annoyance being caused to the owners or occupiers of other premises adjoining the highway and to protect the apparatus of statutory undertakers, and the operators of electronic communications code networks or driver information systems (section 142(5)).

Withdrawal of licence

A highway authority may serve notice on the licensee withdrawing a licence on the expiration of a period of not less than 7 days beginning with the date of service of the notice on the licensee, if any condition of the licence is contravened by the licensee or on the expiration of a period of not less than 3 months beginning with that date, if the authority consider the withdrawal of the licence is necessary to exercise their functions as a highway authority (section 142(6)).

Where a licence expires or is withdrawn or surrendered, the highway authority may remove all or any of the trees, shrubs, plants or grass to which the licence relates and reinstate the highway and may recover the expenses reasonably incurred by them in so doing from the last licensee. However, if it is satisfied that the last licensee can, within such reasonable time as they may specify, remove such trees, shrubs, plants or grass or such of them as they may specify and reinstate the highway, may authorise him to do so at his own expense (section 142(7)).

The licensee and the person who immediately before the expiration, withdrawal or surrender of a licence was the licensee or, if that person has died, his personal representatives must indemnify the highway authority against any claim in respect of injury, damage or loss arising out of:

(a) the planting or presence in a highway of trees, shrubs, plants or grass to which the licence relates, or

FC 3.23 Licence to plant trees, shrubs, etc. – Section 142 Highways Act 1980

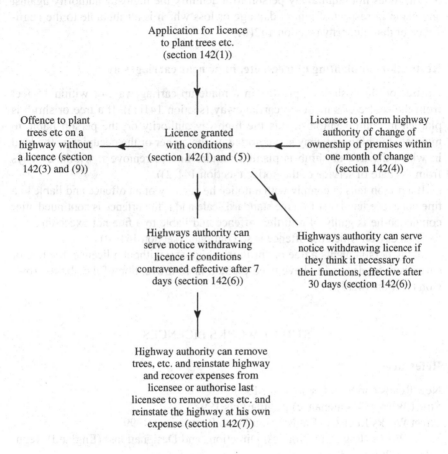

Application for licence
to plant trees etc.
(section 142(1))

Offence to plant
trees etc on a
highway without
a licence (section
142(3) and (9))

Licence granted
with conditions
(section 142(1) and (5))

Licensee to inform highway
authority of change of
ownership of premises within
one month of change
(section 142(4))

Highways authority can
serve notice withdrawing
licence if conditions
contravened effective after 7
days (section 142(6))

Highways authority can serve
notice withdrawing licence if
they think it necessary for
their functions, effective after
30 days (section 142(6))

Highway authority can remove
trees, etc. and reinstate highway
and recover expenses from
licensee or authorise last
licensee to remove trees etc. and
reinstate the highway at his own
expense (section 142(7))

(b) the execution by any person of any works authorised by the licence or by the highway authority, or

(c) the execution by or on behalf of the highway authority of any works,

but this does not require any person to indemnify the highway authority against any claim in respect of injury, damage or loss which is attributable to the negligence of that authority (section 142(8)).

Restriction on planting of trees, etc. in or near carriageway

No tree or shrub shall be planted in a made-up carriageway, or within 15 feet from the centre of a made-up carriageway (section 141(1)). If a tree or shrub is planted in contravention of this the highway authority or, the person liable to maintain the highway, may serve notice on the owner or the occupier of the land in which the tree or shrub is planted require him to remove it within 21 days from the date of service of the notice (section 141(2)).

If a person fails to comply with a notice he is guilty of an offence and liable to a fine not exceeding level 1 on the standard scale and if the offence is continued after conviction he is guilty of a further offence and liable to a fine not exceeding 50p for each day on which the offence is so continued (section 141(3)).

If any person plants a tree or shrub in a highway without a licence, the tree or shrub is to be deemed to have been planted in contravention of the above provision (section 142(9)).

STREET WORKS LICENCES

References

New Roads and Street Works Act 1991.
Street Works (Maintenance) Regulations 1992.
Street Works Register (Registration Fees) Regulations 1999.
Street Works (Registers, Notices, Directions and Designations) (England) Regulations 2007.

Extent

This procedure applies to the United Kingdom.

Scope

The street authority may grant a licence (a 'street works licence') permitting a person:

(a) to place, or to retain, apparatus in the street, and

(b) to inspect, maintain, adjust, repair, alter or renew the apparatus, change its position or remove it,

and to execute any works required for or incidental to such works (including, breaking up or opening the street, or any sewer, drain or tunnel under it, or tunnelling or boring under the street) (section 50(1)).The works referred to above are those required for or incidental to works include reinstatement of the street, and where an undertaker has failed to comply with his duties with respect to reinstatement of the street, any remedial works (section 50(1)).

A street works licence authorises the licensee to execute the works permitted by the licence without obtaining any consent which would otherwise be required to be given:

(a) by any other relevant authority in its capacity as such, or
(b) by any person in his capacity as the owner of apparatus affected by the works (section 50(2)).

A street works licence does not dispense the licensee from obtaining any other consent, licence or permission which may be required; and it does not authorise the installation of apparatus for the use of which the licence of the Secretary of State is required, unless and until that licence has been granted (section 50(3)).

Schedule 3 (below) has effect with respect to the grant of street works licences, the attachment of conditions and other matters (section 50(4)).

A street works licence may be granted:

(a) to a person on terms permitting or prohibiting its assignment, or
(b) to the owner of land and his successors in title (section 50(5)).

Street works licences (schedule 3) (FC 3.24)

Before granting a street works licence the street authority must give not less than 10 working days' notice to each of the following:

(a) where the works are likely to affect a public sewer, to the sewer authority;
(b) where the works are to be executed in a part of a street which is carried or crossed by a bridge, or crosses or is crossed by any other property vested in or held by a transport authority, to that authority;
(c) where in any other case the part of the street in which the works are to be executed is carried or crossed by a bridge, to the bridge authority;
(d) to any person who has given advance notice of certain works (section 54) of his intention to execute street works which are likely to be affected by the works to which the licence relates; and
(e) to any other person having apparatus in the street which is likely to be affected by the works

but a failure to do so does not affect the validity of the licence (schedule 3 para. 1).

The street authority may charge a reasonable fee for legal or other expenses in connection with the grant of a street works licence and an annual fee of a reasonable amount for administering the licence from the licensee.

This does not apply to the authority's own land (schedule 3 para. 2).

FC 3.24 Street works licences – New Roads and Street Works Act 1991

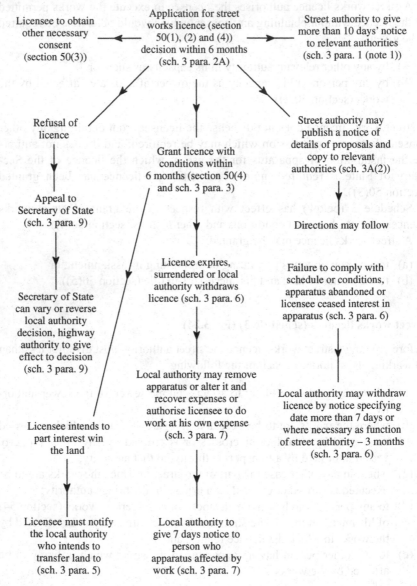

Licensee to obtain other necessary consent (section 50(3))

Application for street works licence (section 50(1), (2) and (4)) decision within 6 months (sch. 3 para. 2A)

Street authority to give more than 10 days' notice to relevant authorities (sch. 3 para. 1 (note 1))

Refusal of licence

Grant licence with conditions within 6 months (section 50(4) and sch. 3 para. 3)

Street authority may publish a notice of details of proposals and copy to relevant authorities (sch. 3A(2))

Appeal to Secretary of State (sch. 3 para. 9)

Directions may follow

Secretary of State can vary or reverse local authority decision, highway authority to give effect to decision (sch. 3 para. 9)

Licence expires, surrendered or local authority withdraws licence (sch. 3 para. 6)

Failure to comply with schedule or conditions, apparatus abandoned or licensee ceased interest in apparatus (sch. 3 para. 6)

Local authority may remove apparatus or alter it and recover expenses or authorise licensee to do work at his own expense (sch. 3 para. 7)

Licensee intends to part interest win the land

Local authority may withdraw licence by notice specifying date more than 7 days or where necessary as function of street authority – 3 months (sch. 3 para. 6)

Licensee must notify the local authority who intends to transfer land to (sch. 3 para. 5)

Local authority to give 7 days notice to person who apparatus affected by work (sch. 3 para. 7)

Notes
1. Relevant authorities – highway authority, sewer authority, bridge authority, transport authority, apparatus owner.
2. Licensee proposing to cease using or abandoning the apparatus, etc. must give local authority 6 weeks' notice before acting (sch. 3 para. 5).
3. Offence to do street works without a licence (section 51(1)).

Decisions on applications to install facilities must be determined within 6 months (schedule 3 para. 2A)

A street authority may attach conditions to a licence in the interests of safety, to minimise the inconvenience to persons using the street (having regard, in particular, to the needs of people with a disability), or to protect the structure of the street and the integrity of apparatus in it (schedule 3 para. 3).

Where assignment of a licence is permitted, a condition may be attached requiring the consent of the street authority to any assignment (schedule 3 para. 4).

Where the licensee proposes to cease using or abandon the apparatus, or to part with his interest in the apparatus, he must give the street authority at least 6 weeks' notice before doing so.

Where the licensee under a licence granted to the owner of land and his successors in title proposes to part with his interest in the land, he must, before doing so, notify the street authority stating to whom the benefit of the licence is to be transferred.

A person who fails to comply with this obligation commits an offence and is liable on summary conviction to a fine not exceeding level 4 on the standard scale (schedule 3 para. 5).

The street authority may by notice in writing served on the licensee withdraw a street works licence:

(a) if the licensee fails to comply with the schedule or any condition of the licence;

(b) if the authority become aware that the licensee:

   (i) has ceased to use or has abandoned the apparatus, or intends to do so, or

   (ii) has parted with or intends to part with his interest in the apparatus in a case where assignment of the licence is prohibited; or

(c) if the authority consider the withdrawal of the licence is necessary in exercise of their functions as street authority.

The withdrawal takes effect at the end of the period specified in the notice.

The period shall not be less than 7 working days in the case of a withdrawal, and not be less than 3 months in the case of a withdrawal necessary in exercise of their functions as street authority (schedule 3 para. 6).

Expiry, withdrawal or surrender

Where a street works licence expires or is withdrawn or surrendered, the street authority may remove the apparatus to which the licence relates or alter it and reinstate the street, and may recover the expenses from the former licensee. If they are satisfied that the former licensee can, within such reasonable time which they may specify, remove the apparatus or alter it and reinstate the street, they may authorise him to do so at his own expense.

The street authority or the former licensee, must give not less than 7 working days' notice to any person whose apparatus is likely to be affected and must satisfy their requirements about the method of executing the works and the supervision of the works by them (schedule 3 para. 7).

The licensee named on a street works licence must indemnify the street authority against any claim in respect of injury, damage or loss arising out of the placing or presence in the street of apparatus to which the licence relates, or the execution by any person of any works authorised by the licence and the former licensee must indemnify the street authority against any claim in respect of injury, damage or loss arising out of the execution by the authority or the licensee of any works above (schedule 3 para. 8).

Appeal

Where the apparatus referred to in an application for a street works licence is made to a local highway authority is to be placed or retained on a line crossing the street, and not along the line of the street, a person aggrieved by:

(a) the refusal of the authority to grant him a licence,
(b) their refusal to grant a licence except on terms prohibiting its assignment, or
(c) any terms or conditions of the licence granted to him,

may appeal to the Secretary of State.

Where on an appeal the Secretary of State reverses or varies the decision of the local highway authority, it is the duty of that authority to give effect to his decision (schedule 3 para. 9).

Restriction on works following substantial street works

The street authority may publish a notice:

(a) specifying the nature and location of the proposed works and the date on which it is proposed to begin them;
(b) stating that the authority propose to issue a direction imposing a restriction on street works;
(c) stating the duration of the proposed restriction and the part of the highway to which it relates;
(d) requiring any other undertakers who propose to execute street works in that part of the highway, and who have not already done so, to notify the authority of their proposed works within the period specified in the notice.

A notice must be published in the prescribed form and manner; and comply with the requirements of its form and content.

A copy of a notice must be given to each of the following:

(a) where there is a public sewer in the part of the highway, to the sewer authority;
(b) where that part of the highway is carried or crossed by a bridge vested in a transport authority, or crosses or is crossed by any other property held by or used for the purposes of a transport authority, to that authority;
(c) where in any other case that part of the highway is carried or crossed by a bridge, to the bridge authority;

   (d) any person who has given notice of his intention to execute street works in that part of the highway;

   (e) any person who has apparatus in that part of the highway (schedule 3A(2)).

After the expiry of the notice period the street authority may issue directions about the date from which the undertaker is proposing to execute and who has given notice to execute the substantial street works.

Offences

Where a direction is given to an undertaker as respects the date on which he may begin to execute the works proposed by him, and he begins to execute those works before that date, he is guilty of an offence.

After the expiry of the notice period, any undertaker who, before completion of the works executes any other street works in the part of the highway, commits an offence.

This does not apply where an undertaker executes emergency works.

A person guilty of an offence is liable on summary conviction to a fine not exceeding level 5 on the standard scale (schedule 3A(3)).

Directions (FC 3.25)

After the expiry of the notice period and before completion of the works, the authority may give a direction restricting the execution of street works in the part of the highway for such period following completion of the works specified in the direction.

A direction must be given in the prescribed manner and comply with such requirements as to its form and content.

The street authority must send a copy of any direction to the persons specified above.

A direction ceases to have effect if the works are not completed within the period prescribed.

A direction may be revoked at any time by the authority which gave it.

Where a direction ceases to have effect or is revoked the street authority must notify the persons specified above.

If the street authority decides not to give a direction, it must notify the persons specified above (schedule 3A(4)).

Where a direction is in force, an undertaker cannot execute street works in the part of the highway to which the restriction relates during the period specified in the direction.

This does not apply where an undertaker executes emergency works or where an undertaker executes works with the consent of the street authority. The consent of the street authority must not be unreasonably withheld. This may be settled by arbitration or by a person specified by the Secretary of State on appeal by the undertaker.

An undertaker who executes works during the period of the direction commits an offence and is liable on summary conviction to a fine not exceeding level 5 on the standard scale.

An undertaker convicted of an offence is liable to reimburse the street authority any costs reasonably incurred by them in reinstating the highway (schedule 3A(5)).

An undertaker is taken not to have failed to fulfil any statutory duty to afford a supply or service if, or to the extent that, his failure is attributable to a restriction imposed under this Schedule (schedule 3A(6)).

Prohibition of unauthorised street works

It is an offence for a person other than the street authority:

(a) to place apparatus in a street, or
(b) to break up or open a street, or a sewer, drain or tunnel under it, or to tunnel or bore under a street, for the purpose of placing, inspecting, maintaining, adjusting, repairing, altering or renewing apparatus, or of changing the position of apparatus or removing it,

otherwise than following a statutory right or under a street works licence (section 51(1)).

A person committing such an offence is liable on summary conviction to a fine not exceeding level 5 on the standard scale (section 51(2)).

This does not apply to works for road purposes or to emergency works of any description (section 51(3)).

If a person commits such an offence, the street authority may:

(a) in the case of an offence direct him to remove the apparatus in respect of which the offence was committed, and
(b) in any case, direct him to take such steps as appear to them necessary to reinstate the street or any sewer, drain or tunnel under it.

If he fails to comply with the direction, the authority may remove the apparatus or carry out the necessary works and recover from him the costs reasonably incurred by them in doing so (section 51(4)).

The street works register

A street authority must keep a register for the street works executed or proposed to be executed in the street (section 53(1)). The register must contain prescribed information in the form and manner prescribed (section 53(2)).

The authority must make the register available for inspection, at all reasonable hours and free of charge by any person having authority to execute works of any description in the street, or otherwise appearing to the authority to have a sufficient interest, and by any other person. This does not apply to restricted information (section 53(3)).

The Street Works (Registers, Notices, Directions and Designations) (England) Regulations 2007 prescribes the form and content of the register. They also designate the fees payable by the undertaker for registrations.

FC 3.25 Street licences – directions – Schedule 3A New Roads and Street Works Act 1991

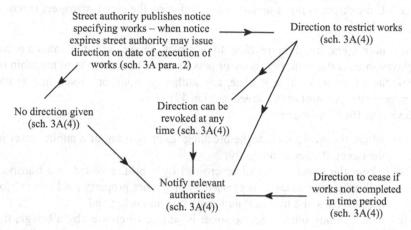

Street authority publishes notice specifying works – when notice expires street authority may issue direction on date of execution of works (sch. 3A para. 2)

Direction to restrict works (sch. 3A(4))

No direction given (sch. 3A(4))

Direction can be revoked at any time (sch. 3A(4))

Notify relevant authorities (sch. 3A(4))

Direction to cease if works not completed in time period (sch. 3A(4))

Notes

1. Relevant authorities – highway authority, sewer authority, bridge authority, transport authority, apparatus owner.
2. Offence to start works before expiry of publication notice and to carry out works not specified (sch. 3A (3)).
3. Offence to carry out works during the time of the direction (sch. 3A (5)).

Definitions

street authority means:

    (a) if the street is a maintainable highway, the highway authority, and

    (b) if the street is not a maintainable highway, the street managers (section 49).

street managers, used in relation to a street which is not a maintainable highway, means the authority, body or person liable to the public to maintain or repair the street or, if there is none, any authority, body or person having the management or control of the street (section 49).

relevant authorities means:

    (a) where the works include the breaking up or opening of a public sewer in the street, the sewer authority;

    (b) where the street is carried or crossed by a bridge vested in a transport authority, or crosses or is crossed by any other property held or used for the purposes of a transport authority, that authority; and

    (c) where in any other case the street is carried or crossed by a bridge, the bridge authority (section 49).

emergency works means works whose execution at the time when they are executed is required in order to put an end to, or to prevent the occurrence of, circumstances then existing or imminent (or which the person responsible for the works believes on reasonable grounds to be existing or imminent) which are likely to cause danger to persons or property (section 52(1)).

TABLES AND CHAIRS, A-BOARDS AND PAVEMENT DISPLAY LICENCES

Reference

Highways Act 1980.

Extent

This procedure applies in England and Wales.

Scope

This procedure applies to licences for placing tables, etc. onto pavements and the prevention of obstructions on pavements.

Execution of works and use of objects, etc. by persons other than councils (FC 3.26)

A council can grant a person permission to place objects or structures on, in or over such a highway, for the purpose of enhancing the amenity of the highway

FC 3.26 Tables and chairs licence – Highways Act 1980

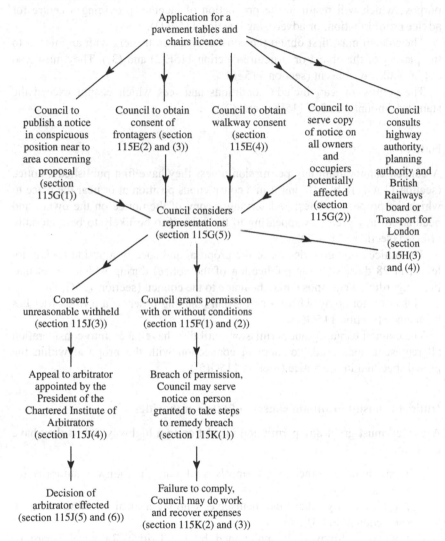

Application for a
pavement tables and
chairs licence

Council to
publish a notice
in conspicuous
position near to
area concerning
proposal
(section
115G(1))

Council to obtain
consent of
frontagers (section
115E(2) and (3))

Council to obtain
walkway consent
(section
115E(4))

Council to
serve copy
of notice on
all owners
and
occupiers
potentially
affected
(section
115G(2))

Council
consults
highway
authority,
planning
authority and
British
Railways
board or
Transport for
London
(section
115H(3)
and (4))

Council considers
representations
(section 115G(5))

Consent
unreasonable withheld
(section 115J(3))

Council grants permission
with or without conditions
(section 115F(1) and (2))

Appeal to arbitrator
appointed by the
President of the
Chartered Institute of
Arbitrators
(section 115J(4))

Breach of permission,
Council may serve
notice on person
granted to take steps
to remedy breach
(section 115K(1))

Decision of
arbitrator effected
(section 115J(5) and (6))

Failure to comply,
Council may do work
and recover expenses
(section 115K(2) and (3))

Note

If consultee consent not given within 28 days of request refusal deemed to be unreasonable (section 115J(3)).

and its immediate surroundings; or of providing a service for the benefit of the public or a section of the public or to provide, maintain and operate facilities for recreation or refreshment or both.

This also applies to using objects or structures on, in or over a highway for a purpose which will result in the production of income; providing a centre for advice or information; or advertising (section 115E(1)).

The council must first obtain the consent of the frontagers with an interest to the placing of the object or structure (section 115E(2) and (3)). They must also obtain walkway consent (section 115E(4)).

The permission can include conditions and fees which cannot exceed the standard amount (section 115F(1) and (2)).

Notices

A council cannot grant any permission unless they have first published a notice (section 115G(1)), by affixing it in a conspicuous position at or near the place to which the proposal relates; and serving a copy of the notice on the owner and occupier of any premises appearing to the council to be likely to be materially affected (section 115G(2)).

The notice must give details of the proposal and specify a period (being not less than 28 days after the publication of the notice) during which representations regarding the proposal may be made to the council (section 115G(3)).

This does not apply where a pedestrian planning order or a traffic order has been made (section 115G(4)).

The council cannot grant permission until they have taken into consideration all representations made to them in connection with the proposal within the period specified in the notice (section 115G(5)).

Duties to consult or obtain consent of other authorities

A council must grant any permission in relation to a highway unless they have consulted:

(i) any authority other than themselves who are the highway authority for the highway;

(ii) any authority other than themselves who are a local planning authority (section 115H(3)); and

(iii) where a highway is maintained by the British Railways Board or Transport for London or any of its subsidiaries (section 115H(4)).

Consents not to be unreasonably withheld

Consent is not to be unreasonably withheld but may be given subject to any reasonable conditions (section 115J(1)).

It may be reasonable for consent to be given for a specified period of time or subject to the payment of a reasonable sum (section 115J(2)).

Consent is to be treated as unreasonably withheld if:

(a) the council have served a notice asking for consent on the person whose consent is required; and

(b) he fails within 28 days of the service of the notice to give the council notice of his consent or his refusal to give it (section 115J(3)).

Any question whether consent is unreasonably withheld or is given subject to reasonable conditions must be referred to and determined by an arbitrator to be appointed, in default of agreement, by the President of the Chartered Institute of Arbitrators (section 115J(4)).

If the arbitrator determines that consent or condition has been unreasonably withheld or unreasonable, but it appears to him that there are conditions subject to which it would be reasonable to give it, he may direct that it can be treated as having been given subject to those conditions (section 115J(5) and (6)).

The expenses and remuneration of the arbitrator must be paid by the council seeking the consent (section 115J(7)) and where the arbitration concerns the consent of the British Railways Board or Transport for London, or any of its subsidiaries, the arbitrator may give such directions he thinks fit about the payment of his expenses and remuneration (section 115J(8)).

Failure to comply with terms of permission

If it appears to a council that a person to whom they have granted has committed any breach of the terms of that permission, they may serve a notice on him requiring him to take such steps to remedy the breach specified in the notice within such time specified (section 115K(1)).

If a person on whom a notice is served fails to comply with the notice, the council may take the steps themselves (section 115K(2)).

Where a council have incurred expenses in the exercise of the power above those expenses, together with interest at such reasonable rate as the council may determine from the date of service of a notice of demand for the expenses, may be recovered by the council from the person on whom the notice was served (section 115K(3)).

Definitions

walkway consent means the consent of any person:

  (i) who is an occupier of the building in which the walkway subsists;

  (ii) whose agreement would be needed for the creation of the walkway if it did not already exist;

  (iii) who is an owner or occupier of premises adjoining the walkway; and

  (iv) who is the owner of the land on, under or above which the walkway subsists.

the standard amount means in relation to permission to use an object or structure or operating facilities for recreation or refreshment or both provided by a council, the aggregate of the cost of providing it; and of such charges as will reimburse the council their reasonable expenses in connection with granting the permission (section 115F(3)).

HOUSE TO HOUSE COLLECTIONS

References

House To House Collections Act 1939.
House To House Collections Regulations 1947.

Extent

These provisions apply to England, Scotland and Wales. The regulations do not apply to Scotland.

Charitable collections from house to house to be licensed

No collection for a charitable purpose can be made unless a licence for its promotion has been issued (section 1(1)).

If a person promotes a collection for a charitable purpose, and it is made without a licence he is guilty of an offence (section 1(2)).

If a person acts as a collector in any locality in a collection for a charitable purpose without a licence he is guilty of an offence (section 1(3)).

If the chief officer of police is satisfied a collection is being made for a charitable purpose is also local and likely to be completed within a short period of time, he may grant a certificate in the prescribed form, and, where a certificate is granted, the provisions of this Act, except sections 5, 6 and 8, do not apply (section 1(4)).

Licences

Where a person who is promoting, or proposes to promote, a collection in any locality for a charitable purpose makes an application to the licensing authority in the prescribed manner specifying the purpose of the collection and the locality within which the collection is to be made, and furnishes them with the prescribed information, the authority must grant a licence authorising him to promote a collection within that locality (section 2(1)).

A licence is for 12 months but the licensing authority has the discretion to make it shorter or longer but no longer than 18 months (section 2(2)).

A licensing authority can refuse to grant a licence, or, where a licence has been granted, can revoke it, if it appears to the authority:

(a) that the total amount likely to be applied for charitable purposes as the result of the collection is inadequate in proportion to the value of the proceeds likely to be received;

(b) that remuneration which is excessive in relation to the total amount above is likely to be, or has been, retained or received out of the proceeds of the collection by any person;

(c) that the grant of a licence would be likely to facilitate the commission of an offence under section 3 of the Vagrancy Act 1824, or that an offence under that section has been committed in connection with the collection;

(d) that the applicant or the holder of the licence is not a fit and proper person to hold a licence because he has committed offences specified in the Schedule to this Act, or has others relating to fraudulent or dishonest acts or;

(e) that the applicant or the holder of the licence, in promoting a collection in respect of which a licence has been granted to him, has failed to exercise due diligence to secure that persons authorised by him to act as collectors for the purposes of the collection were fit and proper persons; or

(f) that the applicant or holder of the licence has refused or neglected to give information the authority required for the licence application (section 2(3)).

When a licensing authority refuse to grant a licence or revoke a licence which has been granted, they must give written notice to the applicant or holder of the licence stating the grounds and informing him of the right of appeal, and the applicant or holder of the licence may appeal to the Minister for the Cabinet Office against the refusal or revocation of the licence and the decision of the Minister for the Cabinet Office is final (section 2(4)).

The appeal may be brought within 14 days from the date on which notice is given (section 2(5)).

If the Minister for the Cabinet Office decides that the appeal is allowed, the licensing authority must issue a licence or cancel the revocation (section 2(6)).

The form of the application and the badges and certificates are described in regulation 4 of the House to House Collections Regulations 1947 and in schedules 1 to 4.

Exemptions in the case of collections over wide areas

Where the Minister for the Cabinet Office is satisfied that a person pursues a charitable purpose throughout the whole of England or a substantial part of it and wants to promote collections, the Minister can by order direct that he is exempt from the provisions of subsection (2) of section 1 of this Act for all collections in localities described in the order (section 3(1)).

Any order made under this section may be revoked or varied by a subsequent order made by the Minister for the Cabinet Office (section 3(2)).

Regulations

The Minister for the Cabinet Office may make regulations for prescribing anything needed for such collections where licences have been granted and carried out by, and the conduct of promoters and collectors in relation to such collections (section 4(1)). This may include prescribed badges and certificates of authority, the age of collectors, for preventing annoyance to the occupants of houses visited by collectors, the prescribed information for expenses, proceeds and application of the proceeds of collections, and requiring the information furnished to be vouched and authenticated (section 4(2)).

Any person who contravenes or fails to comply with the provisions of a regulation is guilty of an offence (section 4(3)).

Unauthorised use of badges, etc.

If any person, displays or uses a prescribed badge or a prescribed certificate of authority, not made under this Act, or one which resembles a prescribed badge or a prescribed certificate of authority calculated to deceive, he is guilty of an offence (section 5).

Collector to give name, etc. to police on demand

A police constable may require any person whom he believes a collector for the purposes of a collection for a charitable purpose to declare his name and address and to sign his name, and if any person fails to comply with this requirement, he is guilty of an offence (section 6).

Delegation of functions

The functions conferred on a chief officer of police may be delegated to a police inspector (section 7).

Penalties

Any promoter operating without a licence is liable, on summary conviction, to imprisonment for a term not exceeding 6 months or to a fine not exceeding level 3 on the standard scale, or both (section 8(1)).

Any collector operating without a licence is liable, on summary conviction, to a fine not exceeding level 2 on the standard scale or imprisonment for a term not exceeding 3 months, or both (section 8(2)).

Any person contravening regulations is liable on summary conviction, to a fine not exceeding level 1 on the standard scale (section 8(3)).

Any person using unauthorised badges or certificates is liable, on summary conviction, to imprisonment for a term not exceeding 6 months or to a fine not exceeding level 3 on the standard scale, or both (section 8(4)).

Any person failing to give his name and address or sign his name is liable, on summary conviction, to a fine not exceeding level 1 on the standard scale (section 8(5)).

If any person when giving information knowingly or recklessly makes a statement false in a material particular, he is guilty of an offence, and liable, on summary conviction, to imprisonment for a term not exceeding 6 months or to a fine not exceeding level 3 on the standard scale, or both (section 8(6)).

Where an offence committed by a corporation is proved to have been committed with the consent or connivance of, or to be attributable to any culpable neglect of duty on the part of, any director, manager, secretary, or other officer of the corporation, he, as well as the corporation, is deemed to be guilty of that offence and liable to be proceeded against and punished accordingly (section 8(7)).

Schedule 1

Offences to which paragraph (d) of subsection (3) of section 2 applies

Sections 47 to 56 of the Offences against the Person Act 1861.
Robbery, burglary, and blackmail.
In Scotland involving personal violence or lewd, indecent, or libidinous conduct, or dishonest appropriation of property.
Under the Street Collections Regulation (Scotland) Act 1915.
Section 5 of the Police, Factories, etc. (Miscellaneous Provisions) Act 1916.

PLEASURE BOAT LICENSING

Reference

Public Health Acts Amendment Act 1907 section 94.

Extent

These provisions apply in England, Scotland, Northern Ireland and Wales.

Scope

A local authority may grant licences for pleasure boats and pleasure vessels (with terms and conditions) to be let for hire or to be used for carrying passengers for hire. Licences also apply to persons in charge of or navigating such boats and vessels the local authority may charge an annual fee for each type of licence.

The period of the licence is determined by the local authority and the licence can be suspended or revoked where they think it is necessary or desirable in the interests of the public.

It is an offence to let for hire pleasure boats not licenced under these provisions and is liable to a penalty not exceeding level 3 on the standard scale. However, it is a defence where the failure to comply is due to a reasonable excuse for the failure.

Any person aggrieved by the withholding, suspension or revocation of any licence may appeal to a court held after the expiration of 2 clear days after such withholding, suspension, or revocation but the aggrieved person must give 24 hours' written notice of an appeal with details of the grounds of appeal. The court has the power to make an order as they see fit and to award costs. Costs are recoverable summarily as a civil debt. These provisions do not apply to pleasure boats, etc. on any inland waterway owned or managed by Canal & River Trust or licensed under other relevant regulations.

Definition

let for hire means let for hire to the public (section 94).

POISONS LICENCE

References

Poisons Act 1972.
Poison Rules 1982 (as amended).
Poisons List Order 1982 (as amended).

Extent

This procedure applies to England, Scotland and Wales.

Scope

The retailing of poisons is controlled by the Poisons Act 1972. The Poisons List Order contains a list of poisons covered by the Act.

The Active Ingredients approved for use in the UK are specified under Parts I and II of the Poisons List, as follows.

Part I

Aluminium phosphide
Arsenic; its compounds, other than those specified in Part II of this List
Barium, salts of, other than barium sulphate and the salts of barium specified in
    Part II of this List
Bromomethane
Chloropicrin
Fluoroacetic acid; its salts; fluoroacetamide
Hydrogen cyanide; metal cyanides, other than ferrocyanides and ferricyanides
Lead acetates; compounds of lead with acids from fixed oils
Magnesium phosphide
Mercury, compounds of, the following:
    nitrates of mercury; oxides of mercury; mercuric cyanide oxides; mercuric thio-
    cyanate; ammonium mercuric chlorides; potassium mercuric iodides; organic
    compounds of mercury which contain a methyl (CH_3) group directly linked to
    the mercury atom
Oxalic acid
Phenols (phenol; phenolic isomers of the following – cresols, xylenols, monoethyl-
    phenols) except in substances containing less than 60 per cent, weight in weight,
    of phenols; compounds of phenols with a metal, except in substances containing
    less than the equivalent of 60 per cent, weight in weight, of phenols
Phosphorus, yellow
Strychnine; its salts; its quaternary compounds
Thallium, salts of

Part II

Aldicarb
Alpha-chloralose

Ammonia

Arsenic, compounds of, the following:
 calcium arsenites; copper acetoarsenite; copper arsenates; copper arsenites; lead arsenates

Barium, salts of, the following:
 barium carbonate; barium silicofluoride

Carbofuran

Cycloheximide

Dinitrocresols (DNOC); their compounds with a metal or a base

Dinoseb; its compounds with a metal or a base

Dinoterb

Drazoxolon; its salts

Endosulfan

Endothal; its salts

Endrin

Fentin, compounds of

Formaldehyde

Formic acid

Hydrochloric acid

Hydrofluoric acid; alkali metal bifluorides; ammonium bifluoride; alkali metal fluorides; ammonium fluoride; sodium silicofluoride

Mercuric chloride; mercuric iodide; organic compounds of mercury except compounds which contain a methyl (CH_3) group directly linked to the mercury atom

Metallic oxalates

Methomyl

Nicotine; its salts; its quaternary compounds

Nitrobenzene

Oxamyl

Paraquat, salts of

Phenols (as defined in Part I of this List) in substances containing less than 60 per cent, weight in weight, of phenols; compounds of phenols with a metal in substances containing less than the equivalent of 60 per cent, weight in weight, of phenols

Phosphoric acid

Phosphorus compounds, the following:
 azinphos-methyl, chlorfenvinphos, demephion, demeton-S-methyl, demeton-S-methyl sulphone, dialifos, dichlorvos, dioxathion, disulfoton, fonofos, mecarbam, mephosfolan, methidathion, mevinphos, omethoate, oxydemeton-methyl, parathion, phenkapton, phorate, phosphamidon, pirimiphos-ethyl, quinalphos, thiometon, thionazin, triazophos, vamidothion

Potassium hydroxide

Sodium hydroxide

Sodium nitrite

Thiofanox

Zinc phosphide

FC 3.27 Poisons licences – Poisons Act 1972

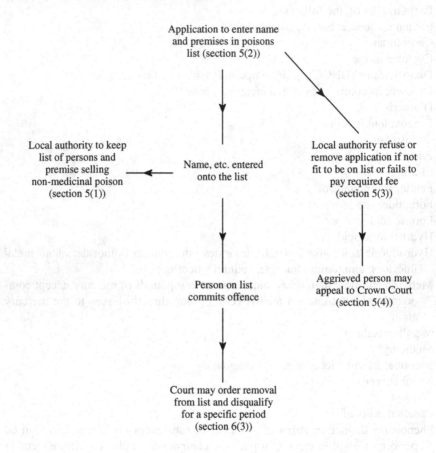

Application to enter name
and premises in poisons
list (section 5(2))

Local authority to keep
list of persons and
premise selling
non-medicinal poison
(section 5(1))

Name, etc. entered
onto the list

Local authority refuse or
remove application if not
fit to be on list or fails to
pay required fee
(section 5(3))

Person on list
commits offence

Aggrieved person may
appeal to Crown Court
(section 5(4))

Court may order removal
from list and disqualify
for a specific period
(section 6(3))

Lists of persons entitled to sell poisons in Part II of Poisons List (FC 3.27)

Every local authority must keep a list of persons and premises entitled to sell non-medicinal poisons included in Part II of the Poisons List (section 5(1)).

The local authority must enter in the list the name of any person who, having premises in the area of the authority, makes an application to the authority in the prescribed form to have his name and premises entered in the list (section 5(2)).

A local authority's list must include particulars of the premises in respect of which the name of any person is entered in the list; be on a prescribed form; and be open at all reasonable times to the inspection of any person without fee (section 6(1)).

A person whose name is on a local authority's list must pay reasonable fees as the authority may determine in respect of the entry of his name in the list; the alteration about the premises in respect of which his name is entered, and the retention of his name on the list in any subsequent year after the year his name is first entered in it (section 6(2)).

A local authority may refuse to enter in, or may remove from, the list the name of any person who fails to pay any fees determined by the authority, or who in the opinion of the authority is, for any sufficient reason relating either to him personally or to his premises, not fit to be on the list (section 5(3)).

If any person is aggrieved by the refusal of a local authority to enter his name in the list or by the removal of his name from the list, he may appeal against the refusal or removal to the Crown Court (section 5(4)).

If any person whose name is on a local authority's list is convicted before any court of any offence which, in the opinion of the court, renders him unfit to have his name on the list, the court may, as part of the sentence, order his name to be removed from the list and direct that he shall, for a period specified in the order, be disqualified for having his name entered in any local authority's list (section 6(3)).

Offences

The following is not lawful:

1. For a person to sell any non-medicinal poison included in Part I of the Poisons List, unless:
   (i)   he is a person lawfully conducting a retail pharmacy business, and
   (ii) the sale is in a registered pharmacy, and
   (iii) the sale is by, or under the supervision of, a pharmacist.
2. For a person to sell any non-medicinal poison included in Part II of the Poisons List, unless:
   (i)   he is a person lawfully conducting a retail pharmacy business and the sale is in a registered pharmacy, or
   (ii) his name is on a local authority's list of the premises on which the poison is sold.
3. For a person to sell any non-medicinal poison included in Part I or in Part II of the Poisons List, unless the container of the poison is labelled in the prescribed manner:

(i)   with the name of the poison, and

(ii) in the case of a preparation which contains a poison, with the prescribed particulars of the proportion which the poison contained in the preparation bears to the total ingredients, and

(iii) with the word 'poison' or other prescribed indication of the character of the article, and

(iv) with the name of the seller of the poison and the address of the premises on which it is sold (section 3(1)).

4. To sell any non-medicinal poison included in Part I of the Poisons List to any person unless that person is either:

(i)   certified in writing in the prescribed manner by a person authorised by the Poisons Rule to give a certificate, or

(ii) known by the seller or by a pharmacist in the employment of the seller at the premises where the sale is done, to be a person to whom the poison may properly be sold.

The seller of any poison must not deliver it until he recorded the delivery in a book stating the date of the sale, the name and address of the purchaser and of the person by whom the certificate required above was given, the name and quantity of the article sold, and the purposes for which it is stated by the purchaser to be required, and the purchaser has signed the entry (section 3(2)).

5. For a non-medicinal poison to be exposed for sale in, or to be offered for sale in an automatic machine (section 3(3)).

6. For any person whose name is entered in a local authority's list to use in connection with his business any title, emblem or description reasonably calculated to suggest that he is entitled to sell any poison which he is not entitled to sell.

If any person acts in contravention above he shall be liable on summary conviction, in respect of each offence, to a fine not exceeding level 2 on the standard scale and, in the case of a continuing offence, to a further fine not exceeding five pounds for every day subsequent to the day on which he is convicted of the offence (section 6(4)).

Exclusion of sales by wholesale and certain other sales

Except as provided by the Poisons Rules, the above does not apply to:

(a)   the sale of poisons in wholesale dealing;

(b)   the sale of poisons to be exported to purchasers outside the United Kingdom;

(c)   the sale of an article to a doctor, dentist, veterinary surgeon or veterinary practitioner for the purpose of his profession;

(d)   the sale of an article for use in or in connection with any hospital, infirmary, dispensary or similar institution approved by an order, whether general or special, of the Secretary of State;

(e)   the sale of an article by a person carrying on a business in the course of which poisons are regularly sold either by way of wholesale dealing or for use by the purchasers in their trade or business (section 4).

Poisons Rules

The Secretary of State has made rules about the following matters:

(a) the sale, whether wholesale or retail, or the supply of non-medicinal poisons, by or to any persons or classes of persons;

(b) the storage, transport and labelling of non-medicinal poisons;

(c) the containers in which non-medicinal poisons may be sold or supplied;

(d) the addition to non-medicinal poisons of specified ingredients for the purpose of rendering them readily distinguishable as non-medicinal poisons;

(e) the compounding of non-medicinal poisons, and the supply of non-medicinal poisons on and in accordance with a prescription duly given by a doctor, a dentist, a veterinary surgeon or a veterinary practitioner;

(f) the period for which any books required to be kept are to be preserved;

(g) the period for which any certificate is to remain in force (section 7).

Penalties

A person who acts in contravention of or fails to comply with the Act (other than section 6(4)) or with the Poisons Rules is, on summary conviction, liable in respect of each offence to a fine not exceeding level 4 on the standard scale, and, in the case of a continuing offence, to a further fine not exceeding ten pounds for every day subsequent to the day on which he is convicted of the offence (section 8(1)).

In the case of proceedings against a person for or in connection with the sale, exposure for sale or supply of a non-medicinal poison effected by an employee it is not a defence that the employee acted without the authority of the employer, and any material fact known to the employee is deemed to have been known to the employer (section 8(2)).

Proceedings for an offence can be commenced at any time within the period of 12 months after the date of the commission of the offence or, in the case of proceedings instituted by, or by the direction of, the Secretary of State, either within that period or within the period of 3 months after the date on which evidence sufficient in the opinion of the Secretary of State to justify a prosecution for the offence comes to his knowledge (section 8(3)).

Inspection and enforcement

An inspector appointed by the General Pharmaceutical Council has power of entry at all reasonable times to enter any registered pharmacy and where he has reasonable cause to suspect that a breach of the law has been committed in relation to poisons.

He can also carry out examination and inquiry and do other things (including the taking, on payment, of samples) as may be necessary for ascertaining whether the Act and Rules are being complied with (section 9(4)).

It is the duty of every local authority to inspect and take all reasonable steps to secure compliance with the Act and with the Poisons Rules relating to substances included in Part II of the Poisons List, and to appoint inspectors (section 9(5)).

A local authority may, with the consent of the General Pharmaceutical Council, appoint an inspector appointed by the General Pharmaceutical Council to be also an inspector for the local authority (section 9(5)).

An inspector appointed by the local authority has power at all reasonable times to enter any premises on which any person whose name is entered in a local authority's list carries on business and where the inspector has reasonable cause to suspect that a breach of the law has been committed in respect of any substances included in Part II of the Poisons List. In either case he has the power to examine and inquire and do other things (including the taking, on payment, of samples) necessary for the purposes of the inspection (section 9(6)).

An inspector appointed by a local authority in England or Wales has power, with the consent of the local authority, to institute proceedings under this Act before a court of summary jurisdiction in the name of the authority, and to conduct any proceedings (section 9(7)).

It is an offence to wilfully delay or obstruct an inspector in the exercise of any of these powers or to refuse to allow the taking of any sample, or fails without reasonable excuse to give any information which he is required to give. A person on summary conviction is liable to a fine not exceeding level 2 on the standard scale (section 9(8)).

The power of entry does not apply to a doctor's, dentist's, veterinary surgeon's or veterinary practitioner premises (section 9(9)).

Definitions

non-medicinal poison means a substance which is included in Part I or Part II of the Poisons List and is neither a medicinal product or a veterinary medicinal product (section 11(1)).

SEX ESTABLISHMENT LICENCES

Reference

Local Government (Miscellaneous Provisions) Act 1982 section 2 and schedule 3.

Extent

These provisions apply to England and Wales.

Scope

A local authority may resolve that the provisions of the control of sex establishments in schedule 3 of the act may apply to their area.

Once the resolution has been passed the provisions come into force 1 month after the resolution is passed.

The local authority must publish notice that they have passed a resolution in 2 consecutive weeks in a local newspaper circulating in their area at least 28 days before the day of the provisions coming into force (section 2).

Provisions introduced in 2011 bring the licensing of lap dancing premises and similar venues in line with other sex establishments (sex shops and sex cinemas).

The provisions allow the licensing authority to prescribe a wider range of conditions on the licences of sexual entertainment venues than those available under the Licensing Act 2003, and allow local people to oppose an application for a sex establishment licence if they have legitimate concerns that a sexual entertainment venue would be inappropriate, given the character of an area.

Licences must be renewed at least annually and local people have the opportunity to raise objections (if any) at that time.

Licensing provisions (FC 3.28)

Sex establishments must only operate under a licence granted by a licensing authority (paragraph 6). An application to operate a sex establishment must be made to the appropriate authority. An application must be made in writing and must include:

(a) the full name of the applicant or body corporate;
(b) his permanent address or address of registered or principal office, the full names and private addresses of the directors or other persons responsible for its management;
(c) his age;
(d) the full address of the premises; and
(e) where an application relates to a vehicle, vessel or stall where it is to be used as a sex establishment (paragraph 10).

The authority may grant to any applicant, and from time to time renew, a licence (with terms, conditions and restrictions) to use any premises, vehicle, vessel or stall specified in it for a sex establishment (paragraph 8).

The licence remains in force for 1 year (paragraph 8) and the authority may, where they think fit, transfer the licence to any other person on application (paragraph 9).

The applicant must give public notice of the application by publishing an advertisement in a local newspaper circulating in the appropriate authority's area not later than 7 days after the date of the application. For premises, notice must be displayed for 21 days beginning with the date of the application on or near the premises and in a place where the notice can conveniently be read by the public. The notice must be prescribed by the authority.

A copy of an application for the grant, renewal or transfer of a licence must be sent to the chief officer of police within 7 days.

Any person objecting to an application for the grant, renewal or transfer of a licence must give notice in writing of his objection to the authority, stating the grounds of the objection within 28 days of the date of the application. The authority then must give notice in writing of the general terms of the objection to the applicant. The authority must not, without the consent of the person making the objection, reveal the objector's name or address to the applicant.

The authority must have regard to any observations made by the chief officer of police and any objections of which notice has been sent to them before determining the application.

FC 3.28 Sex establishments licences – Local Government (Miscellaneous Provisions) Act 1982 Section 2 and Schedule 3

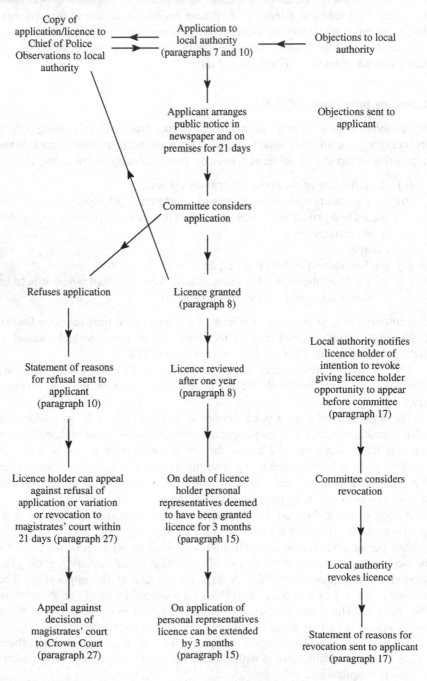

The authority must give an opportunity of being heard by a committee of the authority:

(a) before refusing to grant a licence, to the applicant;

(b) before refusing to renew a licence, to the holder; and

(c) before refusing to transfer a licence, to the holder and the person to be transferred to.

Where the appropriate authority refuse to grant, renew or transfer a licence, they must give the applicant a statement in writing of the reasons for their decision (paragraph 10 and 17).

Licences remain in force where an application has been made for its renewal or transfer until the withdrawal of the application or its determination (paragraph 11).

Waiver

The authority may waive the requirement of a licence in any case where they consider that to require a licence would be unreasonable or inappropriate. Where the authority grant an application for a waiver, they must give the applicant for the waiver, notice that they have granted his application. The authority may at any time terminate the waiver notice by giving 28 days' notice to the person in receipt of the waiver notice (paragraph 7).

Granting licences

Licences cannot be granted:

(a) to a person under the age of 18; or

(b) to a person who is for the time being disqualified due to revocation of a licence; or

(c) to a person, other than a body corporate, who is not resident in a European Economic Area state or was not so resident throughout the period of 6 months immediately preceding the date when the application was made; or

(d) to a body corporate which is not incorporated in a European Economic Area state; or

(e) to a person who has, within a period of 12 months immediately preceding the date when the application was made, been refused the grant or renewal of a licence for the premises, vehicle, vessel or stall in respect of which the application is made, unless the refusal has been reversed on appeal.

Refusal

The authority may refuse an application for the grant or renewal of a licence if:

(a) the applicant is unsuitable to hold the licence by reason of having been convicted of an offence or for any other reason;

(b) the licence were to be granted, renewed or transferred the business to which it relates would be managed by or carried on for the benefit of a

person, other than the applicant, who would be refused the grant, renewal
or transfer of such a licence if he made the application himself;

(c) that the number of sex establishments, or of sex establishments of a
particular kind, in the relevant locality at the time the application is
determined is equal to or exceeds the number which the authority
consider is appropriate for that locality (which may be nil); or

(d) that the grant or renewal of the licence would be inappropriate, having
regard:
  (i) to the character of the relevant locality; or
  (ii) to the use to which any premises in the vicinity are put; or
  (iii) to the layout, character or condition of the premises, vehicle, vessel
       or stall in respect of which the application is made (paragraph 12).

Where the holder of a licence granted dies, the licence shall be deemed to have
been granted to his personal representatives and will remain in force for 3 months
and then expire. However, the authority may on the application of the representatives, extend or further extend the period of 3 months if the authority are satisfied
that the extension is necessary for the purpose of winding up the deceased's estate
and that no other circumstances make it undesirable (paragraph 15).

The authority may, at the written request of the holder of a licence, cancel the
licence (paragraph 16).

Revocation

The authority may, after giving the holder of a licence an opportunity of appearing before and being heard by them, revoke the licence. Where a licence is
revoked, the authority must, if required to do so by the person who held it, give
him a statement in writing of the reasons for their decision within 7 days. Where
a licence is revoked, its holder shall be disqualified from holding or obtaining a
licence in the area of the appropriate authority for a period of 12 months beginning with the date of revocation (paragraph 17).

Variation

The licence holder may apply to the authority to vary the terms, conditions or
restrictions on application. The authority may vary or refuse the application (paragraph 18).

Charging

The authority may charge a reasonable fee for determining the licence (paragraph 19).

Offences

It is an offence for:

(a) a person to knowingly use, or knowingly cause or permit the use of, any
premises, vehicle, vessel or stall without a licence;

(b) a licence holder of a sex establishment, employs in the business of the establishment any person known to him to be disqualified from holding such a licence;

(c) a licence holder or servant or agent of the holder without reasonable excuse knowingly contravenes, or without reasonable excuse knowingly permits the contravention of, a term, condition or restriction specified in the licence (paragraph 20);

(d) any person who, in connection with an application for the grant, renewal or transfer of a licence to make a false statement which he knows to be false in any material respect or which he does not believe to be true (paragraph 21); or

(e) a licence holder without reasonable excuse knowingly permits a person under 18 years of age to enter the establishment; or employs a person known to him to be under 18 years of age in the business of the establishment (paragraph 23).

A person guilty of an offence is liable on summary conviction to a fine not exceeding £20,000 or level 3 on the standard scale (paragraph 22).

Powers of entry

An authorised officer of a local authority may, at any reasonable time, enter and inspect any sex establishment in respect of which a licence is for the time being in force, with a view to seeing:

(a) whether the terms, conditions or restrictions on or subject to which the licence is held are complied with;

(b) whether any person employed in the business of the establishment is disqualified from holding a licence;

(c) whether any person under 18 years of age is in the establishment; and

(d) whether any person under that age is employed in the business of the establishment.

An authorised officer may enter and inspect a sex establishment if he has reason to suspect that an offence has been, is being, or is about to be committed, if he has a warrant granted by a Justice of the Peace. The authorised officer must produce his authority if required to do so by the occupier of the premises or the person in charge of the vehicle, vessel or stall in relation to which the power is exercised.

Any person who without reasonable excuse refuses to permit an authorised officer of a local authority to exercise any such power shall be guilty of an offence and shall for every such refusal be liable on summary conviction to a fine not exceeding level 5 on the standard scale (paragraph 25).

Seizure

A person acting under the authority of a warrant may seize and remove anything found on the premises concerned that the person reasonably believes should be forfeited. The person who, immediately before the seizure, had custody or control of anything seized may request any authorised officer who seized it to provide a record of what was seized. The authorised officer must provide the record within a reasonable time of the request being made (paragraph 25A).

Appeals

Where an application or request for variation is refused or a licence holder is aggrieved by any term, condition or restriction on the licence or the licence is revoked, the applicant or licence holder can appeal to a magistrates' court within 21 days of the relevant date.

An appeal against the decision of a magistrates' court may be brought to the Crown Court. The decision of the Crown Court is final and either court can make such order as it thinks fit. The authority must carry out the order of the court.

Where a licence is revoked or an application for the renewal of a licence is refused, the licence remains in force until the time for bringing an appeal has expired or until the determination or abandonment of the appeal. Where the licence holder makes an application to vary the licence the authority imposes a term, condition or restriction other than one specified in the application, the licence is free of it until the appeal is determined (paragraph 27).

Definitions

sex establishment means a sex cinema, sex shop or sexual entertainment venue.

sex cinema is any premises, vehicle, vessel or stall used to a significant degree for the exhibition of moving pictures, by whatever means produced, which:

- are concerned primarily with the portrayal of, or primarily deal with or relate to, or are intended to stimulate or encourage sexual activity; or acts of force or restraint which are associated with sexual activity; or
- are concerned primarily with the portrayal of, or primarily deal with or relate to, genital organs or urinary or excretory functions.

A sex cinema does not include a dwelling-house to which the public is not admitted.

A premises shall not be treated as a sex cinema only if they are licensed under Section 1 of the Cinemas Act 1985, of their use for a purpose for which a licence is required; or of their use for an exhibition to which section 6 of that Act (certain non-commercial exhibitions) applies given by an exempted organisation within the meaning of section 6(6) of that Act.

sex shop means any premises, vehicle, vessel or stall used for a business which consists to a significant degree of selling, hiring, exchanging, lending, displaying or demonstrating:

- sex articles; or
- other things intended for use in connection with, or for the purpose of stimulating or encouraging sexual activity; or acts of force or restraint which are associated with sexual activity.

No premises shall be treated as a sex shop by reason only of their use for the exhibition of moving pictures by whatever means produced.

sexual entertainment venue means any premises at which relevant entertainment is provided before a live audience for financial gain of an organiser. For the provisions to apply it is not necessary to the organiser to receive financial gain directly or indirectly from the performance or display of nudity.

relevant entertainment means any live performance or any live display of nudity which is of such a nature that, ignoring financial gain, it must reasonably be assumed to be provided solely or principally for the purpose of sexually stimulating any member of the audience (whether by verbal or other means). An audience includes an audience of one.

exempt premises – the following are not defined as sexual entertainment venues:

(a) sex cinemas and sex shops;
(b) premises at which the provision of relevant entertainment is such that:
  (i) there have not been more than 11 occasions on which relevant entertainment has been so provided which fall (wholly or partly) within the period of 12 months;
  (ii) no occasion has lasted for more than 24 hours; and
  (iii) no occasion has begun within the period of 1 month beginning with the end of any previous occasion on which relevant entertainment has been so provided.

SAFETY CERTIFICATES FOR STANDS AT SPORTS GROUNDS

Reference

Fire Safety and Safety of Places of Sport Act 1987.

Extent

These provisions apply to England, Scotland and Wales.

Scope

A safety certificate is required for stands in a sports ground which provides covered accommodation in stands for spectators, and is not a designated sports ground. This applies to covered accommodation for 500 or more spectators to view activities at the ground, but one certificate may be issued for several stands (section 26(1) and (2)).

These are referred to as regulated stands (section 26(5)).

The local authority determines (using relevant criteria and guidance) whether any, and if so, which of the stands at a sports ground in their area is a regulated stand, and to issue safety certificates (section 26(6) to (8)).

The final determination of a local authority that a stand at a sports ground is a regulated stand is subject to appeal (section 26(9)).

FC 3.29 Safety certificates for stands at sports grounds – determination

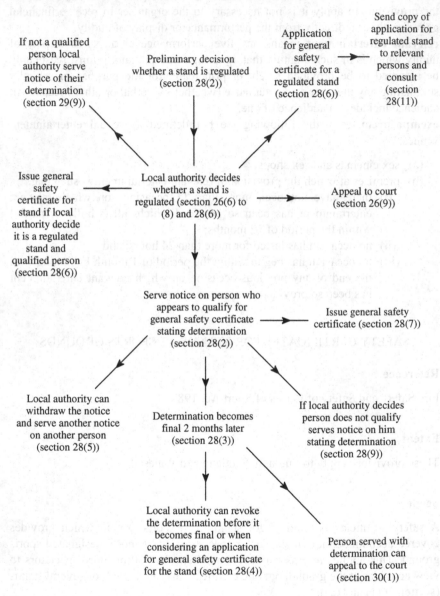

If not a qualified person local authority serve notice of their determination (section 29(9))

Preliminary decision whether a stand is regulated (section 28(2))

Application for general safety certificate for a regulated stand (section 28(6))

Send copy of application for regulated stand to relevant persons and consult (section 28(11))

Issue general safety certificate for stand if local authority decide it is a regulated stand and qualified person (section 28(6))

Local authority decides whether a stand is regulated (section 26(6) to (8) and 28(6))

Appeal to court (section 26(9))

Serve notice on person who appears to qualify for general safety certificate stating determination (section 28(2))

Issue general safety certificate (section 28(7))

Local authority can withdraw the notice and serve another notice on another person (section 28(5))

Determination becomes final 2 months later (section 28(3))

If local authority decides person does not qualify serves notice on him stating determination (section 28(9))

Local authority can revoke the determination before it becomes final or when considering an application for general safety certificate for the stand (section 28(4))

Person served with determination can appeal to the court (section 30(1))

Note
Relevant persons – chief of police, fire and rescue service and building authority.

A safety certificate for a regulated stand at a sports ground may be either a certificate for the use of the stand for viewing an activity or a number of activities specified in the certificate during an indefinite period starting with a specified date or on an occasion or series of specified occasions (section 26(10)).

Contents of safety certificates for stands (FC 3.29)

A safety certificate for a regulated stand must contain terms and conditions (appropriate to the activity) the local authority considers necessary or expedient to secure reasonable safety in the stand when it is in use for viewing the specified activity or activities at the ground (section 27(1) and (6)). This also applies to the amendment or replacement of a safety certificate (section 29(3)).

The chief officer of police determines the extent of the facilities for the provision of their services (section 27(3)).

A safety certificate may include a condition that the following records must be kept:

(a) records of the number of spectators accommodated in covered accommodation in the stand; and
(b) records relating to the maintenance of safety in the stand (section 27(4)).

A general safety certificate must contain a plan of the stand to which it applies and the area in the immediate vicinity of it, and the terms and conditions or those in any special safety certificate issued for the stand must be framed by reference to that plan (section 27(5)).

Issue of certificates

The following persons qualify for the issue of a safety certificate for a regulated stand at a sports ground:

(a) the person who qualifies for the issue of a general safety certificate is the person who is responsible for the management of the ground; and
(b) the person who qualifies for the issue of a special safety certificate for viewing an activity from the stand on any occasion is the person who is responsible for organising that activity (section 28(1)).

The local authority make a preliminary determination whether or not that stand is a regulated stand and, if they determine that it is, they must serve a notice on the person who appears to them to qualify for the issue of a general safety certificate stating their determination and the effects of it (section 28(2)). This becomes final 2 months after the date of the notice (section 28(3)).

A local authority can revoke their determination that a stand at a sports ground is a regulated stand at any time before it becomes final, or on considering an application for a general safety certificate for the stand, whether the determination has or has not become final (section 28(4)).

A local authority can, at any time before a determination becomes final, withdraw the notice of it and serve a further notice on another person, but if they do so the period of 2 months is treated as beginning with the date of the further notice (section 28(5)).

If a local authority receive an application for a general safety certificate for a regulated stand at a sports ground in their area, it is their duty:

(a) if they have not already done so, to determine whether the stand is a regulated stand and; if they determine that it is, to determine whether the applicant is the person who qualifies for the issue of the general safety certificate for it;

(b) if they have made a determination that the stand is a regulated stand and do not decide to revoke it, to determine whether the applicant is the person who qualifies for the issue of the general safety certificate for it;

and a determination made under paragraph (a) above that a stand is a regulated stand is, when made, a final determination (section 28(6)).

If the local authority determine that the applicant is the person who qualifies for the issue of the general safety certificate above they must (if no such certificate is in operation) issue to him such a certificate (section 28(7)). Similar duties apply for an application for a special safety certificate (section 28(8)).

The local authority must, if they determine that an applicant for a safety certificate does not qualify for the issue of the certificate, serve on him a notice stating their determination (section 28(9)).

The local authority must send a copy of an application for a safety certificate for a regulated stand at a sports ground to the chief officer of police, the fire and rescue authority, and the building authority for the area in which the sports ground is situated, and consult them about the terms and conditions to be included in the certificate (section 28(11)).

The local authority may, by notice, require an applicant for a safety certificate to furnish them within such reasonable time specified in the notice with information and plans of the ground they consider necessary for the purpose of discharging their functions for the issue of safety certificates for the regulated stands at the ground (section 28(11)).

If an applicant for a safety certificate fails to comply with this requirement within the time specified, or within such further time as they may allow, he is deemed to have withdrawn his application (section 28(12)).

Amendment, cancellation, etc. of certificates (FC 3.30)

The local authority who have issued a safety certificate for a regulated stand at a sports ground:

(a) must, if at any time it appears to them that the stand in respect of which it was issued is not or has ceased to be a regulated stand, revoke their previous determination and, by notice to its holder, cancel the certificate;

(b) may, in any case where it appears appropriate to them to do so, amend the certificate by notice to its holder; or

(c) may replace the certificate (section 29(1)).

The local authority must, if it appears to them that a safety certificate would require a person to contravene any provision of the Regulatory Reform (Fire

Safety) Order 2005 or regulations made under it, amend the safety certificate by notice in writing to its holder (section 29(1A)).

A safety certificate may be cancelled, amended or replaced either on the application of the holder or without such an application (section 29(2)).

A notice amending a general safety certificate must specify the date on which the amendment to which it relates is to come into operation, and the date specified may be a date later than the date of issue of the notice (section 29(4)).

If the local authority receive an application for the transfer of a safety certificate for a regulated stand at a sports ground from the holder to some other person it is their duty to determine whether that person would, if he made an application for the purpose, qualify for the issue of the certificate; and if they determine that he would, they may transfer the certificate to him and in any case notify him of their determination (section 29(5)). An application may be made either by the holder of the safety certificate or by the person to whom it is proposed that it should be transferred (section 29(6)).

The local authority must send a copy of an application for the transfer of a safety certificate for a regulated stand at a sports ground to the chief officer of police, the fire and rescue authority, and the building authority for the area in which the sports ground is situated (section 29(7)).

The local authority must consult the chief officer of police, the fire and rescue authority, and the building authority about any proposal to amend, replace or transfer a safety certificate (section 29(8)).

The holder of a safety certificate may surrender it to the local authority, then it ceases to have effect (section 29(9)).

The local authority may cancel a safety certificate if the holder dies or (if a body corporate) is dissolved (section 29(10)).

Appeals

A person who has been served with a notice of a determination of a local authority that any stand at a sports ground is a regulated stand may appeal against the determination to the court (section 30(1)).

Any person who has been served with a notice of the determination of a local authority that he does not or, in the case of an application for a transfer, would not qualify for the issue of the certificate may appeal against the determination to the court (section 30(2)).

An applicant for a special safety certificate for a regulated stand at a sports ground may also appeal to the court against a refusal of his application on grounds other than a determination that he does not qualify for the issue of the certificate (section 30(3)).

An interested party may appeal to the court against:

(a) the inclusion of anything in, or the omission of anything from, a safety certificate for a regulated stand at a sports ground; or

(b) the refusal of the local authority to amend or replace a safety certificate for a regulated stand at a sports ground (section 30(4)).

FC 3.30 Stands at sports grounds – amendment, etc.

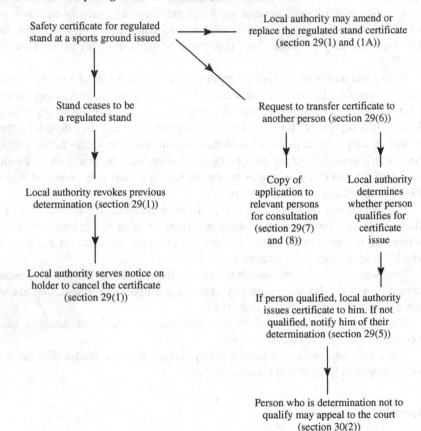

Safety certificate for regulated stand at a sports ground issued

Local authority may amend or replace the regulated stand certificate (section 29(1) and (1A))

Stand ceases to be a regulated stand

Request to transfer certificate to another person (section 29(6))

Local authority revokes previous determination (section 29(1))

Copy of application to relevant persons for consultation (section 29(7) and (8))

Local authority determines whether person qualifies for certificate issue

Local authority serves notice on holder to cancel the certificate (section 29(1))

If person qualified, local authority issues certificate to him. If not qualified, notify him of their determination (section 29(5))

Person who is determination not to qualify may appeal to the court (section 30(2))

Notes
1. A safety certificate may be cancelled, amended or replaced either on the application of the holder or without such an application (section 29(2)).
2. Appeals – see section 30.

An appeal to the magistrates' court in England and Wales is by way of complaint for an order (section 30(6)). An appeal to the court in Scotland is by summary application (section 30(7)).

If a local authority serve on any applicant for a safety certificate a notice of a determination of theirs that he does not qualify for the issue of the certificate, he is deemed to have withdrawn his application on the expiry of the period within which an appeal must be brought (section 30(9)).

If the appeal is withdrawn or the court upholds the authority's determination, the appellant is deemed to have withdrawn his application on the date of the withdrawal of his appeal or of the court's order on the appeal (section 30(10)).

Where an appeal is brought against the inclusion of any term or condition in a safety certificate (whether it was included in the certificate originally or only on its amendment or replacement), the operation of that term or condition is suspended until the court has determined the appeal (section 30(11)).

In England and Wales, the local authority and any interested party may appeal to the Crown Court against an order (section 30(12)). In Scotland this applies but that person cannot be a party to the proceedings on the application (section 30(13)).

Alterations and extensions

If while a general safety certificate for a regulated stand at a sports ground is in operation it is proposed to alter or extend the stand or its installations, and the alteration or extension is likely to affect the safety of persons in the stand, the holder of the certificate must, before the carrying out of the proposals is begun, give notice of the proposals to the local authority (section 32(1) and (2)).

Enforcement

It is the duty of every local authority to enforce this act and to arrange for the periodical inspection of sports grounds at which there are regulated stands (section 34(1)). Local authorities must act in accordance with guidance the Secretary of State may give them (section 34(2)).

Powers of entry and inspection

A person authorised by the local authority, the chief officer of police, the fire and rescue authority, or the building authority may, on production if required of his authority, enter a sports ground at any reasonable time, and inspect the stands and make inquiries relating to them he considers necessary. In particular, he may examine records of the number of spectators accommodated, and the maintenance of safety, in the regulated stands at the ground, and take copies of records (section 35).

Offences

If spectators are admitted to a regulated stand at a sports ground on an occasion when no safety certificate which covers their use of the stand is in operation for it, or any term or condition of a safety certificate for a regulated stand at a sports ground is contravened, any responsible person and, if a safety certificate is in operation, the holder of the certificate, is guilty of an offence (section 36(1)). No offence is committed if the determination that the stand is a regulated stand is not a final one, or an application has been made for a general safety certificate for the stand and has not been withdrawn or deemed to have been withdrawn (section 36(2)).

A person guilty of an offence above is liable on summary conviction, to a fine not exceeding the statutory maximum; or on conviction on indictment, to a fine or to imprisonment for a term not exceeding 2 years or both (section 36(4)).

Where any person is charged with an offence above it is a defence to prove that the spectators were admitted or the contravention of the certificate in question took place without his consent; and that he took all reasonable precautions and exercised all due diligence to avoid the commission of such an offence by himself or any person under his control (section 36(5)).

Where any person is charged as a responsible person with an offence above it is a defence to prove that he did not know that the stand is a regulated stand (section 36(6)).

Any person who:

(a) gives information for the purpose of procuring a safety certificate or the cancellation, amendment, replacement or transfer of a safety certificate, knowingly or recklessly makes a false statement or knowingly or recklessly produces, furnishes, signs or otherwise makes use of a document containing a false statement; or

(b) fails to give a notice concerning amendments, alterations, etc.; or

(c) intentionally obstructs officers or without reasonable excuse refuses, neglects or otherwise fails to answer any question asked by any person in the exercise of such powers,

is guilty of an offence and liable on summary conviction to a fine not exceeding level 5 on the standard scale (section 36(7)).

Where an offence has been committed by a body corporate is proved to have been committed with the consent or connivance of, or to be attributable to any neglect on the part of, a director, manager, secretary or other similar officer of the body corporate, or any person who was purporting to act in that capacity, he, as well as the body corporate, is guilty of that offence and is liable to be proceeded against and punished accordingly (section 36(8)). Where the affairs of a body corporate are managed by its members, this applies to the acts and defaults of a member in connection with his functions of management as if he were a director of the body corporate (section 36(9)).

Service of documents

Any notice or other document required or authorised under this act to be served on any person may be served on him either by delivering it to him or by leaving it at his proper address or by sending it by post (section 38(1)).

Any notice or other document required or authorised to be served on a body corporate or a firm must be served if it is served on the secretary or clerk of that body or a partner of that firm (section 38(2)).

The proper address of a person, in the case of a secretary or clerk of a body corporate is that of the registered or principal office of that body, in the case of a partner of a firm shall be that of the principal office of the firm, and in any other case is the last known address of the person to be served (section 38(3)).

Definitions

responsible person means the person who is concerned in the management of the sports ground or of the regulated stand in question or in the organisation of any activity taking place at the ground at the time when an offence is alleged to have been committed (section 36(3)).
stand means an artificial structure (not merely temporary) which provides accommodation for spectators and is wholly or partly covered by a roof (section 26(11)).

SAFETY CERTIFICATES FOR SPORTS GROUNDS

References

Safety at Sports Grounds Act 1975.
Fire Safety and Safety of Places of Sport Act 1987.
Safety of Sports Grounds Regulations 1987.
Guide to Safety at Sports Grounds.

Extent

The Act applies in England, Scotland and Wales. There are slight variations for Scotland and Wales.

Introduction

Smaller sports grounds with covered stands with a capacity of 500 or more spectators require a safety certificate under the Fire Safety and Safety of Places of Sport Act 1987.

A safety certificate may be either:

- a general safety certificate which covers the use of the stand for viewing an activity, or a number of activities, specified in the certificate for an indefinite period which starts on a specified date; or
- a special safety certificate which covers the use of the stand for viewing a certain specified activity or activities on a certain specified occasion or occasions.

The application will be evaluated in line with recommendations made in the 'Guide to Safety at Sports Grounds'. The guide lays down detailed advice on matters such as:

- safe capacities
- management responsibilities
- circulation of spectators
- barrier design and testing
- spectator accommodation
- fire safety
- communications
- electrical and mechanical services
- first aid and medical provision
- media provision.

Safety certificates for large sports stadia (FC 3.31)

The Secretary of State may by order designate any sports ground requiring a safety certificate which in his opinion has accommodation for more than 10,000 spectators (this number may be changed by order of the Secretary of State and for different classes of sports ground) (section 1(1), (1A) and (1B)).

The Secretary of State may estimate how many spectators a sports ground has accommodation for; and may require any person concerned with the organisation or management of a sports ground to furnish him within a reasonable time with such information he considers necessary to make such an estimate (section 1(2)).

Contents of safety certificates

A safety certificate will have terms and conditions the local authority considers necessary to secure reasonable safety at the sports ground when it is in use for specified activity or activities, and the terms and conditions may involve alterations or additions to the sports ground (section 2(1)).

A condition for services can be reserved for the determination by the chief officer of police (section 2(2A)).

A safety certificate may include a condition to keep records of the attendance of spectators at the sports ground and the maintenance of safety at the sports ground (section 2(3)).

A general safety certificate must have a plan of the sports ground with the terms and conditions in the certificate and any special safety certificate issued for the sports ground must refer to that plan (section 2(4)). A safety certificate may include different terms and conditions in relation to different activities (section 2(5)).

These requirements apply to the amendment or replacement of a safety certificate (section 4(3)).

A safety certificate has no effect to the extent that it would require a person to contravene any provision of the Regulatory Reform (Fire Safety) Order 2005 or regulations made under it (section 4A and 2(2B)).

FC 3.31 Safety certificates for sports grounds – Safety at Sports Grounds Act 1975

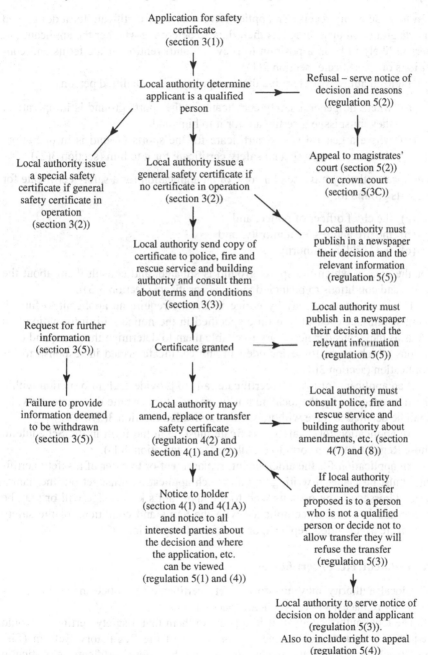

Application for safety certificate (section 3(1))

Local authority determine applicant is a qualified person

Refusal – serve notice of decision and reasons (regulation 5(2))

Appeal to magistrates' court (section 5(2)) or crown court (section 5(3C))

Local authority issue a special safety certificate if general safety certificate in operation (section 3(2))

Local authority issue a general safety certificate if no certificate in operation (section 3(2))

Local authority must publish in a newspaper their decision and the relevant information (regulation 5(5))

Local authority send copy of certificate to police, fire and rescue service and building authority and consult them about terms and conditions (section 3(3))

Local authority must publish in a newspaper their decision and the relevant information (regulation 5(5))

Request for further information (section 3(5))

Certificate granted

Failure to provide information deemed to be withdrawn (section 3(5))

Local authority may amend, replace or transfer safety certificate (regulation 4(2) and section 4(1) and (2))

Local authority must consult police, fire and rescue service and building authority about amendments, etc. (section 4(7) and (8))

Notice to holder (section 4(1) and 4(1A)) and notice to all interested parties about the decision and where the application, etc. can be viewed (regulation 5(1) and (4))

If local authority determined transfer proposed is to a person who is not a qualified person or decide not to allow transfer they will refuse the transfer (regulation 5(3))

Local authority to serve notice of decision on holder and applicant (regulation 5(3)). Also to include right to appeal (regulation 5(4))

Note
Relevant parties – police, fire and rescue service and building authority.

Applications for certificates

If a local authority receive an application for a safety certificate for a designated sports ground in their area, it is their duty to determine whether the applicant is a person likely to be in a position to prevent contravention of the terms and conditions of a certificate (section 3(1)).

If a local authority determine that an applicant is a qualified person:

(a) where no general safety certificate for the sports ground is in operation, they must issue a certificate for it to him; and

(b) where a general safety certificate for the sports ground is in operation, they may issue a special safety certificate for it to him (section 3(2)).

The local authority must send a copy of an application for a safety certificate for a sports ground to:

(a) the chief officer of police, and

(b) the fire and rescue authority, and

(c) the building authority,

for the area in which the sports ground is situated, and consult them about the terms and conditions to be included in the certificate (section 3(3)).

The local authority may by notice in writing require an applicant to furnish them within such reasonable time specified in the notice with information and plans as they consider necessary to enable them to determine the terms and conditions which ought to be included in any certificate issued in response to his application (section 3(4)).

If an applicant for a safety certificate fails to provide such information within the time specified by the local authority, or such further time they may allow, he shall be deemed to have withdrawn his application (section 3(5)).

An application for a safety certificate must be in the form in the Schedule to these Regulations or a form to the like effect (regulation 4(1)).

An application for the amendment, replacement or transfer of a safety certificate must be made in writing and any such application must set out the names and addresses of any persons who to the applicant's knowledge will or may be concerned in ensuring compliance with the terms and conditions of the safety certificate as amended, replaced or transferred (regulation 4(2)).

Amendment, etc. of certificates

The local authority may amend a safety certificate by notice in writing to its holder or replace a safety certificate (section 4(1)).

The local authority must, if it appears to them that a safety certificate would require a person to contravene any provision of the Regulatory Reform (Fire Safety) Order 2005 or its regulations, amend the safety certificate by notice in writing to its holder; but this does not require the local authority to take any action unless they are aware of any such inconsistency between a safety certificate and the Order (section 4(1A)).

A safety certificate may be amended or replaced either on the application of the holder or without an application (section 4(2)).

A notice under amending a general safety certificate must specify the date on which the amendment comes into operation, and the date specified may be a date later than the issue date (section 4(4)).

If the local authority receive an application for the transfer of a safety certificate from the holder to some other person, they must determine whether that person is a qualified person; and if that is the case, they may transfer the certificate to him (section 4(5)). An application may be made either by the holder of a safety certificate or by the transferee (section 4(6)).

The local authority must send a copy of an application for the transfer of a safety certificate for a sports ground to the chief officer of police, the fire and rescue authority and the building authority for the area in which the sports ground is situated (section 4(7)).

The local authority must consult the chief officer of police, the fire and rescue authority and the building authority about any proposal to amend, replace or transfer a safety certificate (section 4(8)).

The holder of a safety certificate may surrender it to the local authority, and it then ceases to have effect (section 4(9)).

The local authority may cancel a safety certificate if the holder dies or (if a body corporate) is dissolved (section 4(10)).

Notices by local authority

As soon as practicable after a local authority have decided to issue, replace, amend or refuse to amend or replace a safety certificate they must serve on every interested party a notice in writing of their decision setting out, stating that a copy of the safety certificate and a copy of any application in respect of which the local authority's decision was taken is available for inspection at a place and at the times specified in the notice and, in the case of a refusal, with their reasons for it (regulation 5(1) and (4)).

As soon as may be after the decision above, the local authority must publish in a newspaper circulating in the locality of the sports ground to which the safety certificate relates a notice setting out that decision and the relevant information (regulation 5(5)).

Where on an application for a special safety certificate a local authority have determined to refuse that application on grounds other than the person is not a qualified person, they must after that refusal serve on the applicant notice in writing of their decision, together with their reasons for it (regulation 5(2)).

Where on an application for the transfer of a safety certificate a local authority:

(a) determine that the person to whom it is proposed to transfer the certificate is not a qualified person, they must, in addition to the notice notifying the applicant of its determination, serve a copy of that notice on the holder of the certificate;

(b) determine that the person to whom it is proposed to transfer the certificate is a qualified person but decide not to transfer the certificate, they must serve on that person and the holder of the certificate notice in writing of their decision together with their reasons for it (regulation 5(3)).

Appeals

Where a local authority determine an applicant not to be a qualified person, they must serve a notice in writing on that person of their determination, and that person may appeal against the determination to the court (section 5(1)).

An applicant for a special safety certificate may also appeal to the court against a refusal of his application on grounds other than a determination that he is not a qualified person (section 5(2)).

An **interested party** may appeal to the court against:

(a) the inclusion of anything in, or the omission of anything from, a safety certificate; or
(b) the refusal of the local authority to amend or replace a safety certificate,

but not against the inclusion in a safety certificate of anything required to be included in it by the Sports Grounds Safety Authority under section 13(2) of the Football Spectators Act 1989 (section 5(3)).

An appeal to the court in England and Wales is by way of complaint for an order, and the Magistrates' Courts Act 1980 shall apply to the proceedings. (section 5(3A)) and the local authority or any interested party may appeal to the Crown Court against an order (section 5(3C)).

An appeal to court in Scotland is by summary application (section 5(3B)) and the local authority or any interested party may appeal against an order even if that person was not party to the proceedings on the application (section 5(3D)).

Interested party includes:

(a) the holder of a safety certificate;
(b) any other person who is or may be concerned in ensuring compliance with the terms and conditions of a safety certificate;
(c) the chief officer of police;
(d) the fire and rescue authority; and
(e) the building authority (section 5(5)).

If a local authority serve a notice determining a person not to be a qualified person, he is deemed to have withdrawn his application on the expiry of the period an appeal against the authority's determination may be brought (section 7(1)). This does not apply where an appeal is brought before the expiry of that period, but if the appeal is withdrawn or the court upholds the authority's determination, the appellant is deemed to have withdrawn his application on the date of the withdrawal of his appeal or of the court's determination (section 7(2)).

Where an appeal is brought against the inclusion of any term or condition in a safety certificate, the operation of that term or condition is suspended until the court has determined the appeal (section 7(3)).

An appeal must be brought in the case of:

(a) a general safety certificate, not later than 28 days, and
(b) a special safety certificate, not later than 7 days,

after the relevant date (regulation 6(1)).

Relevant date means:

(a) in the case of a person to whom a safety certificate is issued, the date of the receipt by him of that certificate;

(b) in the case of a person served with a notice amending a safety certificate; determining a person not to be a qualified person notice in writing of their determination, or refusal and transfer of a certificate, the date of the receipt by him of that notice; and

(c) in the case of any other person, the date of the publication of the notice setting out that decision and the associated information (regulation 6(2)).

Alterations and extensions

If while a general safety certificate is in operation it is proposed to alter or extend the sports ground or any of its installations, and the alteration or extension is likely to affect the safety of persons at the sports ground, the holder of the certificate must, before the carrying out of the proposals is begun, give notice of the proposals to the local authority (section 8(1)). This particularly applies when it is proposed to alter the entrances to or exits from a sports ground (including any means of escape in case of fire or other emergency) (section 8(2)).

Fees

A local authority may determine the fee to be charged in respect of an application for the issue, amendment, replacement or transfer of a safety certificate but the fee must not exceed an amount commensurate with the work actually and reasonably done by or on behalf of the local authority for the application (regulation 8).

Prohibition notices (FC 3.32)

If the local authority are of the opinion that the admission of spectators to a sports ground or any part of a sports ground involves or will involve a risk to them so serious that, until steps have been taken to reduce it to a reasonable level, admission of spectators to the ground or that part of the ground ought to be prohibited or restricted, the authority may serve a notice on **relevant persons** (section 10(1)).

A prohibition notice must:

(a) state that the local authority are of that opinion;

(b) specify the matters which in their opinion give or, will give rise to that risk; and

(c) direct that no, or no more than a specified number of, spectators shall be admitted to, or to a specified part of, the sports ground until the specified matters have been remedied (section 10(2)).

A prohibition notice may prohibit or restrict the admission of spectators generally or on a specified occasion (section 10(3)) or steps to be taken to reduce the risk to a reasonable level and these may require alterations or additions to the ground or things to be done or omitted which would contravene the terms or conditions of a safety certificate for the ground or for any stand at the ground (section 10(4)).

No prohibition notice can include directions which would require the provision of the services at the sports ground of any members of a police force unless the chief officer of police of the force has consented to their inclusion (section 10(5)).

Relevant persons – a prohibition notice must be served on the persons below and in the circumstances specified:

(a) if a general safety certificate is in operation for the ground, on the holder of it;

(b) if the prohibition or restriction applies to an occasion in respect of which a special safety certificate for the ground is in operation, on the holder of it;

(c) if no safety certificate is in operation for the ground, on the person who appears to the local authority to be responsible for the management of the ground;

(d) if the prohibition or restriction applies to an occasion and no safety certificate is in operation for the ground, on each person who appears to the local authority to be responsible for organising an activity at the ground on that occasion;

(e) if a general safety certificate is in operation for a stand at the ground, on the holder of it;

(f) if the prohibition or restriction applies to an occasion in respect of which a special safety certificate for a stand at the ground is in operation, on the holder of it,

but the prohibition notice served on any person above is not affected by a failure to serve another person required to be served with such a notice (section 10(6)).

A prohibition or restriction contained in a prohibition notice takes effect immediately it is served if the authority are of the opinion, and stated in the notice, that the risk to spectators is or, will be imminent, and in any other case shall take effect at the end of a period specified in the notice (section 10(7)).

A copy of any prohibition notice must be sent by the local authority to the chief officer of police, the fire and rescue authority and the building authority (section 10(8)).

The local authority who have served a prohibition notice may amend the prohibition notice by notice served on the relevant persons and copies shall be sent to the officer and authorities above (section 10(9)).

A notice amending a prohibition notice must specify the date on which the amendment is to come into operation (section 10(10)).

The local authority may withdraw a prohibition notice or notice of amendment at any time (section 10(11)).

Appeals against prohibition notices

Any person aggrieved by a prohibition notice or any amendment of a prohibition notice may appeal to the court against the notice (section 10A(1) and (2)).

In England and Wales an appeal to the court is by way of complaint for an order, and the Magistrates' Courts Act 1980 applies to the proceedings (section 10A(3)). Any of the following persons may appeal to the Crown Court against an order:

(a) any person aggrieved by the notice;
(b) the local authority;
(c) the chief officer of police;
(d) the fire and rescue authority; and
(e) the building authority (section 10A(7)).

In Scotland an appeal to the court is by summary application (section 10A(4)). Any of the following persons may appeal against an order:

(a) any person aggrieved by the notice;
(b) the local authority;
(c) the chief officer of police; and
(d) the building authority,

even if that person was not party to the proceedings on the application (section 10A(8)).

The court may either cancel or affirm the notice or, in the case of an appeal against an amendment, annul or affirm the amendment and, if it affirms the notice or the notice as amended, may do so either in its original form or as amended, or with such modifications of the notice as the court may think fit (section 10A(5)).

The bringing of the appeal does not suspend the operation of the notice or the notice as amended (section 10A(6)).

An appeal by an aggrieved person against a prohibition notice must be brought not later than 21 days after the day on which the notice was served on him (regulation 7(1)).

An appeal by an aggrieved person against an amendment to a prohibition notice must be not later than 21 days after the day on which the notice amending the prohibition notice was served on him (regulation 7(2)).

Enforcement

The local authority has duty to enforce the provisions of the Act including guidance and arrange for the inspection of designated sports grounds every 12 months (section 10B)).

Powers of entry and inspection

A person authorised by the local authority; the chief officer of police; the fire and rescue authority; the building authority; or the Secretary of State, may, on production if required of his authority, enter a sports ground at any reasonable time, and inspect it and make necessary inquiries, and in particular may examine records of attendance at the ground and records relating to the maintenance of safety at the ground, and take copies of those records (section 11).

FC 3.32 Sports ground licences – Prohibition notices

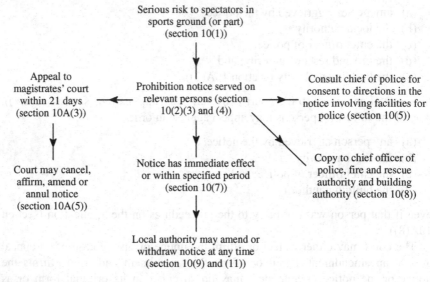

Serious risk to spectators in
sports ground (or part)
(section 10(1))

Appeal to
magistrates' court
within 21 days
(section 10A(3))

Prohibition notice served on
relevant persons (section
10(2)(3) and (4))

Consult chief of police for
consent to directions in the
notice involving facilities for
police (section 10(5))

Court may cancel,
affirm, amend or
annul notice
(section 10A(5))

Notice has immediate effect
or within specified period
(section 10(7))

Copy to chief officer of
police, fire and rescue
authority and building
authority (section 10(8))

Local authority may amend or
withdraw notice at any time
(section 10(9) and (11))

Note
Relevant persons and circumstances – see section 10(6).

Offences

Any responsible person and, if a safety certificate is in operation, the holder of the certificate, is guilty of an offence if:

(a) spectators are admitted to a designated sports ground after the operational date but at a time when no application for a general safety certificate has been made or such an application has been made but has been withdrawn, or is deemed to have been withdrawn; or

(b) when a general safety certificate is in operation in respect of a sports ground spectators are admitted to the sports ground on an occasion when it is used for an activity to which neither the general safety certificate nor a special safety certificate relates; or

(c) spectators are admitted to a designated sports ground on an occasion when, following the surrender or cancellation of a safety certificate, no safety certificate is in operation for that sports ground; or

(d) any term or condition of a safety certificate is contravened otherwise than in relation to a prohibition notice; or

(e) spectators are admitted to a sports ground in contravention of a prohibition notice (section 12(1)).

responsible person means a person who is concerned in the management of the sports ground in question or the organisation of any activity taking place there at the time when an offence is alleged to have been committed (section 12(2)).

A person guilty of an offence above is liable on summary conviction, to a fine of not more than the prescribed sum; or on conviction on indictment, to imprisonment for not more than 2 years or a fine or to both (section 12(3)).

Where any person is charged with an offence above it is a defence to prove:

(a) that the spectators were admitted or the contravention of the certificate or prohibition notice in question took place without his consent; and

(b) that he took all reasonable precautions and exercised all due diligence to avoid the commission of such an offence by himself or any person under his control (section 12(4)).

Any person is guilty of an offence and liable on summary conviction to a fine not exceeding level 5 on the standard scale, who:

(a) without reasonable excuse, refuses, neglects or otherwise fails to comply with a requirement above within the time specified by the Secretary of State; or

(b) in purporting to carry out such a requirement, or for the purpose of procuring a safety certificate or the amendment, replacement or transfer of a safety certificate, knowingly or recklessly makes a false statement or knowingly or recklessly produces, furnishes, signs or otherwise makes use of a document containing a false statement; or

(c) fails to give an alteration or extension notice; or

(d) intentionally obstructs any person in the exercise of powers, or without reasonable excuse refuses, neglects or otherwise fails to answer any question asked by any person in the exercise of those powers (section 12(6)).

Where an offence has been committed by a body corporate and is proved to have been committed with the consent or connivance of, or to be attributable to any neglect on the part of, a director, manager, secretary or other similar officer of the body corporate, or any person who was purporting to act in any such capacity, he, as well as the body corporate, is guilty of that offence and is liable to be proceeded against and punished accordingly (section 12(7)). Where the affairs of a body corporate are managed by its members, this applies in relation to the acts and defaults of a member in connection with his functions of management as if he were a director of the body corporate (section 12(8)).

Service of documents

Any notice or other document required or authorised under the Act to be served on any person may be served on him either by delivering it to him or by leaving it at his proper address or by sending it by post (section 14(1)).

Any notice or other document so required or authorised to be served on a body corporate or a firm shall be duly served if it is served on the secretary or clerk of that body or a partner of that firm (section 14(2)).

Application to Crown

Safety certificate requirements bind the Crown, but the Secretary of State substitutes for the local authority (section 16(1)).

Officers have no right of entry to premises occupied by the Crown (section 16(2)).

Sports Grounds Safety Authority

The Sports Grounds Safety Authority was established by the Sports Grounds Safety Authority Act 2011 (section 1). It may provide relevant advice to a Minister of the Crown, and if requested by a Minister of the Crown, must provide relevant advice to that Minister (section 2). The Authority may provide advice relating to safety at sports grounds in England or Wales to local authorities or other bodies or persons (section 3). It may also give advice to other bodies and persons outside England and Wales (section 4).

Definitions

qualified person means an applicant for a safety certificate for a designated sports ground who is likely to be in a position to prevent contravention of the terms and conditions of a certificate.

sports ground means any place where sports or other competitive activities take place in the open air and where accommodation has been provided for spectators, consisting of artificial structures or of natural structures artificially modified for the purpose.

safety certificate means either:

   (a) a certificate issued by the local authority for the area in which a sports ground is situated in respect of the use of the sports ground for an activity or a number of activities specified in the certificate during an indefinite period commencing with a date specified; or

(b) a certificate issued by that authority in respect of the use of the sports ground for an activity or a number of activities specified in the certificate on an occasion or series of occasions specified (section 1(3)).

STREET COLLECTIONS

References

Police, Factories, &c. (Miscellaneous Provisions) Act 1916.
Charitable Collections (Transitional Provisions) Order 1974 (articles below).

Extent

These provisions apply to England and Wales.

Scope

A licensing authority may, by a resolution, adopt the model street collection regulations. The resolution must send a copy to the Secretary of State (there is no need for confirmation) and insert an advertisement in two newspapers circulating within the area with a copy of the resolution and stating that a copy of the regulations will be provided free of charge to any person on application to the authority (section 5).

Model Street Collection Regulations

A promoter must apply in writing at least 1 month before the date for a proposed collection, to obtain a permit to collect in a street and public place (other than a collection taken at a meeting in the open air (articles 2 and 3). Collections must be made on the day and between the hours stated in the permit (article 4). The collection may be limited to streets or public places as the licensing authority thinks fit (article 5).

The promoter must provide written authority to a collector before collection is carried out (article 6).

The licensing authority may waive the requirement for a collector to remain stationary if in connection with a procession (article 10).

Payment for services connected with the collection may be approved by the licensing authority.

Within 1 month after the date of any collection the person to whom a permit has been granted must provide the licensing authority with:

(a) a statement showing the amount received and the expenses and payments in connection with the collection certified by that person and a qualified accountant;

(b) a list of the collector;

(c) a list of the amounts contained in each collecting box;

(d) a description of the proper application of the proceeds of the collection and

(e) publish in a newspaper or newspapers a statement showing the name of the person to whom the permit was been granted, the area to which the permit relates, the name of the charity or fund to benefit, the date of the collection the amount collected, and the amount of the expenses and payments in connection with the collection.

The licensing authority may extend the period of 1 month (article 16).

It is an offence to contravene any of the regulations and if guilty will be liable on summary conviction to a fine not exceeding level 1 on the standard scale (article 18).

Definitions

collection means a collection of money or a sale of articles for the benefit of charitable or other purposes.

promoter means a person who causes others to act as collectors.

permit means a permit for a collection.

collecting box means a box or other receptacle for the reception of money from contributors (article 1).

STREET TRADING

Reference

Local Government (Miscellaneous Provisions) Act 1982 section 3 and schedule 4.

Adoption

The procedures only operate through a specific resolution of adoption by the local authority (section 3). There are no requirements about the advertisement, etc. of any intention to adopt these provisions.

Extent

This procedure applies in England and Wales but does not apply in Greater London (for which see the London Local Authorities Act 1990) or in Scotland and Northern Ireland.

Designation (FC 3.33)

Following adoption, the local authority may designate streets or parts of streets to be either:

(a) prohibited streets in which street trading is not allowed;

(b) licence streets in which licences to trade are required; or

(c) consent streets where the consent requirements operate.

FC 3.33 Street trading: designation of – Schedule 4 of Local Government (Miscellaneous Provisions) Act 1982 (LG(MP)A)

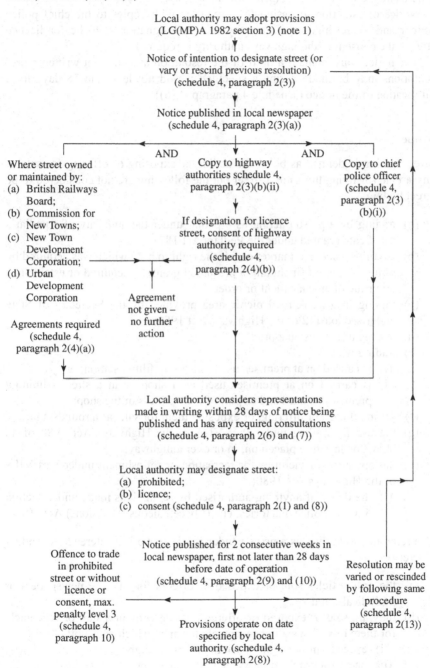

Local authority may adopt provisions
(LG(MP)A 1982 section 3) (note 1)

Notice of intention to designate street (or
vary or rescind previous resolution)
(schedule 4, paragraph 2(3))

Notice published in local newspaper
(schedule 4, paragraph 2(3)(a))

AND — AND

Where street owned
or maintained by:
(a) British Railways
    Board;
(b) Commission for
    New Towns;
(c) New Town
    Development
    Corporation;
(d) Urban
    Development
    Corporation

Agreements required
(schedule 4,
paragraph 2(4)(a))

Copy to highway
authorities schedule 4,
paragraph 2(3)(b)(ii)

If designation for licence
street, consent of highway
authority required
(schedule 4,
paragraph 2(4)(b))

Agreement
not given –
no further
action

Copy to chief
police officer
(schedule 4,
paragraph 2(3)
(b)(i))

Local authority considers representations
made in writing within 28 days of notice being
published and has any required consultations
(schedule 4, paragraph 2(6) and (7))

Local authority may designate street:
(a) prohibited;
(b) licence;
(c) consent (schedule 4, paragraph 2(1) and (8))

Offence to trade
in prohibited
street or without
licence or
consent, max.
penalty level 3
(schedule 4,
paragraph 10)

Notice published for 2 consecutive weeks in
local newspaper, first not later than 28 days
before date of operation
(schedule 4, paragraph 2(9) and (10))

Resolution may be
varied or rescinded
by following same
procedure
(schedule 4,
paragraph 2(13))

Provisions operate on date
specified by local
authority (schedule 4,
paragraph 2(8))

Note
1. No advertisement, publicity, etc. required at this stage.

These designations may be rescinded or changed from one type to another at any time using the full procedure (schedule 4 paragraph 2).

Notice of intention to designate is required with copies to the chief police officer and to the highways authority. If the designation is to be for licence streets, the consent of the highways authority is required.

The notice must contain a draft of the resolution and say that written representations may be made during a specified period not less than 28 days from publication of the notice (schedule 4 paragraph 2(6)).

Scope

Street trading is defined as being the selling or exposing or offering for sale of any article (or living thing) in a street but the following are not considered to be street trading:

(a) trading by a person acting as a pedlar under the authority of a pedlar's certificate granted under the Pedlars Act 1871;

(b) anything done in a market or fair the right to hold which was acquired by virtue of a grant (including a presumed grant) or acquired or established by virtue of an enactment or order;

(c) trading in a trunk road picnic area provided by the Secretary of State under section 112 of the Highways Act 1980;

(d) trading as a news vendor;

(e) trading which:
   (i) is carried on at premises used as a petrol filling station; or
   (ii) is carried on at premises used as a shop or in a street adjoining premises so used and as part of the business of the shop;

(f) selling things, or offering, or exposing them for sale, as a roundsman;

(g) the use for trading under Part VIIA of the Highways Act 1980 of an object or structure placed on, in or over a highway;

(h) the operation of facilities for recreation or refreshment under Part VIIA of the Highways Act 1980;
   (i) the doing of anything authorised by regulations made under section 5 of the Police, Factories, etc. (Miscellaneous Provisions) Act 1916.

The reference to trading as a news vendor in (d) above is a reference to trading where:

(a) the only articles sold or exposed or offered for sale are newspapers or periodicals; and

(b) they are sold or exposed or offered for sale without a stall or receptacle for them or with a stall or receptacle for them which does not:
   (i) exceed 1 m in length or width or 2 m in height;
   (ii) occupy a ground area exceeding 0.25 m^2; or
   (iii) stand on the carriageway of a street (schedule 4 paragraph 1(2) and (3)).

Licences (FC 3.35)

Applications must be in writing and must give:

(a) full name and address of applicant;
(b) street, day and times of proposed trading;
(c) description of articles, stalls or containers;
(d) any other particulars required by the Council which can include two photographs of the applicant (schedule 4 paragraph 3(2) and (3)).

Unless one or more of the following grounds of refusal are applicable, the local authority must grant the licence and may even grant it if the grounds of refusal are available:

(a) that there is not enough space in the street for the applicant to engage in the trading in which he desires to engage without causing undue interference or inconvenience to persons using the street;
(b) that there are already enough traders trading in the street from shops or otherwise in the goods in which the applicant desires to trade;
(c) that the applicant desires to trade on fewer days than the minimum number specified in a resolution under schedule 4 paragraph 2(11);
(d) that the applicant is unsuitable to hold the licence by reason of having been convicted of an offence or for any other reason;
(e) that the applicant has at any time been granted a street trading licence by the council and has persistently refused or neglected to pay fees due to them for it or charges due to them under schedule 4 paragraph 9(6), for services rendered by them to him in his capacity as licence holder;
(f) that the applicant has at any time been granted a street trading consent by the council and has persistently refused or neglected to pay fees due to them for it;
(g) that the applicant has without reasonable excuse failed to avail himself to a reasonable extent of a previous street trading licence (schedule 4 paragraph 3(6)).

Also licences must be refused if the applicant is under 17 years of age or the location is covered by a control order (road-side sales) under section 7 Local Government (Miscellaneous Provisions) Act 1976.

The licence issued must state:

(a) the street in which and days and times between which the holder is able to trade; and
(b) the description of articles in which he may trade and may also state a particular location for trading (schedule 4 paragraph 4(1)).

These are known as the principal terms of the licence and a breach may result in prosecution (schedule 4 paragraph 4(1)–(3)).

In addition, subsidiary terms may be applied at the discretion of the local authority and these may include:

(a) specifying the size and type of any stall or container which the licence holder may use for trading;

(b) requiring that any stall or container so used shall carry the name of the licence holder or the number of his licence or both; and

(c) prohibiting the leaving of refuse by the licence holder or restricting the amount of refuse which he may leave or the places in which he may leave it (schedule 4 paragraph 4(5)).

Licences may be revoked if:

(a) owing to circumstances which have arisen since the grant or renewal of the licence, there is not enough space in the street for the licence holder to engage in the trading permitted by the licence without causing undue interference or inconvenience to persons using the street;

(b) the licence holder is unsuitable to hold the licence by reason of having been convicted of an offence or for any other reason;

(c) since the grant or renewal of the licence, the licence holder has persistently refused or neglected to pay fees due to the council for it or charges due to them under schedule 4 paragraph 9(6), for services rendered by them to him in his capacity as licence holder; or

(d) since the grant or renewal of the licence, the licence holder has without reasonable excuse failed to avail himself of the licence to a reasonable extent (paragraph 5(1)),

and the local authority may also vary the principal terms by altering the days or times of trading or restricting the type of goods sold, subject to notice and appeal (schedule 4 paragraph 5(2)).

The licence street provisions are most appropriate for the formalised street market situations and imply a positive will to promote trading on behalf of the local authority to the extent that refusal powers are limited.

Consents (FC 3.34)

There are no specified particulars for consent applications although they must be in written form (schedule 4 paragraph 7(1)). Unless the applicant is under 17 years of age or the location is covered by a control order under section 7 of the Local Government (Miscellaneous Provisions) Act 1976 (roadside sales) in which situations refusal is mandatory, consents are entirely at the discretion of the local authority as it sees fit. If consent is given the local authority may attach conditions at its discretion including those to prevent obstruction and nuisance or annoyance.

Specific consent to trade from a vehicle or portable stall is required. Consent procedures are most applicable where trading is to be itinerant or infrequent and there is no appeal against refusals or conditions to be applied. There is no requirement in these provisions to hear applicants but in the case of *R.* v. *Bristol City Council, ex parte Pearce and Another* (1984), it was indicated that the local authority should tell applicants of the contents of any objections and give them an opportunity to comment (schedule 4 paragraph 7).

FC 3.34 Street trading: consents – Schedule 4 of Local Government (Miscellaneous Provisions) Act 1982 (LG(MP)A)

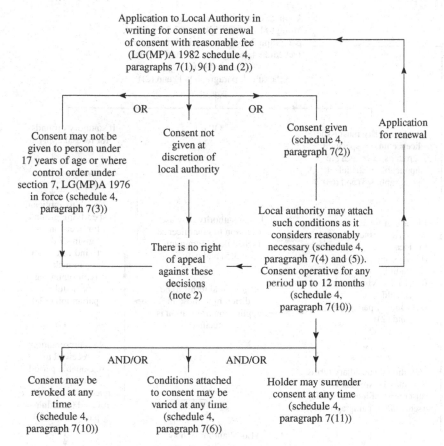

Application to Local Authority in writing for consent or renewal of consent with reasonable fee (LG(MP)A 1982 schedule 4, paragraphs 7(1), 9(1) and (2))

OR — OR

Application for renewal

Consent may not be given to person under 17 years of age or where control order under section 7, LG(MP)A 1976 in force (schedule 4, paragraph 7(3))

Consent not given at discretion of local authority

Consent given (schedule 4, paragraph 7(2))

Local authority may attach such conditions as it considers reasonably necessary (schedule 4, paragraph 7(4) and (5)). Consent operative for any period up to 12 months (schedule 4, paragraph 7(10))

There is no right of appeal against these decisions (note 2)

AND/OR — AND/OR

Consent may be revoked at any time (schedule 4, paragraph 7(10))

Conditions attached to consent may be varied at any time (schedule 4, paragraph 7(6))

Holder may surrender consent at any time (schedule 4, paragraph 7(11))

Notes
1. The maximum penalty for contravening a condition of consent is level 3 (schedule 4, paragraph 10(4)).
2. Although no appeal procedure exists, in the case of *R.* v. *Bristol City Council, ex parte Pearce and Another* (1984) the judge commented that the local authority should tell the applicant of the content of their objections and give him an opportunity to comment.

FC 3.35 Street trading: licences – Schedule 4 of Local Government (Miscellaneous Provisions) Act 1982 (LG(MP)A)

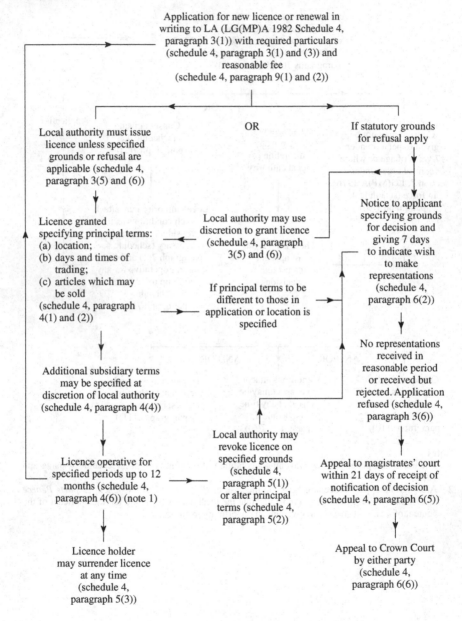

Application for new licence or renewal in writing to LA (LG(MP)A 1982 Schedule 4, paragraph 3(1)) with required particulars (schedule 4, paragraph 3(1) and (3)) and reasonable fee (schedule 4, paragraph 9(1) and (2))

OR

Local authority must issue licence unless specified grounds or refusal are applicable (schedule 4, paragraph 3(5) and (6))

If statutory grounds for refusal apply

Licence granted specifying principal terms:
(a) location;
(b) days and times of trading;
(c) articles which may be sold
(schedule 4, paragraph 4(1) and (2))

Local authority may use discretion to grant licence (schedule 4, paragraph 3(5) and (6))

Notice to applicant specifying grounds for decision and giving 7 days to indicate wish to make representations (schedule 4, paragraph 6(2))

If principal terms to be different to those in application or location is specified

Additional subsidiary terms may be specified at discretion of local authority (schedule 4, paragraph 4(4))

No representations received in reasonable period or received but rejected. Application refused (schedule 4, paragraph 3(6))

Local authority may revoke licence on specified grounds (schedule 4, paragraph 5(1)) or alter principal terms (schedule 4, paragraph 5(2))

Licence operative for specified periods up to 12 months (schedule 4, paragraph 4(6)) (note 1)

Appeal to magistrates' court within 21 days of receipt of notification of decision (schedule 4, paragraph 6(5))

Licence holder may surrender licence at any time (schedule 4, paragraph 5(3))

Appeal to Crown Court by either party (schedule 4, paragraph 6(6))

Note

1. The maximum penalty for breaching the principal terms of the licence is level 3. There is no offence committed in breaching the subsidiary terms but this could be taken into account in any consideration of renewal or revocation (schedule 4, paragraph 10(4)).

Fees

The local authority may charge such fees as they consider reasonable for both licences and consents with different fee levels being possible for different types of licence/consents, periods, streets and articles. Fees may be paid in instalments and the initial application need only be accompanied by a deposit at the discretion of the local authority. Fees are returnable on surrender, revocation and refusal.

Separate and additional charges may be levied on licence holders for the collection of refuse, street sweeping and any other services rendered by the local authority (schedule 4 paragraph 9).

Definition

street includes:

(a) any road, footway, beach or other area to which the public have access without payment; and

(b) a service area as defined in section 329 of the Highways Act 1980,

and also includes any part of a street (schedule 4 paragraph 1(1)).

TAXI LICENCES – HACKNEY CARRIAGES IN LONDON

References

London Hackney Carriages Act 1831.
London Hackney Carriages Act 1843.
London Hackney Carriages Act 1850.
London Hackney Carriages Act 1853.
Metropolitan Public Carriage Act 1869.
London Cab Order 1934 (LCO 1934).

Extent

These procedures apply in London and the United Kingdom.

Scope

These procedures cover the licensing of hackney carriage licensing in London. Transport for London (TfL) licenses London taxis and taxi drivers under the Metropolitan Public Carriage Act 1869.

Grant of hackney carriage licences (FC 3.36)

TfL has the function of licensing hackney carriages to ply for hire (section 6(1) MPCA 1869).

A licence may be granted with conditions, in a prescribed form, subject to revocation or suspension (section 6(2) MPCA 1869).

A licence is in force for 1 year, if not revoked or suspended (section 6(4) MPCA 1869).

Fees are determined by TfL and payable:

(a) by the applicant for a licence, on making the application;

(b) by the applicant on making the application for the taking or re-taking of any test or examination, or any part of a test or examination, concerning fitness; and

(c) by any person granted a licence, on the grant of the licence (section 6(5) MPCA 1869).

Different amounts may be determined for different purposes or different cases (section 6(7) MPCA 1869).

TfL may remit (cancel) or refund the whole or part of a fee (section 6(8) MPCA 1869).

A London cab order may be made for the transfer of a licence to the surviving spouse or surviving civil partner or to any child of full age of any person to whom such a licence has been granted who may die during the period of the licence leaving a surviving spouse or surviving civil partner or child of full age (section 6(9) MPCA 1869).

Standings for hackney carriages to be appointed

TfL can appoint standings for hackney carriages at places they think convenient in any street, thoroughfare, or place of public resort within the metropolitan police district, and alter them at their discretion. They can make regulations concerning their boundaries, and the number of carriages to be allowed at any standing, and the times at and during which they may stand and ply for hire at any standing.

They can also make regulations for enforcing order at every standing, and for removing any person who unnecessarily loiters or remains at or about any standing.

TfL must advertise all the orders and regulations made by it in the London Gazette, and a copy signed by a person authorised by TfL, to be hung up for public inspection in the offices of TfL, and at each of the magistrates' courts acting for an area falling wholly within an inner London borough; and that copy is regarded as evidence in the courts as if it were the original (section 4 London Hackney Carriages Act 1850).

Carriage inspection and powers to suspend licences, and recall plate

TfL can inspect all hackney carriages, as often as necessary, and if any carriage, at any time is in a condition unfit for public use, TfL must give notice in writing to the proprietor, and the notice must be personally served on the proprietor or delivered at his usual place of residence, and may be personally served on the driver of the carriage, and if, after notice has been served on the proprietor or driver, the carriage is used or let to hire as a hackney carriage, whilst in a condition unfit for public use, TfL can suspend, for a time they deem proper, the licence of the proprietor of the carriage (section 2(1) London Hackney Carriage Act 1853).

Applications for cab licences

Every application for a cab licence must be made in a form, and include declarations and information as TfL require (article 5(1) LCO 1934).

Where the cab is jointly owned or owned by a partnership firm or a limited liability company, the application must be made in the name of one of the joint owners or by the senior partner of the firm, or the Secretary, Manager or other duly authorised officer of the company, and that person is deemed to be the applicant for the licence, and the licence if granted must be issued to him (article 5(2) LCO 1934).

TfL may require applicants to provide different information depending on whether or not the applicant has previously held or currently holds a cab licence or cab drivers licence (article 5(3) LCO 1934).

Grant of cab licences

TfL must grant a cab licence if it is satisfied that:

(a) the applicant is a fit and proper person to hold a cab licence;
(b) the vehicle in respect of which the application is made conforms to the conditions of fitness laid down by TfL; and
(c) the requirements as to liability to third parties in article 8 are met (article 7 LCO 1934).

Presentation of motor cabs for licensing

The applicant for licence for a motor cab must present the vehicle for inspection and testing by TfL within a period and place TfL may require by notice (article 10 LCO 1934).

At the time of the examination of the motor cab by the Public Carriage Examiner, he must be given an application form and the certificate below (article 3(2) LCO 1996).

The certificate must relate to the cab and be in the prescribed form in the Schedule to this Order and signed on, or during the 28 day period ending immediately before, the day on which the certificate is handed to the Public Carriage Examiner by either an authorised examiner or a person authorised by an authorised examiner to carry out such an examination on his behalf. The signature must be accompanied by an embossment by a stamp of the authorised examiner by or on whose behalf the certificate was signed; and the statement made in paragraph 2 of the certificate must have been true at the time the certificate was signed (article 3(3) LCO 1996).

If the Public Carriage Examiner has reasonable grounds for suspecting that the motor cab does not comply with regulation 61 of the Road Vehicles (Construction and Use) Regulations 1986 he may direct that the motor cab be presented again at the place at which the examination was carried out with a certificate which satisfies the requirements above and which is signed after the direction is given (article 3(4) LCO 1996).

FC 3.36 Licensing of hackney carriages in London – Metropolitan Public Carriage Act 1869 (MPCA1869) and London Cab Order 1934 (LCO 1934)

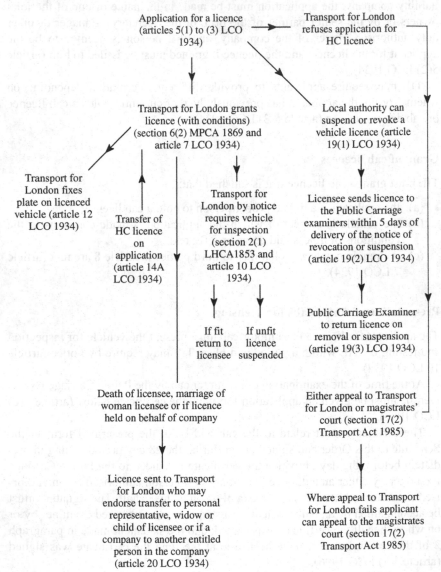

Application for a licence (articles 5(1) to (3) LCO 1934)

Transport for London refuses application for HC licence

Transport for London grants licence (with conditions) (section 6(2) MPCA 1869 and article 7 LCO 1934)

Local authority can suspend or revoke a vehicle licence (article 19(1) LCO 1934)

Transport for London fixes plate on licenced vehicle (article 12 LCO 1934)

Transfer of HC licence on application (article 14A LCO 1934)

Transport for London by notice requires vehicle for inspection (section 2(1) LHCA1853 and article 10 LCO 1934)

Licensee sends licence to the Public Carriage examiners within 5 days of delivery of the notice of revocation or suspension (article 19(2) LCO 1934)

If fit return to licensee

If unfit licence suspended

Public Carriage Examiner to return licence on removal or suspension (article 19(3) LCO 1934)

Death of licensee, marriage of woman licensee or if licence held on behalf of company

Either appeal to Transport for London or magistrates' court (section 17(2) Transport Act 1985)

Licence sent to Transport for London who may endorse transfer to personal representative, widow or child of licensee or if a company to another entitled person in the company (article 20 LCO 1934)

Where appeal to Transport for London fails applicant can appeal to the magistrates court (section 17(2) Transport Act 1985)

Notes
1. LCO 1934 – London Cab Order 1934.
2. MPCA 1869 – Metropolitan Public Carriage Act 1869.
3. LHCA 1853 – London Hackney Carriage Act 1853.
4. Appeals – section 300 to 302 Public Health act 1936 apply (Section 77(1)) Local Government (Miscellaneous Provisions) Act 1976.

Affixing of plates, etc.

On granting a cab licence TfL must fix to the cab:

(a) the plates and notices described in Schedule B to the Order in the positions required by the Schedule; and

(b) notices or marks as TfL may direct.

The plates remain the property of TfL (article 12 LCO 1934).

Form of cab licence and conditions to be complied with

Every cab licence must be in the form contained in Schedule C to the Order, and is in force for 1 year unless revoked or suspended, and granted subject to the specified provisions and subject to the conditions in article 14 (article 14 LCO 1934).

Transfer of cab licence

TfL must transfer a licence from a previous vehicle owner to a new owner on application made if the application is made in a form, and include declarations and information as TfL require; and the new owner satisfies TfL that he is a fit and proper person to hold a cab licence (article 14A LCO 1934).

Revocation or suspension of cab licences

A cab licence can be revoked or suspended by TfL on any of the following grounds:

(a) if the licence has been obtained by any misrepresentation, fraud or concealment of any material circumstances; or

(b) if TfL, by reason of any new circumstance arising or coming to its knowledge after the grant of a licence, or by reason of the condition of the cab, is satisfied that a licence could not properly be granted to the licensee if he were an applicant for a new licence; or

(c) if the licensee fails to comply with any of the provisions or conditions subject to which the licence has been granted,

provided that in a case where more than one licence granted to the same licensee becomes liable to revocation or suspension. TfL, if it is of the opinion that it would be contrary to the public interest to revoke or suspend all of those licences, may revoke or suspend only one or more of them as it may think fit (article 19(1) LCO 1934).

In the event of the revocation or suspension of a cab licence, the licensee must, within 5 days of a notice having been delivered to him personally or sent to him by registered post or by the recorded delivery service at the address mentioned in or last endorsed on the licence, send or deliver the licence to the Public Carriage Examiner at the appointed passing station for cancellation or for retention during the time of suspension. If required in the notice he must bring or send the cab to which the licence relates to that passing station in order that the plates fixed to the cab may be removed; and if the licensee fails to do this, he is guilty of a breach of this Order (article 19(2) LCO 1934).

On the removal of a suspension of a cab licence which has not expired the Public Carriage Examiner must return the licence to the licensee and refix the plates, if removed (article 19(3) LCO 1934).

Transfer of cab licences on death, etc.

In the event of the death of any licensee during the term of his cab licence, the licence must be sent or delivered to the office of TfL, who may by endorsement transfer the licence to the personal representative of the deceased person, or to his widow or child, if the representative or widow or child is of full age and satisfies TfL of his or her fitness to hold the licence.

In the event of the marriage of a woman licensee during the period of her licence, the licence may be transferred to her husband in a similar way to above.

In the case of a licence held on behalf of a firm or company, the licence may be transferred from the licensee to any other person who would be entitled to apply for a licence on behalf of the firm or company in a similar way to above (article 20 LCO 1934).

Cab licences to be surrendered on expiry and plates removed

The owner of a cab, or, where the owner is a firm or company, the person holding the licence for the cab on its behalf, must within 3 days after the expiration of the period of the licence, deliver up the licence and the plates fixed to the cab to TfL or a Public Carriage Examiner, and if he fails so to do he is guilty of a breach of this Order (article 21 LCO 1934).

Responsibility of firm or company

Where a cab licence is held by any person on behalf of a firm or company, both that person and the firm or company, is deemed to be the licensee, and is liable for any breach of this Order or any failure to comply with any of the provisions or conditions subject to which the licence is granted (article 22 LCO 1934).

Appeals

Where the licensing authority has refused to grant, or has suspended or revoked, a licence the applicant for, or holder of, the licence may, before the expiry of the designated period (28 days – London Taxis (Licensing Appeals) Regulations 1986):

(a) require the authority to reconsider his decision; or
(b) appeal to a magistrates' court (section 17(2) Transport Act 1985).

Any call for a reconsideration must be made to the licensing authority in writing (section 17(3) Transport Act 1985).

On any reconsideration the person calling for the decision to be reconsidered is entitled to be heard either in person or by his representative (section 17(4)).

If the person calling for a decision to be reconsidered is dissatisfied with the decision of the licensing authority on reconsideration, he may, before the expiry of the designated period, appeal to a magistrates' court (section 17(5) Transport Act 1985).

On any appeal to it, the court may make such order as it thinks fit; and any order which it makes shall be binding on the licensing authority (section 17(6) Transport Act 1985).

Where a person holds a licence which is in force when he applies for a new licence in substitution for it, the existing licence continues in force until the application for the new licence, or any appeal in relation to that application, is disposed of, but the licensing authority can revoke the existing licence (section 17(7) Transport Act 1985).

Where the licensing authority refuses to grant the new licence the application is not treated as disposed of:

(a) where no call for a reconsideration of the authority's decision is made, until the expiry of the designated period;
(b) where such a reconsideration is called for, until the expiry of the designated period which begins by reference to the decision of the authority on reconsideration (section 17(8)).

Where the licensing authority suspends or revokes a licence, or confirms a decision to do so, he may, if the holder of the licence so requests, direct that his decision does not have effect until the expiry of the designated period (section 17(9) Transport Act 1985).

designated period means a period specified by London cab order (section 17(10)).

Any power to make a London cab order includes power to vary or revoke a previous order (section 17(11) Transport Act 1985).

Driver's licences (FC 3.37)

Hackney carriage to be driven by licensed drivers

TfL licenses drivers of hackney carriages (section 8(1) MPCA 1869).

No hackney carriage can ply for hire unless under the charge of a driver having a licence from TfL (section 8(2) MPCA 1869).

If any hackney carriage plies for hire where the driver has no licence both the person driving the carriage, and the owner of the carriage, unless he proves that the driver acted without his consent is liable to a penalty not exceeding level 3 on the standard scale (section 8(3) MPCA 1869).

TfL may send to the Commissioner of Police of the Metropolis or the Commissioner of Police for the City of London:

(a) details of a person to whom TfL is considering granting a licence, and
(b) request the Commissioner's observations and the Commissioner must respond to the request (section 8(4) MPCA 1869).

A licence may be granted with conditions, in a form, subject to revocation or suspension in such event, and generally be dealt with in such manner which may be prescribed (section 8(5) MPCA 1869). A licence is in force for 3 years, if not revoked or suspended (section 8(7) MPCA 1869).

Fees are payable determined by TfL:

(a) by the applicant for a licence, on making the application;

(b) by the applicant on making the application for the taking or re-taking of any test or examination, or any part of a test or examination, concerning fitness; and

(c) by any person granted a licence, on the grant of the licence (section 8(8) MPCA 1869).

Different amounts may be determined for different purposes or different cases (section 8(10) MPCA 1869).

TfL may remit or refund the whole or part of a fee (section 8(11) MPCA 1869).

Granting driver's licences

TfL can grant a licence to act as driver of hackney carriages to any person. The applicant must produce a certificate to satisfy TfL of his good behaviour and fitness to drive a hackney carriage.

Every licence must specify the number of the licence, the name and address, age, and a description of the person to whom it is granted.

Every licence must have on it the day on which the licence is granted, TfL must place columns for every proprietor employing the driver named in the licence to enter his own name and address, and the days on which the driver starts and finishes work.

If particulars entered or endorsed on any licence are erased or defaced the licence is wholly void.

TfL must at the time of granting any licence, give the driver an abstract of the laws in force relating to the driver, and the penalties to which he is liable for any misconduct, and also a badge marked with his office or employment, and the licence number (section 8 London Hackney Carriages Act 1843).

Persons applying for licences to sign a requisition for the same, etc.

Before any licence is granted, a requisition, in a form TfL appoints for that purpose, and accompanied with a required certificate, must be made and signed by the person applying for the licence; and in every requisition all particulars TfL require must be given.

Every person applying for or attempting to procure any licence and every person referred to, in the requisition who makes or causes to be made any false representation, or who does not truly answer all questions in relation to such application for a licence, wilfully and knowingly makes any misrepresentation, is liable for every offence the sum of level 3 on the standard scale.

FC 3.37 Hackney carriage driver's licences in London

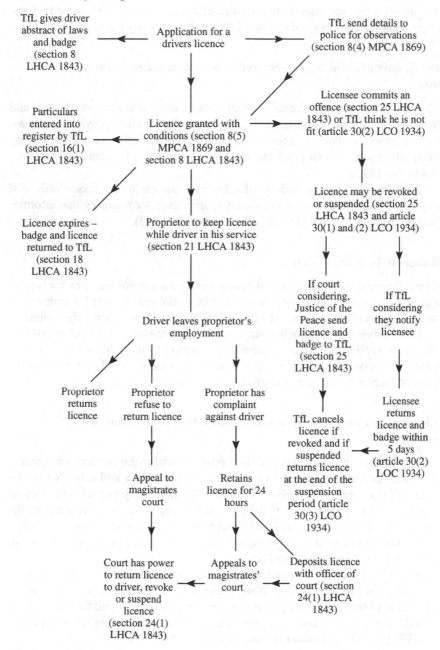

Notes

1. LCO 1934 – London Cab Order 1934.
2. MPCA 1869 – Metropolitan Public Carriage Act 1869.
3. LHCA 1843 – London Hackney Carriage Act 1843.
4. Appeals – section 300 to 302 Public Health act 1936 apply (Section 77(1)) Local Government (Miscellaneous Provisions) Act 1976.

TfL can recover the penalty before any magistrate at any time within 1 calendar month after the commission of the offence, or during the period of the licence improperly obtained (section 140 London Hackney Carriages Act 1843).

Particulars of licences to be entered in a book at the registrar's office

The particulars of every licence granted must be entered in a register by TfL and this is good evidence in court proceedings, and every person applying at all reasonable times is allowed access to a certified copy of the particulars of any licensed person, without payment of a fee (section 16(1) London Hackney Carriages Act 1843).

TfL may disclose the address of a licensed person to any person only if it appears to TfL that the person has a sufficient reason for requiring that information (section 16(2) London Hackney Carriages Act 1843).

Badges to be worn by drivers

Every licensed driver must at all times during his employment, and when he shall be required to attend before any justice of the peace, wear his badge conspicuously on his breast, so that whole of the writing on it is distinctly legible.

Every driver not wearing his badge, or who, when required refuses to produce his badge for inspection, or to permit any person to note the writing on it, is liable for every offence to pay a fine of level 1 on the standard scale (section 17 London Hackney Carriages Act 1843).

Licences and badges to be delivered up on the discontinuance of licences

On the expiration of any licence the person to whom the licence was granted must deliver the licence and badge to TfL within 3 days and a licence holder who wilfully neglects to deliver them to TfL, and every person who uses or wears or detains any badge without having a licence in force relating to the badge, or who for the purpose of deception uses or wears or has any badge resembling or intended to resemble any badge, is liable for every such offence to a fine on level 1 on the standard scale.

TfL, or any person employed by them, can prosecute any person neglecting to deliver up his licence or badge, at any period within 12 calendar months after the expiration of the licence. Any constable or any person employed for that purpose by TfL, can seize and take away any badge, wherever found, in order to deliver it to TfL (section 18 London Hackney Carriages Act 1843).

New badges to be delivered instead of defaced or lost badges

Whenever the writing on any badge becomes obliterated or defaced, so it is not distinctly legible, and whenever any badge is proved to the satisfaction of TfL to have been lost or mislaid, the person to whom the licence relating to any such badge was granted must deliver the badge (if he has it in his possession) and produce the licence to Transport for London, and that person is entitled to have a new badge delivered to him, on payment, to TfL, of a sum of money TfL considers reasonable.

If any badge proved represented to have been lost or mislaid and later found it must be delivered to TfL. Every person into whose possession any such badge refuses or neglects for 3 days to deliver it to TfL, and every person licensed who uses or wears the badge granted to him after the writing has been obliterated, defaced or obscured, so it is not distinctly legible, is liable for every such offence to a fine of level 1 on the standard scale (section 19 London Hackney Carriages Act 1843).

Proprietor to retain the licence of drivers or conductors employed by him, and produce them in case of complaint

Every proprietor of a hackney carriage who permits or employs any licensed person to act as the driver must retain possession of the driver's licence while in his service; and in cases of complaint where the proprietor of a hackney carriage is summoned to produce the driver of the carriage before a justice of the peace he must also produce the driver's licence if at the time of receiving the summons the driver is in his service.

If the driver complained of is found guilty of the offence alleged against him the justice of the peace must endorse the driver's licence the nature of the offence, and the amount of the penalty; and if the proprietor does not retain the driver's licence or refuses or neglects to produce the licence is liable for every offence a fine equivalent to level 1 on the standard scale (section 21 London Hackney Carriages Act 1843).

Proceedings with respect to licences on quitting service

When any licensed driver leaves the service of any proprietor that proprietor must, on demand, return his licence to him.

If the proprietor has any complaint against the driver the proprietor can retain the licence for up to 24 hours, excluding Saturday or Sunday, Christmas Day or Good Friday, and bank holidays, and within that time apply to a magistrates' court for a summons against him.

The proprietor, when applying for the summons, must deposit the licence with the designated officer for the magistrates' court; and if any proprietor who on demand of the licence refuses or neglects to deliver the driver's licence and does not apply for summons within 24 hours, and deposit the licence, or appears to prosecute his complaint, the driver can apply to a magistrates' court for a summons against the proprietor.

On hearing and deciding the case the justice, if he thinks there was no just cause for detaining the licence, or that there has been needless delay on the part of the proprietor in bringing the matter to a hearing, he has power to order the proprietor to pay compensation to the driver as the justice thinks reasonable; and the justice must order that the licence is delivered to the driver, unless any misconduct is proved against him, and if there is the justice may revoke or suspend the licence.

If the proprietor neglects to apply for a summons and deposit the licence, after its demand, any justice of the peace may order compensation to be paid by him to the driver, and no proprietor can retain the licence of his driver beyond the time specified (section 24(1) London Hackney Carriages Act 1843).

Licences may be revoked or suspended

Where any driver is convicted of any offence, a justice of the peace can revoke the driver's licence, and any other licence he holds, or suspend the licence for a time the justice thinks proper. The justice can also require the proprietor or driver, who has the licence and the badge, to deliver them up. Every proprietor or driver who refuses or neglects to deliver up the licence or badge, is liable to a fine on level 1 on the standard scale.

The justice must send the licence and badge to TfL, who must cancel the licence if it has been revoked by the justice, or, if it has been suspended, must, at the end of the time for which it has been suspended, re-deliver the licence, with the badge, to the person to whom it was granted.

A magistrates' court that makes an order revoking or suspending any licence may, if the court thinks fit, suspend the effect of the order pending an appeal against the order (section 25 London Hackney Carriages Act 1843).

Complaints to be made within 28 days

All complaints under the provisions of the London Hackney Carriage Act, 1831, or of this Act, or of the orders and regulations made under either of them, must be made within 28 days after the day on which the cause of complaint arises (section 38 London Hackney Carriages Act 1843).

Requisitions for cab-driver's licences

TfL must furnish on demand to any person applying for a cab-driver's licence the form of requisition under Section 14 of the London Hackney Carriages Act 1843 (article 23 LCO 1934).

Grounds for refusal of cab-driver's licences

TfL has the discretion to refuse to grant a cab-driver's licence:

   (a) if the applicant fails to satisfy TfL that he is of good character and fit to act as a cab-driver, or
   (b) if the applicant has within the previous 3 years of the date of his application held a cab-driver's licence and has, otherwise than by reason of illness or other unavoidable cause, failed to act as a cab-driver during any considerable part of the licence period or, where he held more than one licence in those 3 years (article 25 LCO 1934).

Form of cab-driver's licence and issue of copy of licence

Every cab-driver's licence must be in a form and contain particulars TfL think fit, and unless revoked or suspended, be in force for between 1 and 3 years as TfL may direct. TfL may limit in a manner it thinks fit the types of cabs which the licensee is permitted to drive, and in any case where the licensee has not satisfied TfL that he has an adequate knowledge of the metropolitan area, they may attach a condition prohibiting the licensee from plying for hire with a cab in the area except the area which he has satisfied TfL that he has an adequate knowledge (article 27(1) LCO 1934).

When a cab-driver's licence is issued, a copy of the licence must be issued to the licensee (article 27(2) LCO 1934).

Immediately after the licensee receives his licence and a copy, he must sign both the licence and the copy with his usual signature in the space provided, and if he fails so to do he is guilty of a breach of this Order (article 27(3) LCO 1934).

Production of copy of licence

A cab-driver must keep his copy of his licence in his possession at all times during his employment as a cab-driver or when appearing before a court and must produce it for inspection on demand by any police constable or Public Carriage Examiner or any officer of a court, and if he fails so to do he is guilty of a breach of this Order (article 28 LCO 1934).

Cab-drivers to display driver's badge details and licensed area

A cab-driver must display the driver's badge number and licensed area in the form that TfL thinks fit; and notices or marks as they may direct (article 28A LCO 1934).

Fees for cab-driver's licences and for driving tests

If an applicant for a cab-driver's licence has failed on two occasions to pass the driving test on any type of cab, or if the holder of a cab-drivers licence who has applied to have his licence made available for an additional type of cab has failed on two occasions to pass the driving test on that type, he must, if allowed by TfL to undergo any further driving test on that type, pay the Receiver 12½p for the third and each subsequent driving test (article 26 LCO 1934).

Responsibility of cab-owners for safe custody and return of cab-driver's licences

While a cab-driver's licence is retained by a cab-owner, he must preserve it undamaged and undefaced, and make no mark on it otherwise than where the justice of the peace endorses the licence. If the licence is defaced or is lost, he must forthwith provide TfL with a statement of the circumstances in which it was defaced or lost, and, in the case of a defaced licence, must at the same time send or deliver it to TfL (article 29(1) LCO 1934).

If at any time the licensee is required to send or deliver his licence to TfL the cab-owner must return the licence to the licensee for that purpose (article 29(2) LCO 1934).

If any cab-owner acts in contravention of or fails to comply with any of these provisions, he is guilty of a breach of this Order (article 29(3) LCO 1934).

Revocation or suspension of cab-driver's licences

A cab-driver's licence is liable to revocation or suspension by TfL if it is satisfied, by reason of any circumstances arising or coming to its knowledge after the licence was granted, that the licensee is not a fit person to hold such a licence (article 30(1) LCO 1934).

In the event of the revocation or suspension of a cab-driver's licence TfL notify the licensee, and the licensee must, within 5 days after the notice has been delivered to him personally or sent to him by registered post or by the recorded delivery service at the address mentioned in or last endorsed upon the licence, send or deliver the licence and his copy and his badge to TfL for cancellation or for retention during the time of suspension, and if he fails so to do he is guilty of a breach of this Order (article 30(2) LCO 1934).

On the removal of a suspension of a cab-driver's licence which has not expired TfL must return the licence and the copy and the badge to the licensee (article 30(3) LCO 1934).

Breach of terms or conditions of cab-driver's licence

If the holder of a cab-driver's licence drives a cab of any type which by the terms of his licence he is not permitted to drive (not being a cab which is withdrawn from hire), or plies for hire with a cab or permits the cab to be hired in any part of the metropolitan area in which by a condition attached to his licence he is prohibited from plying for hire with a cab, he is guilty of a breach of this Order (article 31(1) LCO 1934).

Change of cab-driver's address to be notified

If during the period of his licence the holder of a cab-driver's licence changes his address, he must, within 7 days of the change of address, notify it and send or deliver his licence and his copy to TfL, who must endorse the new address on the licence and the copy and return them to the licensee (article 32(1 LCO 1934)).

If a licensee fails to comply with the requirements of this paragraph, he is guilty of a breach of this Order (article 32(2) LCO 1934).

Powers and rules

Regulations as to hackney and stage carriages

TfL may from time to time by London cab order make regulations for all or any of the following purposes:

* regulating the number of persons to be carried in any hackney carriage, how that number is displayed and how hackney carriages are furnished or fitted.

- fixing the stands of hackney carriages, and the persons to attend at the stands. For stands within the city of London the consent of the Court of the Lord Mayor and Aldermen is required for any stand appointed by TfL.
- fixing the rates or fares, as well for time as distance, to be paid for hackney carriages, and for securing the publication of the fares. The Mayor of London may direct TfL on the calculation of those rates or fares.
- forming, in the case of hackney carriages, a table of distances, for the purposes of any fare to be charged by distance, by the preparation of a book, map, or plan, or any combination.
- securing the safe custody and re-delivery of any property accidentally left in hackney carriages and fixing the charges to be paid for it, with power to sell or to be given to the finder in the event of its not being claimed within a certain time (section 9(1) MPCA 1869).

Any power to make a London cab order includes power to vary or revoke a previous order (section 4 MPCA 1869).

Offences and penalties

Penalty on use of unlicensed carriages

If any unlicensed hackney carriage plies for hire, the owner of the carriage is liable to a penalty not exceeding level 4 on the standard scale for every day during which the unlicensed carriage plies. And if any unlicensed hackney carriage is found on any stand, the owner of the carriage is liable to a penalty not exceeding level 4 on the standard scale for each time it is so found. The driver is liable to the same penalty unless he proves that he was ignorant of the fact of the carriage being unlicensed.

Any hackney carriage plying for hire, and any hackney carriage found on any stand without having a distinguishing mark, or other prescribed distinguishing mark is deemed to be an unlicensed carriage (section 7 MPCA 1869).

Penalties for breach of regulations

Where TfL is authorised to make a London cab order, it may annex a penalty not exceeding level 1 on the standard scale for the breach of the order or any regulations made; and any penalties are deemed to be penalties under this Act, and may be enforced accordingly (section 10 MPCA 1869).

Grant of licences by other persons at direction of TfL

Any licence which may be granted by TfL may, if TfL directs, be granted by the persons appointed for the purpose in the direction (section 11 MPCA 1869).

Powers to execute Act

TfL may appoint officers and constables of the metropolitan police force, it thinks fit to perform any duties to execute this Act, and may use fees, etc. raised under the Act for administration it thinks just (section 12 MPCA 1869).

Recovery of penalties

All penalties may be recovered summarily (section 13 MPCA 1869).

Placard etc. may be affixed to lamp post

TfL may attach any placard or signal to any lamp post for the purpose of its duties under this Act (section 14 MPCA 1869).

Hackney carriages standing in any street deemed to be plying for hire; and the driver refusing to go with any person liable to a penalty

Every hackney carriage found standing in any street or place, unless actually hired, is deemed to be plying for hire. The driver of every such hackney carriage which is not actually hired is obliged and compelled to go with any person wanting to hire it.

On hearing any complaint against the driver of any such hackney carriage for any refusal the driver is obliged to give evidence of having been hired at the time of the refusal, and where the driver fails to produce sufficient evidence of having been hired he is liable to a fine on level 1 on the standard scale (section 35 London Hackney Carriages Act 1831). This does not apply where a cab-driver has a condition attached to his licence that he is prohibited from plying for hire with a cab (article 31(2) London Cab Order 1934).

Compensation to be made to drivers improperly summoned for refusing to carry any person

If the driver of any hackney carriage in civil and explicit terms declares to any person, wanting to hire a hackney carriage that it is actually hired and is summoned to answer for an alleged refusal to carry that person in his hackney carriage, and on the hearing of the complaint produces sufficient evidence to prove that the hackney carriage was at the time actually and bona fide hired, and he did not appear to use uncivil language, or that he improperly conducted himself towards the party who he is summoned by, the justice must order the person summoning the driver to give him compensation for his loss of time in attending to make his defence to the complaint as the justice deems reasonable (section 36 London Hackney Carriages Act 1831).

Persons refusing to pay the driver his fare, or for any damage

If any person refuses or omits to pay the driver of any hackney carriage, the due sum to him for the hire of the hackney carriage, or if any person defaces or in any manner injures any hackney carriage, a justice of the peace, on complaint can grant a summons, or, if it appears to him necessary, a warrant, for bringing before him or any other justice the defaulter or defender, and, on proof of the facts made on oath before any such justice, to award reasonable satisfaction to the party complaining for his fare or for his damages and costs, and also a reasonable compensation for his loss of time in attending to make and establish the complaint (section 41 London Hackney Carriages Act 1831).

Deposit to be made for carriages waiting

Where any hackney carriage is hired and taken to any place of public resort, or elsewhere, and the driver is required there to wait with the hackney carriage, the driver can demand and receive from the person hiring and requiring him to wait a reasonable sum as a deposit, over and above the fare to which such driver is entitled for driving him there. The sum deposited must be accounted for by the driver when the hackney carriage is finally discharged.

If any driver receiving a deposit refuses to wait with the hackney carriage at the place where he was required to wait, or if such driver goes away, or permits the hackney carriage to be driven or taken away, without the consent of the person making the deposit, before the agreed time for which the sum was deposited, or if the driver on the final discharge of the hackney carriage refuses to account for the deposit, the driver is liable to a fine on level 1 on the standard scale (section 47 London Hackney Carriages Act 1831).

Penalty for permitting persons to ride without consent of the hirer

If the proprietor or driver of any hackney carriage which is hired permits or suffers any person to ride or be carried in, on, or about the hackney carriage, without the express consent of the person hiring, the proprietor or driver is liable to a fine on level 1 on the standard scale (section 50 London Hackney Carriages Act 1831).

Improperly standing with carriage; refusing to give way to or obstructing any other driver or depriving him of his fare

If any proprietor or driver of any hackney carriage wrongfully, in a forcible or clandestine manner, takes away the fare from any other proprietor or driver who, in the judgment of any justice of the peace before the complaint is heard, appears to be fairly entitled to the fare; every such proprietor, driver, offending is liable to a fine on level 1 on the standard scale (section 51 London Hackney Carriages Act 1831).

Proprietors or drivers misbehaving

If the proprietor or driver of any hackney carriage, or any other person having its care, is, by intoxication, or by wanton and furious driving, or by any other wilful misconduct, injures or endangers any person in his life, limbs, or property, or if any proprietor or driver, uses any abusive or insulting language, or is guilty of other rude behaviour, to or towards any person whatever, or assaults or obstructs any officer of police, constable, watchman or patrol, in the execution of his duty, every proprietor, driver or other person offending is liable to a fine on level 1 on the standard scale (section 56 London Hackney Carriages Act 1831).

Justices empowered to award compensation to drivers for their loss of time in attending to answer complaints which are not substantiated against them

If any driver of a hackney carriage is summoned or brought before any justice of the peace to answer any complaint or information touching or concerning any offence committed or alleged to have been committed by the driver against the provisions of this Act, and the complaint or information is withdrawn or quashed or dismissed, or if the defendant is acquitted of the offence charged against him, the justice can, if he thinks fit, order that the complainant or informant pay the driver compensation for his loss of time in attending the justice that seems reasonable (section 57 London Hackney Carriages Act 1831).

Penalty on persons acting as drivers, etc. without licences and badges

Every person granted a licence and badge who transfers or lends the licence, or permit any other person to use or wear the badge is liable to a fine of five pounds. Every proprietor who knowingly suffers any person not licensed to act as driver of any hackney carriage is liable to a fine of ten pounds (section 10 London Hackney Carriages Act 1843).

No person to act as driver of any carriage without the consent of the proprietor

Every driver authorised by any proprietor to act as driver of any hackney carriage, who allows any other person to act as driver of the hackney carriage, without the consent of the proprietor, and every person, whether licensed or not, who acts as driver of any carriage without the consent of the proprietor, is liable to a fine on level 1 on the standard scale.

Every driver charged with such an offence who, when required by a justice of the peace, refuses to give the name and address of the person allowed to act as driver and the number of the badge without consent of the proprietor, is liable to a further penalty of level 1 on the standard scale.

Any police constable, without any warrant for that purpose, can take into custody any person unlawfully acting as a driver and to take him before any justice of the peace and, if necessary, to take charge of the carriage, and to deposit it in some place of safe custody until it can be applied for by the proprietor (section 27 London Hackney Carriages Act 1843).

Punishment for furious driving, and wilful misbehaviour

Every driver of a hackney carriage, who is guilty of wanton or furious driving, or who by carelessness or wilful misbehaviour causes any hurt or damage to any person or property in any street or highway, and every driver, who during his employment is drunk, or uses any insulting or abusive language, or is guilty of any insulting gesture or any misbehaviour, is liable for every such offence to a fine on level 1 on the standard scale.

A justice considering any such complaint, has the discretion, instead of inflicting a penalty, to commit the offender to prison for up to 2 calendar months (section 28 London Hackney Carriages Act 1843).

Penalty on drivers of hackney carriages for loitering or causing any obstruction, or plying for hire by making any noise, etc.

Every driver of a hackney carriage who plys for hire elsewhere than at some standing or place appointed for that purpose, or who by loitering or by any wilful misbehaviour causes any obstruction in or upon any public street, road or place, and every driver of a hackney carriage, whether hired or unhired, who allows a person other than himself, other than the hirer or a person employed by the hirer, to ride on the driving box, is liable to a fine on level 1 on the standard scale (section 33 London Hackney Carriages Act 1843).

Justices may hear complaints and award penalties

Any justice of the peace can hear and determine all complaints under the provisions of this Act or of the London Hackney Carriage Act, 1831 to judge the payment of any penalty or of any sum of money, or of the orders and regulations made under them, and to order their payment, with or without costs, either immediately, or at a time and place, and by instalments, as he thinks fit (section 39 London Hackney Carriages Act 1843).

Providing for cases where there are more proprietors than one

Where there is more than one proprietor of any hackney carriage it is sufficient, in any information, summons, order, conviction, warrant, or any other proceeding under this Act, or of the London Hackney Carriage Act, 1831 to name one of the proprietors without reference to any other or others of them, and to describe and proceed against him as if he were sole proprietor (section 44 London Hackney Carriages Act 1843).

Distances drivers of hackney carriages required to drive

The driver of every hackney carriage which plys for hire at any place within the limits of this Act can (unless the driver has a reasonable excuse, to be allowed by the justice before whom the matter shall be brought in question) drive the hackney carriage to any place to which he is required by the hirer, more than 6 miles (20 miles if hired at Heathrow airport (London Cab Order 1972)) from the place where the carriage was hired, or for more than 1 hour from the time when hired (section 7 London Hackney Carriage Act 1853).

Number of persons to be carried to be painted or marked on hackney carriage

The driver of any hackney carriage must, if required by the hirer, carry in the carriage the number (or less) of persons painted or marked on it (section 9 London Hackney Carriage Act 1853).

Quantity of luggage to be carried without extra charge

The driver of every hackney carriage must carry in or on the carriage a reasonable quantity of luggage for every person hiring the carriage (section 10 London Hackney Carriage Act 1853).

Advertising vehicles, etc. prohibited

It is unlawful to carry on any carriage, in any thoroughfare or public place to the obstruction or annoyance of the inhabitants or passengers, any picture, placard, notice or advertisement, whether written, printed or painted upon or posted or attached to any part of any carriage, or on any board, or otherwise (section 16 London Hackney Carriage Act 1853).

Drivers and conductors of metropolitan stage carriages and drivers of hackney carriages, liable to penalties for offences

The driver of any hackney carriage, who commits any of the following offences is liable to a penalty not exceeding level 3 on the standard scale, for each offence:

- Every driver of a hackney carriage who demands or takes more than the proper fare, or who refuses to admit and carry in his carriage the number of persons painted or marked on the carriage or specified in the certificate granted by Transport for London, or who refuses to carry a reasonable quantity of luggage for any person hiring or intending to hire the carriage (section 17(1) London Hackney Carriage Act 1853).
- Every driver of a hackney carriage who refuses to drive the carriage to any place up to six miles, or who refuses to drive any carriage for up to 1 hour, if required by any person hiring or intending to hire the carriage, or who drives the carriage at a reasonable and proper speed, not less than 6 miles an hour, except in cases of unavoidable delay, or when required by the hirer to drive at any slower pace (section 17(2) London Hackney Carriage Act 1853).
- Every driver of a hackney carriage who plys for hire any carriage which is at the time unfit for public use (section 17(3) London Hackney Carriage Act 1853).

This does not apply where a cab-driver is prohibited from plying for hire with a cab by a condition attached to his licence (article 31(2) London Cab Order 1934).

Power to police, magistrates or justices of the peace to hear and determine offences

Two justices of the peace can hear and determine all offences against the provisions of this Act, and also all disputes or causes of complaint arising from it (section 18 London Hackney Carriage Act 1853).

Penalty for offences against this Act for which no penalty is appointed

For every offence against the provisions of this Act, for which no special penalty is appointed, the offender is liable to a penalty not exceeding level 1 on the standard scale (section 19 London Hackney Carriage Act 1853).

Power of Assistant Commissioner to vary directions, etc.

Any appointment made or approval or direction given by TfL under this Order may at any time be revoked or varied by it (article 4 LCO 1934).

Disqualification of persons under 21

A cab licence must not be granted to a person under the age of 21 years, and if it is, it is void (article 6 LCO 1934).

Obtaining licence by false statements or by withholding information

If any person in obtaining the grant of a cab licence to himself or to any other person knowingly makes any false statement or withholds any material information, he is guilty of a breach of this Order (article 15 LCO 1934).

Plying for hire without plates, etc., or with plates, etc., defaced

If the owner of a cab, or, where the owner is a firm or company, the person holding the licence in respect of the cab on its behalf, causes or permits the cab to ply for hire:

(a) without any of the plates or notices fixed to the vehicle, or with any such plate or notice so defaced that any figure is illegible, or

(b) without any of the marks placed on the vehicle by the authority of TfL, or with any such mark so obliterated, or indistinct that any material particular is illegible,

he is guilty of a breach of this Order (article 16(1) LCO 1934).

If any person without lawful authority removes, conceals, obliterates or alters any plate notice or mark, he is guilty of a breach of this Order (article 16(2) LCO 1934).

If any plate or notice is lost, or has become defaced, or if any such mark has become obliterated or indistinct, the cab-owner or the licensee must bring or send the cab to the appointed passing station and the Public Carriage Examiner must fix a new plate or notice or place a new mark on the cab (article 16(3) LCO 1934).

If a licensed cab is granted a new licence to take effect after the expiration of the current licence it may ply for hire whilst displaying plates and notices related to that new licence rather than the current licence (article 16(4) LCO 1934).

Defaced or altered cab licences to be void

A cab licence which is defaced or on which there is an unauthorised erasure or alteration of any material particular is void (article 17 LCO 1934).

Possession of defaced, etc., cab licence or plates, etc.

If any person uses or has in his possession without lawful authority any altered or irregular cab licence, or any altered or irregular plate notice or mark, or any counterfeit of any such licence plate notice or mark, he is guilty of a breach of this Order (article 18 LCO 1934).

Disqualification of persons under 21

A cab-driver's licence cannot be granted to a person under the age of 21 years, and if granted is void (article 24 LCO 1934).

Carriage of persons and luggage

No person other than the driver can be carried on any cab in excess of the number of persons which it is licensed to carry, provided that, an infant in arms is not counted as a person and two children under 10 years of age counts as one person (article 33(1) LCO 1934).

No person other than the driver can, without the authority of Transport for London, be carried on the driving box or platform of a motor cab (article 33(2) LCO 1934).

No luggage can be carried on the roof of a motor cab unless the cab is fitted for that purpose and is licensed to carry luggage on the roof (article 33(3) LCO 1934).

The driver of the cab, and, unless he proves that the breach occurred without his knowledge or consent, the owner, is liable for any breach of these provisions (article 33(4) LCO 1934).

Taximeters

An approved taximeter is to be fitted and sealed. The owner of every motor cab must fit the cab with a taximeter of a type approved by TfL (article 35(1) LCO 1934).

Any such meter must be so construed and adjusted that:

(a) after the taximeter has been started at the commencement of the hiring or at such later time as the driver thinks fit, the prescribed fare payable for the hiring is automatically recorded and displayed by the taximeter;

(b) the total up to an amount of not less than 10p of any extra charges payable by a hirer of the cab, can be displayed by the meter (article 35(2) LCO 1934).

Every taximeter must be sealed as TfL may direct and may be marked TfL permit, and no person can place a seal or mark on the taximeter unless he is authorised by TfL (article 35(3) LCO 1934).

Removal of or tampering with taximeters

Where a taximeter is fitted to a cab with a licence, no person can, without the authority of TfL, remove or tamper with the taximeter, or the mechanism by which the taximeter is operated, or break, alter, deface, or otherwise tamper with any seal or mark placed on the taximeter by direction of TfL (article 36 LCO 1934).

Plying for hire without taximeters or with taximeters not sealed, etc.

The owner or driver of a motor cab must not allow the cab to ply for hire:
  (i) if a taximeter is not fitted, or
  (ii) if a taximeter has not been sealed and marked in accordance with the directions of TfL, or
  (iii) if the seal or any mark placed on a taximeter by direction of TfL is broken, altered, defaced, or otherwise tampered with, as soon as practicable and in any event within 24 hours the cab must be brought or sent to the appointed passing station in order that a new seal or mark may be placed on the taximeter (article 37 LCO 1934).

Lighting of taximeters and 'taxi' signs

The owner of every motor cab must provide a lamp placed on the cab to allow the readings on the dial of the taximeter easily legible at all times of the day and night, and must maintain the lamp in proper working order and condition (article 38(1) LCO 1934).

The driver of a motor cab must keep the lamp properly alight throughout any part of a hiring which is during the hours of darkness, and must light the lamp during a hiring at any other time at the request of a hirer so as to enable the hirer to read the dial of the taximeter (article 38(2) LCO 1934).

Where a motor cab is provided with means for illuminating either the flag of the taximeter or a sign bearing the word 'Taxi' fitted with the approval of TfL on the top of the cab, the driver, when plying for hire with the cab during any part of the hours of darkness, must light the flag or the sign or both (article 38(3) LCO 1934).

Starting and stopping the taximeter

The driver of a motor cab must start the taximeter no sooner than when the cab is hired or at such later time as the driver thinks fit (article 39(1) LCO 1934).

The driver of a motor cab must stop the taximeter no later than when the hiring is terminated or at such earlier time as the driver thinks fit (article 39(2) LCO 1934).

Fares for motor cabs

Fares for motor cabs are contained in Article 40. Similar fares apply where only luggage is transported without passengers (article 42 LCO 1934).

Provision of receipt on request

The driver of a motor cab must, if requested by a passenger during or immediately after a journey, provide the passenger with a receipt for the fare paid by him for that journey (article 41 LCO 1934).

Property accidentally left in cabs – care and disposal of property left in cabs

Any person who finds any property accidentally left in a cab must immediately hand it to the cab-driver (article 51(1) LCO 1934).

Immediately after the termination of every hiring of a cab, the cab-driver must carefully search the cab, or, if this is impracticable, must look inside the cab, to ascertain whether any property has been accidentally left, and if he does not carefully search the cab at the termination of the hiring he shall do so as soon as practicable (article 51(2) LCO 1934).

Any cab-driver who finds any property left in the cab or to whom any such property is handed must, within 24 hours, deposit the property at a police station in the metropolitan area, in the state in which it was found by, or handed to, him, and must state the particulars of the finding: provided that if the property is sooner claimed by the owner and satisfactory proof of ownership is given, it must be restored to the owner instead of being deposited at a police station (article 51(3) LCO 1934).

Disposal by Assistant Commissioner of deposited property

If any property found in a cab and deposited at a police station by the cab-driver is not claimed within 3 months from the last day of the month in which the property reaches the police station and proved to the satisfaction of TfL to belong to the claimant, TfL may at its discretion either deliver the property to the cab-driver, or sell or otherwise dispose of the property and pay to the cab-driver a reasonable award as it can give, but the award cannot be less than 5p and:

• for property consisting of coin, paper money, any gold or silver article, jewellery, watch or clock, and not being of greater value than £10, 15p in the £ on the value of the property;
• for property of any other kind and not being of greater value than £10, 12½p in the £ on the value of the property;
• for property above the value of £10, such a sum as TfL deems reasonable:

> Provided that a cab-driver who has failed to satisfy TfL that he has complied fully with the Regulations is not entitled to receive any award, but TfL may, at its discretion, award a sum as it may consider reasonable in all the circumstances.

(article 52(1) LCO 1934)

If any property found in a cab and deposited at a police station by the cab-driver is claimed within 3 months from the last day of the month in which the property is received at the police station, and the claimant proves to the satisfaction of TfL that he is entitled to the property it must be delivered to him on payment to TfL of:

(a) a fee in respect of the cost of collecting, keeping in safe custody, and restoring lost property;

(b) an award to the cab-driver who deposited the property; and

(c) an additional sum (if any) as may be payable below (article 52(2) LCO 1934).

The fee payable above is determined in accordance with the scale in Article 52 (3).

The award to be paid to the cab-driver is determined in accordance with the scale in Article 52 (4).

The value of any property is determined in accordance with the scale in Article 52 (5).

If the property is forwarded to the claimant by post or other means, the cost of postage or other means of conveyance and any other expenses incurred must be paid to TfL by the claimant (article 52(6) LCO 1934).

In the case of any unclaimed property contained in any package, bag or other receptacle, TfL may open it and the contents examined if it deems it necessary to do so for the purpose either of identifying and tracing the owner of the property, or of ascertaining the nature of its contents with a view to securing its safe custody or ascertaining whether the property is of a perishable nature.

In the case of any property which is claimed by any person, TfL may, if the claim of that person to be entitled to the property cannot otherwise be established to the satisfaction of TfL, require the claimant to open the property and to submit the contents to examination for the purpose of establishing his claim to ownership (article 52(7) LCO 1934).

Definitions

hackney carriage means any carriage for the conveyance of passengers which plies for hire and is neither a stage carriage nor a tramcar (MPCA 1869).
matter of fitness means:

(a) any matter which TfL must be satisfied before granting a licence, or

(b) any matter where TfL is not satisfied with respect to the matter, they may refuse to grant a licence (section 6(6) MPCA 1869).

stage carriage means any carriage for the conveyance of passengers which plies for hire in any public street, road or place, and in which the passengers or any of them are charged to pay separate and distinct or at the rate of separate and distinct fares for their respective places or seats (MPCA 1869).
proprietor means every person who, either alone or in partnership with any other person, keeps any hackney carriage, or who is concerned otherwise than as a driver or attendant in employing for hire any hackney carriage (London Hackney Carriages Act 1843).

passing station means any place appointed by TfL as a place where cabs may be examined, and if TfL appoints any passing station for the examination of any particular cab or cabs, that passing station is deemed to be the appointed passing station for that cab or those cabs (article 2(1) London Cab Order 1934).

Public Carriage Examiner means any person appointed by TfL to examine and inspect public carriages for the purposes of the Metropolitan Public Carriage Act 1869 (article 2(1) London Cab Order 1934).

Metropolitan area means the area consisting of the City of London and the Metropolitan Police District (article 2(1) London Cab Order 1934).

cab has the same meaning as hackney carriage has in the Metropolitan Public Carriage Act 1869 (article 2(1) London Cab Order 1934).

motor cab means any mechanically propelled cab (article 2(1) London Cab Order 1934).

horse cab means any cab drawn by animal power, and **horse** includes any animal used to draw a cab (article 2(1) London Cab Order 1934).

cab licence means a licence in pursuance of Section 6 of the Metropolitan Public Carriage Act 1869 and of Part III of this Order in respect of a cab (article 2(1) London Cab Order 1934).

cab-driver's licence means a licence in pursuance of Section 8 of the Metropolitan Public Carriage Act 1869, as amended by Section 39 of the Road Traffic Act 1934, and of Part IV of this Order to drive cabs (article 2(1) London Cab Order 1934).

owner or cab-owner in relation to a cab which is the subject of a hiring agreement or hire purchase agreement means the person in possession of the cab under that agreement (article 2(1) London Cab Order 1934).

licensee means any person to whom a licence is granted (article 2(1) London Cab Order 1934).

PRIVATE HIRE VEHICLE LICENSING IN LONDON

References

Private Hire Vehicles (London) Act 1998.
The Private Hire Vehicles (London) (Operator's Licences) Regulations 2000.
The Private Hire Vehicles (London PHV Driver's Licences) Regulations 2003.
Private Hire Vehicles (London PHV Licences) Regulations 2004.

Extent

These procedures apply in London and the United Kingdom.

Scope

The Private Hire Vehicles (London) Act 1998 (PHVLA 1998) provides for the licensing of minicabs in London and applies to private hire vehicle (PHV) operators, drivers and vehicles.

Vehicle licensing (FC 3.38)

Requirement for private hire vehicle licence

A vehicle cannot be used as a private hire vehicle on a road in London unless it has a private hire vehicle licence (section 6(1) PHVLA 1998).

If a vehicle is used for private hire without a licence the driver and operator of a vehicle are each guilty of an offence (section 6(2) PHVLA 1998). The owner of a vehicle who permits it to be used for private hire without a licence is also guilty of an offence (section 6(3) PHVLA 1998)). It is a defence in proceedings for an offence for the driver or operator to show that he exercised all due diligence to prevent the vehicle being used as such (section 6(4) PHVLA 1998). A person guilty of an offence is liable on summary conviction to a fine not exceeding level 4 on the standard scale (section 6(5) PHVLA 1998).

London PHV licences

The owner of any vehicle constructed or adapted to seat fewer than nine passengers may apply to the licensing authority for a private hire vehicle licence for London for that vehicle (section 7(1) PHVLA 1998).

The licensing authority must grant a London PHV licence for a vehicle if the authority is satisfied:

(a) that the vehicle is suitable in type, size and design for use as a private hire vehicle, is safe, comfortable and in a suitable mechanical condition for that use, and is not of such design and appearance as would lead any person to believe that the vehicle is a London cab;

(b) an insurance policy is in force for the use of the vehicle or that which complies with the requirements of Part VI of the Road Traffic Act 1988; and

(c) that any further prescribed requirements are met (section 7(2) PHVLA 1998).

A London PHV licence cannot be granted in respect of more than one vehicle (section 7(3)) and must be granted subject to prescribed conditions and other conditions the licensing authority may think fit (section 7(4) PHVLA 1998). The licence must be in a form and contain particulars the licensing authority think fit (section 7(5) PHVLA 1998). The licence is to be granted for 1 year or for a shorter period the licensing authority may consider appropriate in the circumstances of the case (section 7(6) PHVLA 1998).

An applicant for a London PHV licence may appeal to a magistrates' court against a decision not to grant a licence or against any condition (other than a prescribed condition) to which the licence is subject (section 7(7) PHVLA 1998).

Obligations of owners of licensed vehicles

The owner must present the vehicle for inspection and testing by or on behalf of the licensing authority within a period and at a place the authority may, by

FC 3.38 Licensing of private hire vehicles in London – Private Hire Vehicles (London) Act 1998 and Private Hire Vehicles (London PHV Licences) Regulations 2004

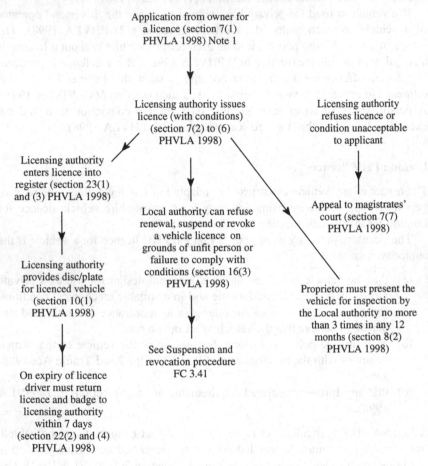

Notes
1. Application must be accompanied with fee, vehicle insurance policy, etc. – see page 313.
2. Appeals – sections 25 and 26 PHVLA 1998.
3. PHVLA 1998 – Private Hire Vehicles (London) Act 1998.

notice, reasonably require. However, it cannot be required to be presented on more than three separate occasions during any one period of 12 months (section 8(2) PHVLA 1998).

The owner must report any accident to the vehicle materially affecting the safety, performance or appearance of the vehicle, or the comfort or convenience of persons carried in the vehicle, to the licensing authority as soon as reasonably practical and in any case within 72 hours of the accident occurring (section 8 (3) PHVLA 1998).

If the ownership of the vehicle changes, the previous owner must notify the licensing authority within 14 days of the change with the name and address of the new owner (section 8 (4) PHVLA 1998).

A person who, without reasonable excuse, contravenes these provisions is guilty of an offence and liable on summary conviction to a fine not exceeding level 3 on the standard scale (section 8 (5) PHVLA 1998).

Fitness of licensed vehicles

A constable or authorised officer has power at all reasonable times to inspect and test, to ascertain its fitness, any vehicle with a London PHV licence (section 9(1) PHVLA 1998).

If a constable or authorised officer is not satisfied as to the fitness of such a vehicle he may by notice to the owner of the vehicle:

(a) require the owner to make the vehicle available for further inspection and testing at a reasonable time and place specified in the notice; and

(b) if he thinks fit, suspend the London PHV licence until a time the constable or authorised officer is satisfied as to the fitness of the vehicle (section 9(2) PHVLA 1998). Such a notice must state the grounds on which the licence is being suspended and the suspension takes effect on the day on which it is served on the owner (section 9(3) PHVLA 1998).

A licence suspended remains suspended until a time the constable or authorised officer notifies the owner that the licence is again in force (section 9(4) PHVLA 1998).

If a licence remains suspended at the end of 2 months from the day on which a notice was served on the owner of the vehicle a constable or authorised officer may by notice to the owner direct that the licence is revoked and the revocation takes effect 21 days after the owner is served with that notice (section 9(5) PHVLA 1998).

An owner may appeal against such notices to a magistrates' court (section 9(6) PHVLA 1998).

Identification of licensed vehicles

The licensing authority must issue a disc or plate for each vehicle with a London PHV licence which identifies that vehicle as such a vehicle (section 10(1) PHVLA 1998) and the disc or plate must be exhibited on the vehicle in the pre-scribed manner (section 10(2) PHVLA 1998).

The licensing authority may by notice exempt a vehicle from this requirement when it is being used to provide a service specified in the notice if the authority considers it inappropriate (having regard to that service) to require the disc or plate in question to be exhibited (section 10(3) PHVLA 1998). Failure of the driver and operator of a vehicle to comply with these requirements is an offence (section 10(4)) and the owner of a vehicle who permits it to be used as such is also guilty of an offence (section 10(5) PHVLA 1998).

It is a defence in proceedings for an offence for the driver or operator to show that he exercised all due diligence to prevent the vehicle being used as such (section 10(6) PHVLA 1998).

A person guilty of such an offence is liable on summary conviction to a fine not exceeding level 3 on the standard scale (section 10(7) PHVLA 1998).

A London PHV licence may be suspended or revoked where:

(a) the licensing authority is no longer satisfied that the vehicle to which it relates is fit for use as a private hire vehicle; or
(b) the owner has failed to comply with any condition of the licence or any other obligation imposed on him (section 16(3) PHVLA 1998).

Prohibition of taximeters

No vehicle with a London PHV licence can be equipped with a taximeter (section 11(1) PHVLA 1998). If such a vehicle is equipped with a taximeter, the owner of that vehicle is guilty of an offence and liable on summary conviction to a fine not exceeding level 3 on the standard scale (section 11(2) PHVLA 1998).

Private Hire Vehicles (London PHV Licences) Regulations 2004 (PHVLR 2004)

Further requirements that must be met

In addition to the requirements specified in section 7(2)(a) and (b) of the 1998 Act the requirements specified in Schedule 1 of these regulations must be satisfied before granting a London PHV licence in respect of a vehicle.

Transport for London may exempt a vehicle from any of the requirements of this regulation and Schedule 1 if they are requested by the applicant for a London PHV licence for the vehicle and are satisfied that, having regard to exceptional circumstances, it is reasonable to do so (regulation 3 PHVLR 2004).

Prescribed licence conditions

The conditions specified in Schedule 2 of these regulations apply to a London PHV licence (regulation 4 PHVLR 2004).

Exhibition of identification disc

The identification disc must be exhibited by being affixed to the top of the inside of the front windscreen on the passenger side of the vehicle so that:

(a) on the side which faces outwards, the following particulars are clearly legible:
  (i) the registration mark of the vehicle;
  (ii) the maximum number of passengers which may be carried in the vehicle in accordance with the conditions of the licence for the vehicle;
  (iii) the number of the London PHV licence for the vehicle;
  (iv) the date of the expiry of the licence; and
  (v) a statement that the licence has been issued by the Public Carriage Office of Transport for London; and
(b) on the side which is visible from inside the vehicle, a statement that the vehicle is licensed by the Public Carriage Office of Transport for London is clearly legible (regulation 5 PHVLR 2004).

Fees

The fees for an application, by an applicant for a London PHV licence are specified in regulation 6 PHVLR 2004 and also apply to the renewal of a licence.

Register of licences

In addition to the particulars specified in section 23(1)(a) of the 1998 Act, the register must contain the following particulars for each London PHV licence:

(a) the registration mark of the vehicle to which the licence relates;
(b) an indication as to whether the licence:
  (i) is a London PHV licence or a temporary permit; and
  (ii) is current, suspended or revoked (regulation 7 PHVLR 2004).

Prohibition of certain signs, notices, etc.

No signs or advertising material can be displayed on a private hire vehicle, except:

(a) badges or emblems on the radiator or windscreen which are issued by an organisation providing vehicle repair or recovery services; or concerned with driving skills and qualifications; or
(b) a sign displayed temporarily on a stationary vehicle which contains, and contains only the name and address of a person operating the private hire vehicle, or the name under which he carries on that business and its address; and the name of a passenger for whom a private hire booking has been made (regulation 8 PHVLR 2004).

Operator's licence (FC 3.39)

Requirement for London operator's licence

A London PHV operator's licence is required for the invitation or acceptance of, private hire bookings in London (section 2(1) PHVLA 1998).

It is an offence liable on summary conviction to a fine not exceeding level 4 on the standard scale for a person to contravene this requirement (section 2(2) PHVLA 1998).

London operator's licences

Any person may apply to the licensing authority for a London PHV operator's licence (section 3(1) PHVLA 1998) and an application must state the address of any premises in London which the applicant proposes to use as an operating centre (section 3(2) PHVLA 1998).

The licensing authority must grant a London PHV operator's licence to the applicant if the authority is satisfied that the applicant is a fit and proper person to hold a London PHV operator's licence; and any prescribed requirements are met (section 3(3) PHVLA 1998).

A London PHV operator's licence must be granted subject to conditions prescribed and other conditions as the licensing authority may think fit (section 3(4) PHVLA 1998).

A London PHV operator's licence is for 5 years or a shorter period the licensing authority may consider appropriate in the circumstances of the case (section 3(5) PHVLA 1998).

A London PHV operator's licence must specify the address of any premises in London which the holder of the licence may use as an operating centre and be in a form and contain particulars the licensing authority may think fit (section 3(6) PHVLA 1998).

An applicant for a London PHV operator's licence may appeal to a magistrates' court against a decision not to grant such a licence, not to specify an address proposed in the application as an operating centre or any condition (other than a prescribed condition) to which the licence is subject (section 3(7) PHVLA 1998).

The information which an applicant for a London PHV operator's licence may be required to furnish includes:

(a) any premises in London which he proposes to use as an operating centre;
(b) any convictions recorded against him;
(c) any business activities he has carried on before making the application;
(d) if the applicant is or has been a director or secretary of a company, that company;
(e) if the applicant is a company, information about the directors or secretary of that company;
(f) if the applicant proposes to act as an operator in partnership with any other person, information about that person (section 15(3) PHVLA 1998).

Obligations of London operators

The holder of a London PHV operator's licence must not (in London) accept a private hire booking other than at an operating centre specified in his licence (section 4(1) PHVLA 1998).

FC 3.39 Licensing of operators of private hire vehicles in London – Private Hire Vehicles (London) Act 1998 (PHVLA 1998) and Private Hire Vehicles (London) (Operators' Licences) Regulations 2000 (PHVLOLR 2000)

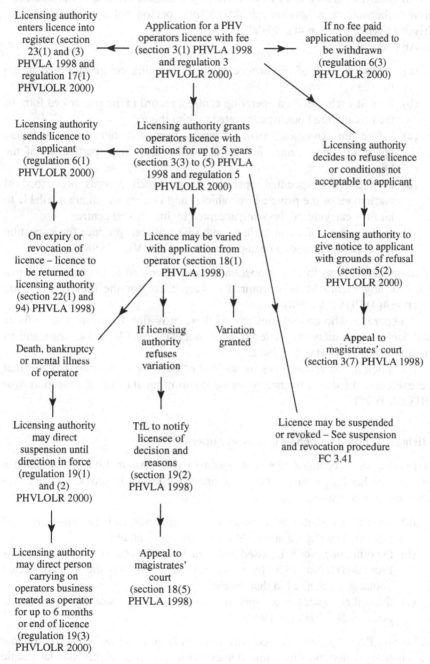

Notes
1. PHVLA 1998 Private Hire Vehicles (London) Act 1998.
2. PHVLOLR 2000 – Private Hire Vehicles (London) (Operators' Licences) Regulations 2000.

An operator must ensure that any vehicle which is provided by him for carrying out a private hire booking accepted by him in London is a vehicle for which a London PHV licence is in force driven by a person holding a London PHV driver's licence; or a London cab driven by a person holding a London cab-driver's licence (section 4(2) PHVLA 1998).

An operator must:

(a) display a copy of his licence at each operating centre specified in the licence;

(b) keep at each specified operating centre a record in the prescribed form of the private hire bookings accepted by him there;

(c) before the commencement of each journey booked at a specified operating centre, enter in the record the prescribed particulars of the booking;

(d) keep at each specified operating centre such records prescribed of particulars of the private hire vehicles and drivers which are available to him for carrying out booking accepted by him at that centre;

(e) at the request of a constable or authorised officer, produce for inspection any record required to be kept (section 4(3) PHVLA 1998).

If an operator ceases to use an operating centre specified in his licence he must preserve any record he was required to keep there for the prescribed period (section 4(4) PHVLA 1998).

An operator who contravenes any of these provision is guilty of an offence and liable on summary conviction to a fine not exceeding level 3 on the standard scale (section 4(5) PHVLA 1998).

It is a defence in proceedings for such an offence for an operator to show that he exercised all due diligence to avoid committing the offence (section 4(6) PHVLA 1998).

Hirings accepted on behalf of another operator

A London PHV operator (the first operator) who has, in London, accepted a private hire booking cannot arrange for another operator to provide a vehicle to carry out that booking as sub-contractor unless:

(a) the other operator is a London PHV operator and the sub-contracted booking is accepted at an operating centre in London;

(b) the other operator is licensed under the Local Government (Miscellaneous Provisions) Act 1976 by the district council and the sub-contracted booking is accepted in that district; or

(c) the other operator accepts the sub-contracted booking in Scotland (section 5(1) PHVLA 1998).

A London PHV operator who contravenes this is guilty of an offence and liable on summary conviction to a fine not exceeding level 3 on the standard scale (section 5(2) PHVLA 1998). It is a defence in proceedings for an offence for an operator to show that he exercised all due diligence to avoid committing such an offence (section 5(3) PHVLA 1998). It is immaterial whether or not sub-contracting is permitted by the contract between the first operator and the person who made the booking (section 5(4) PHVLA 1998).

A contract of hire between a person who made a private hire booking at an operating centre in London and the London PHV operator who accepted the booking remains in force despite the making of arrangements by that operator for another contractor to provide a vehicle to carry out that booking as sub-contractor (section 5(5) PHVLA 1998).

A London PHV operator's licence may be suspended or revoked where:

(a) the licensing authority is no longer satisfied that the licence holder is fit to hold such a licence; or

(b) the licence holder has failed to comply with any condition of the licence or any other obligation imposed on him (section 16(2) PHVLA 1998).

Variation of operator's licence at the request of the operator

The licensing authority may, on the application of a London PHV operator, vary his licence by adding a reference to a new operating centre or removing an existing reference to an operating centre (section 18(1) PHVLA 1998).

An application for the variation of a licence must be made in a form, and include declarations and information, the licensing authority may require (section 18(2) PHVLA 1998).

The licensing authority may require an applicant to furnish further information as he may consider necessary for dealing with the application (section 18(3) PHVLA 1998).

The licensing authority must not add a reference to a new operating centre unless the authority is satisfied that the premises in question meet any prescribed requirements (section 18(4) PHVLA 1998).

An applicant for the variation of a London PHV operator's licence may appeal to a magistrates' court against a decision not to add a new operating centre to the licence (section 18(5) PHVLA 1998).

Variation of operator's licence by the licensing authority

The licensing authority may:

(a) suspend the operation of a London PHV operator's licence for any operating centre specified in the licence; or

(b) vary a licence by removing a reference to an operating centre previously specified in the licence,

if the authority is no longer satisfied that the operating centre in question meets any prescribed requirements or for any other reasonable cause (section 19(1) PHVLA 1998).

Where the licensing authority has decided to suspend the operation of a licence or vary a licence the authority must give notice of the decision and the grounds for it to the licence holder and the decision takes effect 21 days after the day on which the licence holder is served with that notice (section 19(2) PHVLA 1998).

If the licensing authority is of the opinion that the interests of public safety require the authority's decision to have immediate effect, and the authority includes a statement of that opinion and the reasons for it in the notice, the authority's decision takes effect when the notice is served on the licence holder (section 19(3) PHVLA 1998).

If a licence is suspended in relation to an operating centre, the premises is not regarded as authorised to accept private hire bookings, until the licensing authority by notice states that the licence is no longer suspended (section 19(4) PHVLA 1998).

The holder of a London PHV operator's licence may appeal to a magistrates' court against such a decision (section 19(5) PHVLA 1998).

The Private Hire Vehicles (London) (Operator's Licences) Regulations 2000 (PHVLOLR 2000)

Applications

Every application must:

(a) be made on a form supplied by the licensing authority and include the information and declarations required by that form;

(b) be signed if made by an individual, by that person, or if made by a firm, by one of the partners of that firm with the authority of the others, or if made by any other body or group of persons, by one or more individual persons authorised for that purpose by the body or group; and

(c) be accompanied by the appropriate fee (regulation 3 PHVLOLR 2000).

Determination of applications

If the licensing authority is satisfied that a licence can be properly granted for 5 years in the terms applied for; and without the need for any additional conditions or varied in the terms applied for, the authority must approve the application and give the applicant notice of the decision (regulation 5(1) PHVLOLR 2000).

If the authority is not satisfied and decides:

(a) to approve the application with different terms than applied for;

(b) in the case of an application for the grant of a licence, to approve the application on the basis that additional conditions are attached to the licence, or the licence is granted for a shorter period than 5 years; or

(c) to refuse the application,

the authority must give the applicant notice of the decision and the grounds for it (regulation 5(2) PHVLOLR 2000).

Grant and variation

Where the decision has been made to approve an application the licensing authority must, provided that any appropriate fee is received within the period of 28 days commencing on the date below, grant or vary the licence, and send the licence or any replacement licence to the applicant (regulation 6(1) PHVLOLR 2000).

The date above is the date of service of the notice; or where an appeal is brought against that decision, the date of disposal or withdrawal of that appeal (regulation 6(2) PHVLOLR 2000).

If any appropriate fee is not received by the licensing authority the approval will lapse, the application will be deemed to have been withdrawn and the licensing authority is entitled to retain the fee (regulation 6(3) PHVLOLR 2000).

Fees

Fees are specified in regulations 4 and 7. Where the calculation of any fee would result in the amount payable including a fraction of a pound then the amount payable is adjusted downwards to the nearest pound (regulation 7(6) PHVLOLR 2000).

Refund of fees

Where the licensing authority is satisfied that:

(a) an operator has ceased to operate from every operating centre specified in his licence, other than by reason of the suspension or revocation of that licence;

(b) that operator has transferred some or all of his undertaking as an operator to another person; and

(c) before the date of the transfer the transferee has been granted a new licence in relation to any operating centre specified in the transferor's licence,

the licensing authority must, on receipt of a written request for a refund accompanied by the transferor's licence, refund a proportion of the fee paid for the grant of that licence being an amount calculated below (regulation 8(1) PHVLOLR 2000).

Where a licence has been granted following an election below and before its expiry the operator has been granted a new licence in circumstances where he did not meet the requirements for all of the operating centres specified in that licence, the licensing authority must, upon receipt of a written request for a refund accompanied by the first mentioned licence, refund a proportion of the fee paid for the grant of that licence being an amount calculated below (regulation 8(2) PHVLOLR 2000).

The amount above must be that proportion of the fee which the number of full years remaining on the licence bears to the period for which the licence was granted, the number of full years being calculated from the date of receipt by the licensing authority of both the request for a refund and the licence (regulation 8(3) PHVLOLR 2000).

Where the calculation above would have the result that the amount refundable would include a fraction of a pound then the amount refundable is adjusted downwards to the nearest pound (regulation 8(4) PHVLOLR 2000).

Where a proportion of the fee paid for the grant of a licence is refunded, that licence ceases to have effect (regulation 8(5) PHVLOLR 2000).

Conditions

Every licence must be granted subject to the conditions set out below (regulation 9(1) PHVLOLR 2000).

Where any operating centre specified in the licence which is accessible to members of the public, the operator must maintain an insurance policy against public liability risks for a minimum indemnity of £5,000,000 for any one event (regulation 9(2) PHVLOLR 2000).

The operator must, if required to do so by a person making a private hire booking, agree the fare for the journey booked, or provide an estimate of that fare (regulation 9(3) PHVLOLR 2000).

If, during the period of the licence:

(a) any conviction is recorded:
 (i) where the operator is an individual, against him,
 (ii) where the operator is a firm, against any partner of that firm, or
 (iii) where the operator is another type of body or group of persons, against that body or group or any officer of that body or group;
(b) any information provided in the application for the grant of the licence, or for any variation, changes; or
(c) any driver ceases to be available to the operator for carrying out bookings, due to that driver's unsatisfactory conduct in connection with the driving of a private hire vehicle,

the operator must, within 14 days of the date of such event, give the licensing authority notice containing details of the conviction or change, or, in a case of a driver's unsatisfactory conduct above, the name of the driver and the circumstances of the case (regulation 9(4) PHVLOLR 2000).

No CB (Citizens Band) apparatus can be used in connection with a private hire booking at any operating centre specified in the licence or in any private hire vehicle available for carrying out bookings accepted at any such operating centre (regulation 9(5) PHVLOLR 2000).

The operator must preserve records specified below (regulation 9(6) PHVLOLR 2000).

The operator must establish and maintain a procedure for dealing with complaints and lost property, in connection with any private hire booking accepted by him and keep and preserve such records (regulation 9(7) PHVLOLR 2000).

Where an operator provides a London cab for the purpose of carrying out a private hire booking, any fare payable for the booking must be calculated as if the vehicle was a private hire vehicle unless the fare shown on the taximeter is less (regulation 9(8) PHVLOLR 2000).

Form of record of private hire bookings

The record which an operator is required to keep at each operating centre specified in his licence of the private hire bookings accepted by him there must be kept in writing, or in another form that the information contained in it can easily be reduced to writing (regulation 10 PHVLOLR 2000).

Particulars of private hire bookings

Before the commencement of each journey booked at an operating centre specified in his licence an operator must enter the following particulars of the booking in the record:

(a) the date on which the booking is made and, if different, the date of the proposed journey;

(b) the name of the person for whom the booking is made or other identification of him, or, if more than one person, the name or other identification of one of them;

(c) the agreed time and place of collection, or, if more than one, the agreed time and place of the first;

(d) the main destination specified at the time of the booking;

(e) any fare or estimated fare quoted;

(f) the name of the driver carrying out the booking or other identification of him;

(g) if applicable, the name of the other operator to whom the booking has been sub-contracted, and

(h) the registered number of the vehicle to be used or other means of identifying it as may be adopted (regulation 11 PHVLOLR 2000).

Particulars of private hire vehicles

An operator must keep at each operating centre specified in his licence a record of each private hire vehicle which is available to him for carrying out bookings accepted by him at that centre (regulation 12(1) PHVLOLR 2000).

The particulars are:

(a) the make, model and colour;

(b) the registration mark;

(c) the name and address of the registered keeper;

(d) a copy of any current MOT test certificate;

(e) a copy of the current certificate of insurance or certificate of security;

(f) the date on which the vehicle became available to the operator; and

(g) the date on which the vehicle ceased to be so available (regulation 12(2) PHVLOLR 2000).

Particulars of drivers

An operator must keep at each operating centre specified in his licence a record of each driver who is available to him for carrying out bookings accepted by him at that centre (regulation 13(1) PHVLOLR 2000).

The particulars are:

(a) his surname, forenames, address and date of birth;

(b) his national insurance number;

(c) a photocopy of his driving licence;

   (d) a photograph of him;

   (e) the date on which he became available to the operator; and

   (f) the date on which he ceased to be so available (regulation 13(2) PHVLOLR 2000).

Record of complaints

An operator must keep at each operating centre specified in his licence a record of any complaint made in respect of a private hire booking accepted by him at that centre, and of any other complaint made about his undertaking as an operator at that centre (regulation 14(1) PHVLOLR 2000).

The particulars are:

   (a) the date of the related booking;

   (b) the name of the driver who carried out the booking;

   (c) the registration mark of the vehicle used;

   (d) the name of the complainant and any address, telephone number or other contact details provided by him;

   (e) the nature of the complaint; and

   (f) details of any investigation carried out and subsequent action taken as a result (regulation 14(2) PHVLOLR 2000).

Record of lost property

An operator must keep at each operating centre specified in his licence a record of any lost property found at that centre, or in any private hire vehicle used to carry out a booking accepted by him there (regulation 15(1) PHVLOLR 2000).

These particulars are:

   (a) the date on which it was found;

   (b) the place where it was found and if it was found in a vehicle, the registration mark of that vehicle;

   (c) a description of the item;

   (d) evidence to show that, where practical, an attempt was made to return the item to the owner and whether or not this was successful; and

   (e) in the case of any unclaimed item which has been disposed of, how it was disposed of (regulation 15(2) PHVLOLR 2000).

A similar record must be kept of any property reported to him at that centre as having been lost (regulation 15(3) PHVLOLR 2000).

The particulars are:

   (a) the date of the report;

   (b) the date on which it is alleged to have been lost;

   (c) the place where it is alleged to have been lost;

   (d) a description of the item; and

   (e) evidence to show that, where practical, an attempt was made to find the item (regulation 15(4) PHVLOLR 2000).

Preservation of records

An operator must preserve the particulars of:

(a) each private hire booking for 6 months from the date on which the booking was accepted;

(b) each private hire vehicle and driver for 12 months from the date on which the vehicle or, the driver ceased to be available for carrying out bookings;

(c) each complaint and item of lost property recorded for 6 months from the date on which they were entered in the respective record (regulation 16(1) PHVLOLR 2000).

Where an operator tape-records a private hire booking he must preserve the tape-recording of that conversation for a period of 6 months (regulation 16(2) PHVLOLR 2000).

If an operator ceases to use an operating centre specified in his licence, he must preserve:

(a) private hire bookings records for 6 months; and

(b) private hire vehicles and drivers records for 12 months,

from the date of the last entry (regulation 16(3) PHVLOLR 2000).

Register of licences

The licence register maintained by the licensing authority under the 1998 Act must contain, in addition to the matters set out in section 23(1)(a), further particulars set out below (regulation 17(1) PHVLOLR 2000).

The particulars are:

(a) the address of each operating centre specified in the licence; and

(b) an indication that it is current, suspended or revoked (regulation 17(2) PHVLOLR 2000).

Issue of replacement licences

Where an operator notifies the licensing authority that:

(a) he has adopted, altered or dispensed with a business name;

(b) he has changed his name; or

(c) his licence has been lost, destroyed or defaced,

the licensing authority must issue a replacement licence (regulation 18(1) PHVLOLR 2000).

Except where a licence has been lost or destroyed, no replacement can be issued until the original licence has been returned to the licensing authority (regulation 18(2) PHVLOLR 2000).

Continuance of licence on death, bankruptcy, etc.

Where a licence is granted in the sole name of an individual, in the event of:

(a) the death of that individual;
(b) the bankruptcy of that individual; or
(c) that individual becoming a patient under Part VII of the Mental Health Act 1983,

the licensing authority may direct that the licence is not to be treated as terminated when the individual died but suspended until the date when a direction below comes into force (regulation 19(1) and (2) PHVLOLR 2000).

The licensing authority may direct that a person carrying on the business of the operator is to be treated as if he were the operator specified in the direction for a period not exceeding:

(a) 6 months from the date of the coming into force of that direction; or
(b) if less, the remainder of the period of the licence (regulation 19(3) PHVLOLR 2000).

Definitions

the 1998 Act means the Private Hire Vehicles (London) Act 1998.
licensing authority means the person appointed under section 24(1) of the 1998 Act for the purpose of exercising the functions of the Secretary of State under that Act, or, where no such appointment has been made, the Secretary of State.
operator means a London PHV operator and in relation to a licence means the operator to whom the licence was granted.

Private hire vehicle driver's licence (FC 3.40)

Requirement for private hire vehicle driver's licence

No vehicle can be used as a private hire vehicle on a road in London unless the driver holds a private hire vehicle driver's licence (section 12(1) PHVLA 1998). The driver, operator and owner of a vehicle used where the driver has no such licence are each guilty of an offence (section 12(2) and (3) PHVLA 1998). It is a defence in proceedings against the operator of a vehicle for such an offence to show that he exercised all due diligence to prevent the vehicle being used in such circumstances (section 12(4) PHVLA 1998). A person guilty of an offence is liable on summary conviction to a fine not exceeding level 4 on the standard scale (section 12(5) PHVLA 1998).

London PHV driver's licences

Any person may apply to the licensing authority for a private hire vehicle driver's licence for London (section 13(1) PHVLA 1998).

The licensing authority must grant a London PHV driver's licence to an applicant if the authority is satisfied that:

FC 3.40 Licensing of drivers of private hire vehicles in London – Private Hire Vehicles (London) Act 1998 and Private Hire Vehicles (London PHV Driver's Licences) Regulations 2003

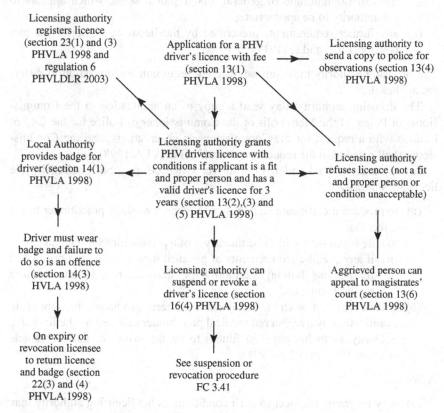

Notes
1. PHVLA 1998 – Private Hire Vehicles (London) Act 1998.
2. PHVLDLR 2003 – Private Hire Vehicles (London PHV Driver's Licences) Regulations 2003.

(a) the applicant is 21, is (and has for at least 3 years been) authorised to drive a motor car and is a fit and proper person to hold a London PHV driver's licence;

(b) the applicant shows to the authority's satisfaction (whether by taking a test or otherwise) that they possess a level of knowledge of London or parts of London, and of general topographical skills, which appears to the authority to be appropriate;

(c) any further requirements prescribed by the licensing authority are met (section 13(2) and (3) PHVLA 1998).

The licensing authority may impose different requirements in relation to different applicants.

The licensing authority may send a copy of an application to the Commissioner of Police of the Metropolis or the Commissioner of Police for the City of London with a request for the Commissioner's observations; and the Commissioner shall respond to the request (section 13(4) PHVLA 1998).

An applicant for a London PHV driver's licence may be required by the licensing authority:

(a) to produce a certificate signed by a registered medical practitioner to the effect that:

  (i) he is physically fit to be the driver of a private hire vehicle; and

  (ii) if any specific requirements of physical fitness have been prescribed for persons holding London PHV licences, that he meets those requirements; and

(b) whether or not such a certificate has been produced, to submit to examination by a registered medical practitioner selected by the licensing authority as to his physical fitness to be the driver of such a vehicle (section 15(4) PHVLA 1998).

A licence:

(a) may be granted subject to such conditions as the licensing authority may think fit;

(b) must be in a form and contain particulars the licensing authority may think fit; and

(c) is granted for 3 years or for a shorter period as the licensing authority may consider appropriate in the circumstances of the particular case (section 13(5) PHVLA 1998).

An applicant may appeal to a magistrates' court against a decision not to grant a London PHV driver's licence or against any condition which the licence is subject (section 13(6) PHVLA 1998).

A person is authorised to drive a motor car if:

(a) he holds a licence granted under Part III of the Road Traffic Act 1988 (other than a provisional licence) authorising him to drive a motor car; or

(b) he is authorised by virtue of section 99A(1) or 109(1) of that Act (Community licences and Northern Ireland licences) to drive a motor car in Great Britain (section 13(7) PHVLA 1998).

Issue of driver's badges

The licensing authority must issue a badge to each person to whom the authority has granted a London PHV driver's licence (section 14(1) PHVLA 1998). The licensing authority may prescribe the form of these badges (section 14(2) PHVLA 1998).

A person issued with a badge must, when he is the driver of a private hire vehicle, wear the badge in such position and plainly and distinctly visible (section 14(3) PHVLA 1998).

The licensing authority may by notice exempt a person from wearing this badge, when he is the driver of a vehicle being used to provide a service specified in the notice if the authority considers it inappropriate (having regard to that service) to require the badge to be worn (section 14(4) PHVLA 1998).

Any person who without reasonable excuse contravenes this requirement is guilty of an offence and liable on summary conviction to a fine not exceeding level 3 on the standard scale (section 14(5) PHVLA 1998).

A London PHV driver's licence may be suspended or revoked where:

(a) the licence holder has, since the grant of the licence, been convicted of an offence involving dishonesty, indecency or violence;

(b) the licensing authority is for any other reason no longer satisfied that the licence holder is fit to hold such a licence; or

(c) the licence holder has failed to comply with any condition of the licence or any other obligation imposed on him (section 16(4) PHVLA 1998).

Private Hire Vehicles (London PHV Driver's Licences) Regulations 2003 (PHVLDLR 2003)

The physical fitness requirement

The physical fitness requirement is that the applicant:

(a) is the holder of a Group 2 licence; or

(b) satisfies Transport for London that he meets the requirements as to physical fitness that he would be required to meet in order to be granted a Group 2 licence.

A 'Group 2 licence' means a licence to drive a motor vehicle granted under Part III of the Road Traffic Act 1988 which is a Group 2 licence as defined by regulation 70 of the Motor Vehicles (Driving Licences) Regulations 1999 (regulation 3 PHVLDLR 2003).

Form of driver's badge

The driver's badge must:

(a) state the name of the licence holder to whom it has been issued;

(b) state the number and expiry date of the holder's licence;

(c) include a photographic image of the holder; and

(d) state on the reverse that it is the property of Transport for London and is to be returned to a specified address in the event of its being found (regulation 4 PHVLDLR 2003).

Fees

Regulation 5 PHVLDLR 2003 specifies the fee for making of an application, by an applicant for a London PHV driver's licence.

Where a London PHV driver's licence ceases to have effect (whether by revocation or otherwise) on the ground that:

(a) the holder of the licence is no longer physically fit to hold such a licence;

(b) the licence is surrendered by the holder; or

(c) the holder dies,

a refund of a proportion of the licence fee is payable to the holder, or holder's personal representatives.

A refund is payable upon receipt of a written request by the holder of the licence accompanied by the licence and the driver's badge issued to the holder.

The amount refundable is equal to that proportion of the licence fee which the number of whole months remaining unexpired of the period for which the licence was granted bears to the whole of that period, rounded up to the nearest whole pound (regulation 5 PHVLDLR 2003).

Register of licences

The register must contain the particulars specified in section 23(1)(a) of the 1998 Act, and an indication in relation to each licence as to whether:

(a) it is a London PHV driver's licence or a driver's temporary permit; and

(b) it is current, suspended or revoked (regulation 6 PHVLDLR 2003).

General provisions

Applications for licences

An application for the grant of a licence must be made in a form, and include declarations and information that the licensing authority may require (section 15(1) PHVLA 1998).

The licensing authority may require an applicant to furnish further information the authority may consider necessary for dealing with the application (section 15(2) PHVLA 1998).

These provisions, as well as those for driver licence and operator's licences, apply to the renewal of a licence as they apply to the grant of a licence (section 15(5) PHVLA 1998).

Power to suspend or revoke licences

The licensing authority may suspend or revoke a licence for any reasonable cause including any ground mentioned below (section 16(1) PHVLA 1998).

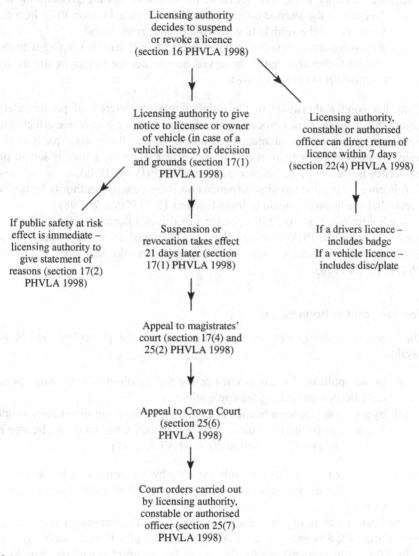

FC 3.41 Suspension and revocation of licences in London – Section 16 and 17 Private Hire Vehicles (London) Act 1998

Licensing authority
decides to suspend
or revoke a licence
(section 16 PHVLA 1998)

Licensing authority to give
notice to licensee or owner
of vehicle (in case of a
vehicle licence) of decision
and grounds (section 17(1)
PHVLA 1998)

Licensing authority,
constable or authorised
officer can direct return of
licence within 7 days
(section 22(4) PHVLA 1998)

If public safety at risk
effect is immediate –
licensing authority to
give statement of
reasons (section 17(2)
PHVLA 1998)

Suspension or
revocation takes effect
21 days later (section
17(1) PHVLA 1998)

If a drivers licence –
includes badge
If a vehicle licence –
includes disc/plate

Appeal to magistrates'
court (section 17(4) and
25(2) PHVLA 1998)

Appeal to Crown Court
(section 25(6)
PHVLA 1998)

Court orders carried out
by licensing authority,
constable or authorised
officer (section 25(7)
PHVLA 1998)

Note
PHVLA 1998 – Private Hire Vehicles (London) Act 1998.

Suspension and revocation under section 16: procedure

Where the licensing authority has decided to suspend or revoke a licence under section 16:

(a) the authority must give notice of the decision and the grounds for the decision to the licence holder or, in the case of a London PHV licence, the owner of the vehicle to which the licence relates; and

(b) the suspension or revocation takes effect after 21 days beginning with the day on which that notice is served on the licence holder or the owner (section 17(1) PHVLA 1998).

If the licensing authority is of the opinion that the interest of public safety require the suspension or revocation of a licence to have immediate effect, and the authority includes a statement of that opinion and the reasons for it in the notice of suspension or revocation, it takes effect when the notice is served on the licence holder or vehicle owner (section 17(2) PHVLA 1998).

A licence suspended remains suspended until the licensing authority by notice directs that the licence is again in force (section 17(3) PHVLA 1998).

The holder of a London PHV operator's or driver's licence, or the owner of a vehicle to which a PHV licence relates, may appeal to a magistrates' court against a decision under section 16 to suspend or revoke that licence (section 17(4) PHVLA 1998).

Fees for grant of licences, etc.

The licensing authority may by regulations provide for prescribed fees to be payable:

(a) by an applicant for any licence or for the variation of a London operator's licence on making the application;

(b) by a person granted a licence or variation, on the grant or variation of the licence and (if the regulations provide) at such times while the licence is in force as prescribed (section 20(1) PHVLA 1998).

Regulations may provide for fees to be payable by instalments, or for fees to be remitted or refunded (in whole or part), in prescribed cases (section 20(2) PHVLA 1998).

The licensing authority may decline to proceed with an application for, or for the variation of, a licence; or the grant or variation of a licence, until any prescribed fee (or instalment) due for the application or grant is paid (section 20(3) PHVLA 1998).

Production of documents

The holder of a London PHV operator's licence or a London PHV driver's licence must at the request of a constable or authorised officer produce his licence for inspection (section 21(1) PHVLA 1998).

The owner of a vehicle with a London PHV licence must at the request of a constable or authorised officer produce the London PHV licence for that vehicle for inspection and the insurance policy certificate or security required in respect of the vehicle by Part VI of the Road Traffic Act 1988 (section 21(2) PHVLA 1998).

A document required to be produced must be produced either forthwith or:

(a) if the request is made by a constable, at any police station within London nominated by the licence holder or vehicle owner when the request is made, or

(b) if the request is made by an authorised officer, at a place the officer may reasonably require,

within 6 days beginning with the day on which the request is made (section 21(3) PHVLA 1998). A person who without reasonable excuse fails to comply is guilty of an offence and liable on summary conviction to a fine not exceeding level 3 on the standard scale (section 21(4) PHVLA 1998).

Return of licences, etc.

The following licences must be returned to the licensing authority by the relevant person after the expiry or revocation of the licence, within 7 days after the day on which the licence expires or the revocation takes effect:

* the holder of a London PHV operator's licence (section 22(1) PHVLA 1998);
* the owner of a vehicle with a London PHV licence (section 22(2) PHVLA 1998); and
* the holder of a London PHV driver's licence (section 22(3) PHVLA 1998).

On the suspension of a licence, the licensing authority, a constable or an authorised officer may by notice direct the holder of the licence, or the owner of the vehicle, to return the licence to the authority, constable or officer within 7 days after the day on which the notice is served on that person. A direction may also direct:

(a) the return by the vehicle owner of the disc or plate which was issued for the vehicle (in the case of a London PHV licence); or

(b) the return by the licence holder of the driver's badge (in the case of a London PHV driver's licence) (section 22(4) PHVLA 1998).

A person who without reasonable excuse fails to comply with these requirements or directions to return a licence, disc, plate or badge is guilty of an offence (section 22(5) PHVLA 1998). A person guilty of an offence is liable on summary conviction to a fine not exceeding level 3 on the standard scale; and in the case of a continuing offence, to a fine not exceeding ten pounds for each day during which an offence continues after conviction (section 22(6) PHVLA 1998).

A constable or authorised officer is entitled to remove and retain the plate or disc from a vehicle to which an expired, suspended or revoked London PHV licence relates following:

(a) a failure to comply the return of the licence from the owner of a vehicle with a London PHV licence or a direction above, or

(b) a suspension or revocation of the licence which has immediate effect (section 22(7) PHVLA 1998).

Register of licences

The licensing authority must maintain a register containing the following particulars for each issued:

(a) the number of the licence, the name of the person to whom it is granted, the date on which it is granted and the expiry date; and

(b) other prescribed particulars (section 23(1) PHVLA 1998).

The register must be available for inspection free of charge by members of the public at a place or places, and during hours, determined by the licensing authority (section 23(2) PHVLA 1998).

The licensing authority must maintain a supplementary register containing, for each issued licence, the address of the person to whom it is granted (section 23(3) PHVLA 1998).

The licensing authority may disclose the address of a licence holder to any person only if it appears to the authority that the person has a sufficient reason for requiring that information (section 23(4) PHVLA 1998).

Delegation of functions by the licensing authority

The functions of the licensing authority (apart from any power to make subordinate legislation) may be exercised by any person appointed by the licensing authority for the purpose to such extent and subject to such conditions as may be specified in the appointment (section 24(1) PHVLA 1998).

An appointment may authorise the person appointed to retain any fees received by him (section 24(2) PHVLA 1998).

It is the duty of a person appointed to comply with any directions given to him by the licensing authority in relation to the exercise of these functions (section 24(3) PHVLA 1998).

Appeals

If the licensing authority has exercised the power to delegate functions, an appeal must be heard by a magistrates' court (section 25(2) PHVLA 1998).

Any appeal is by way of complaint for an order and the Magistrates' Courts Act 1980 applies to the proceedings (section 25(3) PHVLA 1998).

The appellant must bring an appeal within 21 days from the date on which notice of the relevant decision is served on him (section 25(4) PHVLA 1998).

In the case of a decision where an appeal lies, the notice of the decision must state the right of appeal to a magistrates' court and the time within which an appeal may be brought (section 25(5) PHVLA 1998).

An appeal against any decision of a magistrates' court lies to the Crown Court for any party to the proceedings in the magistrates' court (section 25(6) PHVLA 1998).

Where on appeal a court varies or reverses any decision of the licensing authority, a constable or an authorised officer, the order of the court must be carried out by the licensing authority or, a constable or authorised officer (section 25(7) PHVLA 1998).

Effect of appeal

If any decision of the licensing authority against which there is a right of appeal:

(a) involves the execution of any work or the taking of any action;
(b) makes it unlawful for any person to carry on a business which he was lawfully carrying on at the time of the decision,

the decision does not take effect until the time for appealing has expired or (where an appeal is brought) until the appeal is disposed of or withdrawn (section 26(1) PHVLA 1998).

This does not apply in relation to a decision to suspend, vary or revoke a licence if the notice of suspension, variation or revocation directs that, in the interests of public safety, the decision is to have immediate effect (section 26(2) PHVLA 1998).

Obstruction of authorised officers, etc.

A person who wilfully obstructs a constable or authorised officer acting under these provisions is guilty of an offence and liable on summary conviction to a fine not exceeding level 3 on the standard scale (section 27(1) PHVLA 1998).

A person who, without reasonable excuse fails to comply with any requirement properly made to such person by a constable or authorised officer; or fails to give a constable or authorised officer any other assistance or information which he may reasonably require of such person for the purpose of performing his functions, is guilty of an offence and liable on summary conviction to a fine not exceeding level 3 on the standard scale (section 27(2) PHVLA 1998).

A person who makes any statement which he knows to be false in giving any information to an authorised officer or constable is guilty of an offence and liable on summary conviction to a fine not exceeding level 5 on the standard scale (section 27(3) PHVLA 1998).

Penalty for false statements

A person who knowingly or recklessly makes a statement or furnishes information which is false or misleading in any material particular for the purpose of procuring the grant or renewal of a licence, or the variation of an operator's licence, is guilty of an offence and liable on summary conviction to a fine not exceeding level 5 on the standard scale (section 28 PHVLA 1998).

Vehicles used for funerals and weddings

These provisions do not apply to any vehicle whose use as a private hire vehicle is limited to use in connection with funerals or weddings (section 29 PHVLA 1998).

Prohibition of certain signs, notices, etc.

The licensing authority may make regulations prohibiting the display in London on or from vehicles (other than licensed taxis and public service vehicles) of any sign, notice or other feature of a description specified in the regulations (section 30(1) PHVLA 1998).

Before making the regulations the licensing authority must consult bodies appearing to the authority to represent the London cab trade and the private hire vehicle trade in London as the authority considers appropriate (section 30(2) PHVLA 1998).

Any person who drives a vehicle where a prohibition imposed by regulations is contravened, or causes or permits such a prohibition to be contravened in, is guilty of an offence and liable on summary conviction to a fine not exceeding level 4 on the standard scale (section 30(3) PHVLA 1998).

Prohibition of certain advertisements

Where any advertisement:

(a) indicates that vehicles can be hired on application to a specified address in London;
(b) indicates that vehicles can be hired by telephone on a telephone number being the number of premises in London; or
(c) on or near any premises in London, indicates that vehicles can be hired at those premises

it cannot include:

(a) any of the following words: 'taxi', 'taxis', 'cab' or 'cabs', or
(b) any word so closely resembling any of those words as to be likely to be mistaken for it,

(whether alone or as part of another word), unless the vehicles offered for hire are London cabs (section 31(1) and (2) PHVLA 1998).

This does apply to an advertisement which includes the word 'minicab', 'mini-cab' or 'mini cab' (whether singular or plural) (section 31(3) PHVLA 1998).

Any person who issues, or causes to be issued, an advertisement which contravenes this is guilty of an offence and liable on summary conviction to a fine not exceeding level 4 on the standard scale (section 31(4) PHVLA 1998).

It is a defence for a person charged with an offence to prove that:

(a) he is a person whose business it is to publish or arrange for the publication of advertisements;

(b) he received the advertisement in question for publication in the ordinary course of business; and

(c) he did not know and had no reason to suspect that its publication would amount to an offence (section 31(5) PHVLA 1998).

Regulations

The licensing authority may make regulations to deliver the functions of a licensing authority (section 32(1) PHVLA 1998).

Regulations may:

(a) make different provision for different cases;

(b) provide for exemptions from any provision of the regulations; and

(c) contain incidental, consequential, transitional and supplemental provision (section 32(2) PHVLA 1998).

Any power of the licensing authority to make regulations includes power to vary or revoke previous regulations including those made by the Secretary of State by statutory instrument (section 32(4) and (5) PHVLA 1998).

The licensing authority must ensure that any regulations made by the authority are printed and published and a fee may be charged for the sale of those regulations (section 32(6) and (7) PHVLA 1998).

Offences due to fault of other person

Where an offence by any person is due to the act or default of another person, then (whether proceedings are taken against the first mentioned person or not) that other person is guilty of the offence and is liable to be proceeded against and punished accordingly (section 33(1) PHVLA 1998).

Where an offence committed by a body corporate is proved to have been committed with the consent or connivance of, or attributable to any neglect on the part of, any director, manager, secretary or other similar officer of the body corporate (or any person purporting to act in that capacity), he, as well as the body corporate is guilty of the offence is liable to be proceeded against and punished accordingly (section 33(2) PHVLA 1998).

Service of notices

Any notice to be given to any person may be served by post and is properly addressed to a London PHV operator if it is addressed to him at any operating centre of his in London (section 34(1) and (2) PHVLA 1998).

Any notice given to the owner of a vehicle is deemed to have been effectively given if it is given to the person who is for the time being notified to the Secretary of State as the owner of the vehicle (or, if more than one person is currently notified as the owner, if it is given to any of them) (section 34(3) PHVLA 1998).

Definitions

controlled district means any area for which Part II of the 1976 Act is in force by:

(a) a resolution by a district council under section 45 of that Act; or

(b) section 255(4) of the Greater London Authority Act 1999 (section 36 PHVLA 1998).

hackney carriage means a vehicle licensed under section 37 of the Town Police Clauses Act 1847 or any similar enactment (section 36 PHVLA 1998).

licensed taxi means a hackney carriage, a London cab or a taxi licensed under Part II of the 1982 Act (section 36 PHVLA 1998).

licensing authority means Transport for London (section 36 PHVLA 1998).

Delegated authority for day-to-day licensing operations is carried out by the London Taxi and Private Hire Office, formerly the Public Carriage Office (PCO). The Office is responsible for ensuring that taxi drivers and proprietors are of the standard specified by the Mayor of London and that their taxis conform to the specification he sets.

London cab means a vehicle licensed under section 6 of the Metropolitan Public Carriage Act 1869 (section 36 PHVLA 1998).

operating centre means premises at which private hire bookings are accepted by an operator (section 1(5) PHVLA 1998).

operator means a person who makes provision for the invitation or acceptance of, or who accepts, private hire bookings (section 1(1) PHVLA 1998).

owner of a vehicle is taken to be the person by whom it is kept (section 35(1)) and when talking proceedings it is presumed that the owner is the person who is the registered keeper of the vehicle at that time (section 35(2) PHVLA 1998).

However, it is open to the defence to show that the person who was the registered keeper of a vehicle at any particular time was not the person by whom the vehicle was kept at that time; and it is open to the prosecution to prove that the vehicle was kept at that time by some person other than the registered keeper (section 35(3) PHVLA 1998).

private hire vehicle means a vehicle constructed or adapted to seat fewer than nine passengers which is made available with a driver for hire for the purpose of carrying passengers, other than a licensed taxi or a public service vehicle (section 1(1) PHVLA 1998).

private hire vehicle licence means:

(a) a London PHV licence;

(b) if the vehicle is in use for the purposes of a hiring the booking for which was accepted outside London in a controlled district, a licence under the 1976 Act issued by the council for that district; and

(c) if the vehicle is in use for the purposes of a hiring the booking for which was accepted in Scotland, a licence under the Civic Government (Scotland) Act 1982

and it is immaterial that the booking in question is a sub-contracted booking.

This does not apply to a vehicle used for the purposes of a hiring for a journey beginning outside London in an area of England and Wales which is not a controlled district (section 6(7) PHVLA 1998).

taximeter means a device for calculating the fare to be charged in respect of any journey by reference to the distance travelled or time elapsed since the start of the journey (or a combination of both) (section 11(3) PHVLA 1998).

the 1976 Act means the Local Government (Miscellaneous Provisions) Act 1976.

the 1982 Act means the Civic Government (Scotland) Act 1982.

used as a private hire vehicle means it is in use in connection with a hiring for the purpose of carrying one or more passengers or is immediately available to an operator to carry out a private hire booking (section 1(2) PHVLA 1998).

TAXI LICENCES – HACKNEY CARRIAGES OUTSIDE LONDON

References

Town Police Clauses Act 1847 (as amended) (TPCA1847).
Local Government (Miscellaneous Provisions) Act 1976 (as amended) (LG(MP) A1976).
Transport Act 1985.
Local Services (Operation by Licensed Hire Cars) Regulations 2009.

Extent

This procedure applies to England and Wales.

Introduction

In England and Wales, outside London, taxis (hackney carriages) are licensed by district councils under the Town Police Clauses Act 1847 and the Local Government (Miscellaneous Provisions) Act 1976. All taxis and their drivers must be licensed. The licensing conditions that are applied to taxi and PHV(minicab) drivers and the local conditions of vehicle fitness are for each council to decide, so can vary considerably from area to area. Taxis, which are the only vehicles permitted to ply for hire, are normally subject to a stricter regime than the minicabs. The Department for Transport publishes best practice guidance for local licensing authorities.

Extension of taxi licensing in England and Wales

Where the provisions of the Town Police Clauses Act 1847 with respect to hackney carriages and the Town Police Clauses Act 1889 were not in force throughout the whole of the area of a district council in England and Wales whose area lies outside the area to which the Metropolitan Public Carriage Act 1869 applies, they apply throughout that area (section 15(1) Transport Act 1985).

Hackney carriage licences (FC 3.42)

Hackney coaches or carriages of any kind or description adapted to the carriage of persons can be granted a licence to ply for hire within a prescribed distance, or if no distance is prescribed, within five miles from the General Post Office of the city, town, or place (section 37 TPCA1847).

FC 3.42 Licensing of hackney carriages outside London – Town Police Clauses Act 1847 (TPCA1847) and Local Government (Miscellaneous Provisions) Act 1976 (LG(MP)A1976)

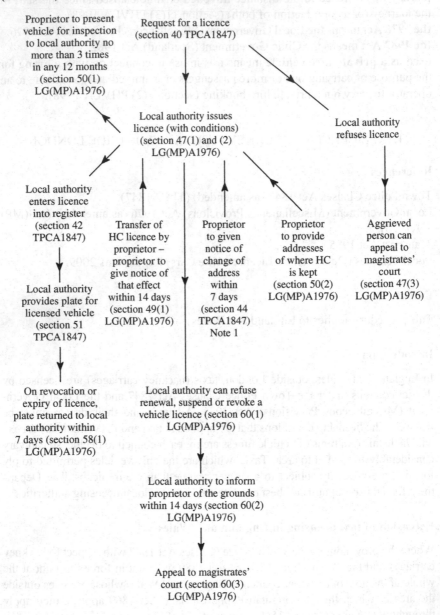

Notes

1. Section 49(2) Local Government (Miscellaneous Provisions) Act 1976 – offence for failure to notify local authority.
2. Section 58(2) Local Government (Miscellaneous Provisions) Act 1976 – offence to fail to return plate or disc under notice.
3. Appeals – section 300 to 302 Public Health Act 1936 apply (Section 77(1) Local Government (Miscellaneous Provisions) Act 1976).

Hackney carriages

A hackney carriage is a wheeled carriage, used in standing or plying for hire in any street within the prescribed distance. It must also have a numbered plate, or a plate resembling or intended to resemble a plate, fixed on it. This does not apply to a licensed stage coach standing or plying for passengers to be carried for hire at separate fares (section 38 TPCA1847).

Persons applying for licence to sign a requisition

Before a licence is granted a requisition for the appropriate licence must be made and signed, by the proprietor or one of the proprietors of the hackney carriage. The requisition must include:

- the name and address of the person applying for the licence;
- every proprietor or part proprietor of the carriage, or person concerned, either solely or in partnership with any other person, in the keeping, employing, or letting to hire of the carriage; and
- any person who, on applying for such licence, states in the requisition the name of any, person who is not a proprietor or part proprietor of the carriage, or who is not concerned in the keeping, employing, or letting to hire of the carriage,

and any person who wilfully omits the name of any person who is a proprietor or part proprietor of the carriage, or who is concerned in the keeping, employing, or letting to hire of the carriage, is liable to a penalty not exceeding level 1 on the standard scale (section 40 TPCA1847).

What must be specified in the licences

Every licence must specify the name and address of every person who is a proprietor or part proprietor of the hackney carriage, in respect of which licence or licences is granted, who is concerned, either solely or in partnership with any other person, in the keeping, employing, or letting to hire of the carriage. The number of the licence must correspond to the number to be painted or marked on the plates to be fixed on the carriage, together with such other particulars as the commissioners think fit (section 41 TPCA1847).

Conditions

A district council may attach conditions it may consider reasonably necessary, to the grant of a licence of a hackney carriage (section 47(1)).

A district council may require any licensed hackney carriage to be of such design or appearance or bear such distinguishing marks as shall clearly identify it as a hackney carriage (section 47(2) LG(MP)A1976).

Any person aggrieved by any conditions attached to the licence may appeal to a magistrates' court (section 47(3)).

Licences to be registered

Every licence must be made out by the clerk of the commissioners, and entered in a register and include offences committed by any proprietor or driver or person attending a carriage and any person can inspect the register at any reasonable time without paying a fee or giving a reward (section 42 TPCA1847).

Licence for 1 year only

Every licence granted is under the common seal of the commissioners and must be signed by two or more of the commissioners, and must not include more than one licensed carriage, and the licence is in force for 1 year from the day of the date of the licence, or until the next general licensing meeting where one is appointed by the commissioners (section 43 TPCA1847).

Notice to be given by proprietors of hackney carriages of any change of address

Any person named in any licence as the proprietor or one of the proprietors, or as being concerned, either solely or in partnership with any person, in the keeping, employing, or letting to hire of any carriage, changes his address, he must inform the commissioners by notice in writing and signed by him and specifying his new address, within 7 days. He must also, at the same time, produce the licence at the office of the commissioners, who must endorse and sign a memorandum specifying the change on the licence.

Any person named in any licence as the proprietor, or one of the proprietors, of any hackney carriage, who changes his place of abode, and neglects or wilfully omits to give notice of such a change, or to produce the licence, is liable to a penalty not exceeding level 1 on the standard scale (section 44 TPCA1847).

Penalty for plying for hire without a licence

If the proprietor or part proprietor of any carriage, or any person concerned above, uses a hackney carriage plying for hire within the prescribed distance without a licence, or during the time when the licence is suspended, or without having the number of the carriage corresponding with the number of the licence openly displayed on the carriage, is liable to a penalty not exceeding level 4 on the standard scale (section 45 TPCA1847).

Number of persons to be carried in a hackney carriage to be painted on the carriage

The number of persons to be carried by a hackney carriage in form, 'To carry persons', must be painted on a plate placed on a conspicuous place on the outside of the carriage, in legible letters, clearly distinguishable from the colour of the background, one inch in length, and of a proportionate breadth; and the driver cannot carry in the carriage a greater number of persons than that number in the carriage (section 51 TPCA1847).

Penalty for neglecting to exhibit the number, or refusal to carry the prescribed number

If the proprietor of any hackney carriage use the carriage without having the number of persons to be carried painted and exhibited on the carriage, or if the driver of the carriage refuses, when required by the hirer, to carry the relevant number of persons, or any less number, the proprietor or driver is liable to a penalty not exceeding level 1 on the standard scale (section 52 TPCA1847).

Transfer of hackney carriages

If the proprietor of a hackney carriage with a licence transfers his interest in the hackney carriage to a person other than the proprietor whose name is specified in the licence, he must within 14 days after the transfer give notice in writing to the district council specifying the name and address of the person to whom the hackney carriage has been transferred (section 49(1) LG(MP)A1976).

If a proprietor without reasonable excuse fails to give notice to a district council he is guilty of an offence (section 49(2) LG(MP)A1976).

Suspension and revocation of vehicle licences

A district council may suspend or revoke, or on appropriate application refuse to renew a vehicle licence on any of the following grounds:

    (a) that the hackney carriage is unfit for use as a hackney carriage or private hire vehicle;

    (b) any offence under, or non-compliance with, the Act of 1847 or this Act by the operator or driver; or

    (c) any other reasonable cause (section 60(1) LG(MP)A1976).

Where a district council suspend, revoke or refuse to renew any licence they must give the proprietor of the vehicle notice of the grounds on which the licence has been suspended or revoked or on which they have refused to renew the licence within 14 days of the suspension, revocation or refusal (section 60(2) LG(MP)A1976).

Any proprietor aggrieved by a decision of a district council may appeal to a magistrates' court (section 60(3) LG(MP)A1976).

Return of identification plate or disc on revocation or expiry of licence, etc.

On the revocation or expiry of a vehicle licence in relation to a hackney carriage; or the suspension of a licence a district council may, by notice, require the proprietor of the licensed hackney carriage to return the plate or disc within 7 days after the service of that notice (section 58(1) LG(MP)A1976).

If any proprietor fails without reasonable excuse to comply with the terms of a notice he is guilty of an offence and liable on summary conviction to a fine not exceeding level 3 on the standard scale and to a daily fine not exceeding ten pounds; and any authorised officer of the council or constable is entitled to remove and retain the plate or disc (section 58(2) LG(MP)A1976).

Fees for vehicle licences (FC 3.43)

A district council may charge fees for the grant of vehicle licences resolved by them from time to time, sufficient to cover:

(a) the reasonable cost of the carrying out by or on behalf of the district council of inspections of hackney carriages to determine whether a licence should be granted or renewed;

(b) the reasonable cost of providing hackney carriage stands; and

(c) any reasonable administrative or other costs in connection with the control and supervision of hackney carriages (section 70(1) LG(MP) A1976).

The fees chargeable must not exceed for a vehicle licence for a hackney carriage, twenty-five pounds or other sums as the council may, from time to time determine (section 70(2) LG(MP)A1976).

If a district council determine that the maximum fees should be varied they must publish, in at least one local newspaper circulating in the district, a notice setting out the variation proposed, drawing attention to where the notice can be and specifying the period over 28 days from the date of the first publication of the notice, within which and how objections to the variation can be made. A copy of the notice must be deposited at the offices of the council which published the notice for 28 days from the date of the first publication and must be open to public inspection without payment at all reasonable hours (section 70(3) LG(MP)A1976).

If no objection to a variation is made within the period specified in the notice or if all objections are withdrawn, the variation comes into operation on the date of the expiration of the period specified in the notice or the date of withdrawal of the objection or, if more than one, of the last objection, whichever date is the later (section 70(4) LG(MP)A1976).

If objection is made and is not withdrawn, the district council must set a further date, within 2 months after the first specified date, on which the variation comes into force with or without modification as decided by the district council after consideration of the objections (section 70(5) LG(MP)A1976).

A district council may remit (return) the whole or part of any fee chargeable for the grant of a licence in any case in which they think it appropriate to do so (section 70(6) LG(MP)A1976).

Driver's licences (FC 3.44)

Drivers required to have a licence

A driver of any licensed hackney carriage must obtain a licence from the commissioners, the licence must be registered by the clerk to the commissioners, and a fee determined by the commissioners must be paid. This personal licence is in force until it is revoked, but it may be suspended (section 46 TPCA1847).

FC 3.43 Hackney carriage licence fees outside London – Section 70 Local Government (Miscellaneous Provisions) Act 1976

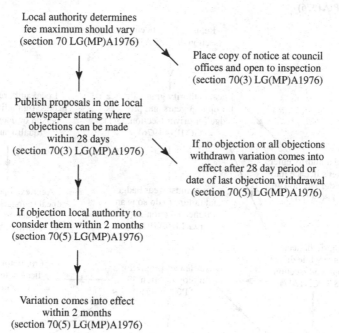

Local authority determines
fee maximum should vary
(section 70 LG(MP)A1976)

Place copy of notice at council
offices and open to inspection
(section 70(3) LG(MP)A1976)

Publish proposals in one local
newspaper stating where
objections can be made
within 28 days
(section 70(3) LG(MP)A1976)

If no objection or all objections
withdrawn variation comes into
effect after 28 day period or
date of last objection withdrawal
(section 70(5) LG(MP)A1976)

If objection local authority to
consider them within 2 months
(section 70(5) LG(MP)A1976)

Variation comes into effect
within 2 months
(section 70(5) LG(MP)A1976)

Note
The setting of fees for private hire vehicles is the same as that for hackney carriages.

FC 3.44 Licensing of drivers of hackney carriages outside London – Town Police Clauses Act 1847 (TPCA1847) and Section 53 Local Government (Miscellaneous Provisions) Act 1976 (LG(MP)A1976)

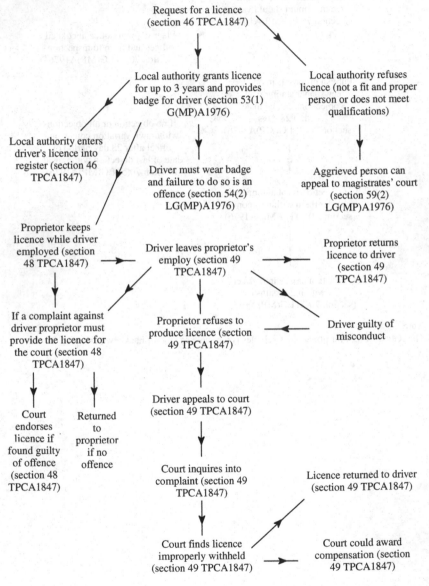

Notes

1. Section 49(2) Local Government (Miscellaneous Provisions) Act 1976 – offence for failure to notify local authority.
2. On second conviction of driver the licence can be suspended or revoked (section 50 TPC Act 1847).

Power to require applicants to submit information for licence applications

A district council may require any applicant for any relevant licence to submit to them information they reasonably consider necessary to enable them to determine whether the licence should be granted and whether conditions should be attached to the licence (section 57(1) LG(MP)A1976).

A district council may require an applicant for a driver's licence in respect of a hackney carriage:

(i) to produce a certificate signed by a registered medical practitioner that he is physically fit to be the driver of a hackney carriage; and

(ii) whether or not such a certificate has been produced, to submit to examination by a registered medical practitioner selected by the district council as to his physical fitness to be the driver of a hackney carriage.

If any person knowingly or recklessly makes a false statement or omits any material particular in giving information, he is guilty of an offence (section 57(3) LG(MP)A1976).

Provisions concerning driver's licences for hackney carriages

Every licence granted by a district council under the 1947 Act to any person to drive a hackney carriage remains in force for 3 years from the date of the licence or for a lesser period as they specify in the licence (section 53(1) LG(MP) A1976).

A district council may charge a fee for the grant of a licence to drive a hackney carriage they consider reasonable with a view to recovering the costs of issue and administration (section 53(2) LG(MP)A1976).

The driver of any hackney carriage licensed by a district council must on request of any authorised officer of the council or of any constable produce for inspection his driver's licence either immediately or:

(a) in the case of a request by an authorised officer, at the principal offices of the council within 5 days beginning with the day following that on which the request is made;

(b) in the case of a request by a constable, within 5 days at any police station in the area of the council and is nominated by the driver when the request is made (section 53(3) LG(MP)A1976).

If any person without reasonable excuse contravenes these provisions, he is guilty of an offence (section 53(4) LG(MP)A1976).

Qualifications for drivers of hackney carriages

A district council must not grant a licence to drive a hackney carriage unless they are satisfied that the applicant is a fit and proper person to hold a driver's licence; or to any person who has been authorised to drive a motor car for at least 12 months, or on the date of the application for a driver's licence is not authorised.

A person is authorised to drive a motor car if he holds a licence granted under the Road Traffic Act 1988 (not a provisional licence) authorising him to drive a motor car (section 59(1) LG(MP)A1976).

Any applicant aggrieved by the refusal of a district council to grant a driver's licence on the ground that he is not a fit and proper person to hold such licence may appeal to a magistrate's court (section 59(2) LG(MP)A1976).

Issue of driver's badges

When granting a driver's licence a district council must issue a driver's badge in such a form prescribed by them (section 54(1) LG(MP)A1976). A driver must at all times, acting within the driver's licence granted to him, wear the badge in such position and manner as to be plainly and distinctly visible. If any person without reasonable excuse contravenes this, he is guilty of an offence (section 54(2) LG(MP)A1976).

Penalty on drivers acting without licence

It is an offence for:

- any person to act as a driver without a licence, or while his licence is suspended,
- to lend a licence or part with a licence, except to the proprietor of the hackney carriage,
- the proprietor of the hackney carriage to employ any person as the driver who has not obtained a licence, or while he is suspended,

and every such driver and proprietor is liable to a penalty not exceeding level 3 on the standard scale (section 47 TPCA1847).

Where the proprietor of hackney carriage permits or employs any licensed person to act as the driver he must keep the driver licences while the driver remains in his employment.

Where there is a complaint and the proprietor of a hackney carriage is summoned to attend before a justice, or to produce the driver, the proprietor must also produce the driver's licence.

If a driver complained of is judged to be guilty of the offence alleged against him, the justice must make an endorsement on the licence of the driver, stating the nature of the offence and the amount of the penalty.

If any proprietor has not kept the driver's licence or he refuses or neglects to produce such a licence the proprietor is liable to a penalty not exceeding level 1 on the standard scale (section 48 TPCA1847).

Drivers leaving

When any driver leaves the service of the proprietor without having been guilty of any misconduct, the proprietor must return the driver's licence.

But if the driver has been guilty of any misconduct, the proprietor must not return the licence, but give him notice of the complaint which he intends to make against him, and summon the driver to appear before a justice to answer the complaint. The justice, having the parties before him, must inquire into and determine the complaint, and if on inquiry it appears that the driver's licence has been improperly withheld, the justice must direct that the licence be given to the driver, and award a sum of money as he thinks proper to be paid by the proprietor to the driver by way of compensation (section 49 TPCA1847).

Licences to be suspended or revoked for misconduct

The commissioners may, on the conviction for the second time of the proprietor or driver of any hackney carriage for any offence with respect to hackney carriages, suspend or revoke, the licence of the proprietor or driver (section 50 TPCA1847).

Suspension and revocation of driver's licences (FC 3.45)

A district council may suspend or revoke or on relevant application refuse to renew the licence of a driver of a hackney carriage on any of the following grounds:

(a) that he has since the grant of the licence been convicted of an offence involving dishonesty, indecency or violence; or been convicted of an offence under or has failed to comply with the provisions of the 1847 Act or of this Act; or

(b) any other reasonable cause (section 61(1) LG(MP)A1976).

Where a district council suspend, revoke or refuse to renew any licence they must give the driver notice of the grounds on which the licence has been suspended or revoked or on which they have refused to renew such licence within 14 days of such suspension, revocation or refusal and the driver must, on demand, return the driver's badge issued to him to the district council.

If any person without reasonable excuse contravenes this section he is guilty of an offence and liable on summary conviction to a fine not exceeding level 1 on the standard scale (section 61(2) LG(MP)A1976).

A suspension or revocation of the licence of a driver takes effect 21 days after the day on which notice is given to the driver (section 61(2A) LG(MP)A1976).

If it appears that the interests of public safety require the suspension or revocation of the licence to have immediate effect, and the notice given to the driver includes such a statement and an explanation why, the suspension or revocation takes effect when the notice is given to the driver (section 61(2B) LG(MP) A1976).

Any driver aggrieved by a decision of a district council may appeal to a magistrates' court (section 61(3) LG(MP)A1976).

Rules and powers

Penalty on driver for refusing to drive

A driver of a hackney carriage standing at any of the stands for hackney carriages appointed by the commissioners, or in any street, who refuses or neglects, without reasonable excuse, to drive the carriage to a place within the prescribed

FC 3.45 Licensing of drivers of hackney carriages outside London (suspension and revocation) Section 61 Local Government (Miscellaneous Provisions) Act 1976 (LG(MP)A1976)

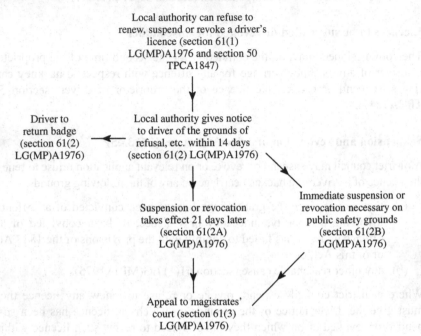

Local authority can refuse to renew, suspend or revoke a driver's licence (section 61(1) LG(MP)A1976 and section 50 TPCA1847)

Driver to return badge (section 61(2) LG(MP)A1976)

Local authority gives notice to driver of the grounds of refusal, etc. within 14 days (section 61(2) LG(MP)A1976)

Immediate suspension or revocation necessary on public safety grounds (section 61(2B) LG(MP)A1976)

Suspension or revocation takes effect 21 days later (section 61(2A) LG(MP)A1976)

Appeal to magistrates' court (section 61(3) LG(MP)A1976)

distance, is liable to a penalty not exceeding level 2 on the standard scale (section 53 TPCA1847).

Penalty for demanding more than the sum agreed

If the proprietor or driver of any hackney carriage, or if any other person on his behalf, agree beforehand with any person hiring the hackney carriage to take for any job a sum less than the fare allowed, the proprietor or driver is liable to a penalty not exceeding level 1 on the standard scale, if he demands more than the agreed fare (section 54 TPCA1847).

Agreement to pay more than the legal fare

No agreement made with the driver, or with any person having or pretending to be the driver for the payment of more than the fare allowed is binding on the person making the agreement. Even if there was such an agreement, the hirer may refuse, after the trip, to pay any sum beyond the fare allowed.

If any person actually pays the driver any sum exceeding the fare to which the driver was entitled, the person paying is entitled, on complaint made against the driver before any justice of the peace to recover back the sum paid beyond the proper fare.

The driver is also liable to a penalty not exceeding the sum of level 3 on the standard scale. If the driver defaults the repayment of the excess fare, or of payment of the penalty, the justice must commit the driver to prison for up to 1 month, unless the excess fare and the penalty is paid sooner (section 55 TPCA1847).

Driver to carry, under an agreement for a discretionary distance, the distance to which hirer is entitled for the fare

If the proprietor or driver of any hackney carriage, or if any other person on his behalf, agrees with any person to carry in a hackney carriage for a distance to be in the discretion of the proprietor or driver, and for a sum agreed, the proprietor or driver is liable to a penalty not exceeding level 1 on the standard scale if the distance is under that to which they were entitled to be carried for the agreed sum (section 56 TPCA1847).

Deposit to be made for carriages required to wait

When any hackney carriage is hired and taken to any place, and the driver is required by the hirer to wait with the hackney carriage, the driver may charge an additional fare for the carriage for the period, as a deposit, during which he is required to wait.

If no fare for time is fixed by bye-laws, then the sum of one shilling and six-pence is chargeable for every half hour during which he is required to wait as a deposit.

If the driver who has received any such deposit refuses to wait, or goes away or permits the hackney carriage to be driven or taken away without the consent of the hirer, before the expiration of the time for which the deposit was made; or if the driver on the final discharge of the hackney carriage refuses to account for the deposit, the driver is liable to a penalty not exceeding level 1 on the standard scale (section 57 TPCA1847).

Overcharge

Every proprietor or driver of any hackney carriage who is convicted of taking as a fare a greater sum than is authorised is liable to a penalty not exceeding level 3 on the standard scale, and a penalty may be recovered before a justice, and on conviction an order may be included for payment of the sum overcharged, over and above the penalty and costs and the overcharge must be returned to the party aggrieved (section 58).

Any proprietor or driver of any hired hackney carriage who permits or suffers any person to be carried in a hackney carriage during the hire, without the express consent of the person hiring the carriage, is liable to a penalty not exceeding level 1 on the standard scale (section 59 TPCA1847).

No person to act as driver of any carriages without the consent of the proprietor

No driver authorised by the proprietor of any hackney carriage can allow any other person to act as driver without the consent of the proprietor.

No person, whether licensed or not, can act as driver of a carriage without the consent of the proprietor; and any person allowing another person to act as driver, and any person acting as driver without consent, is liable to a penalty not exceeding level 1 on the standard scale for every offence (section 60 TPCA1847).

Penalty on drivers misbehaving

If the driver or any other person having or pretending to have the care of any hackney carriage is intoxicated while driving, or if any driver or other person by wanton and furious driving, or by any other wilful misconduct, injure or endanger any person in his life, limbs, or property, he is liable to a penalty not exceeding level 1 on the standard scale (section 61 TPCA1847).

Penalties in case of carriages being unattended at places of public resort

If the driver of any hackney carriage leaves it in any street or at any place of public resort or entertainment, whether it is hired or not, without someone proper to take care of it, any constable may drive the carriage away and deposit it, and the horse or horses harnessed, at some neighbouring livery stable or other place of safe custody, and the driver is liable to a penalty not exceeding level 1 on the standard scale.

If the offender on conviction defaults payment of the penalty, and the expenses of taking and keeping the hackney carriage and horse or horses, together with the harness, they must be sold by order of the justice before conviction is made, and after deducting from the produce of the sale the amount of the penalty, and all costs and expenses above, the surplus (if any) of the produce must be paid to the proprietor of the hackney carriage (section 62 TPCA1847).

Damage done by driver may be recovered from the proprietor

Where any hurt or damage has been caused to any person or property by the driver of any carriage let to hire, the justice before whom the driver has been convicted may direct that the proprietor of the carriage pay a sum, not exceeding five pounds, appearing to the justice a reasonable compensation for the hurt or damage and every proprietor who pays any compensation may recover it from the driver as damages (section 63 TPCA1847).

Improperly standing with carriage; refusing to give way to, or obstructing, any other driver or depriving him of his fare

Any driver of a hackney carriage who:

- allows the carriage to stand for hire across any street or alongside any other hackney carriage, or who refuses to give way, if he conveniently can, to any other carriage;
- obstructs or hinders the driver of any other carriage in taking up or setting down any person into or from such other carriage; or
- wrongfully in a forcible manner prevents or endeavours to prevent the driver of any other hackney carriage from being hired,

is liable to a penalty not exceeding level 1 on the standard scale (section 64 TPCA1847).

Justices empowered to award compensation to drivers for loss of time in attending to answer complaints not substantiated

If any hackney carriage driver is summoned or brought before a justice to answer any complaint or information touching or concerning any offence alleged to have been committed by the driver, and the complaint or information is withdrawn or quashed or dismissed, or if the driver is acquitted of the offence charged against him, the justice, if he thinks fit, may order the complainant or informant to pay to the driver compensation for his loss of time in attending the justice which seems reasonable (section 65 TPCA1847).

Penalty for refusing to pay the fare

If any person refuses to pay fare to the proprietor or driver of any hackney carriage, the fare may, together with costs, be recovered before a justice as a penalty (section 66 TPCA1847).

FC 3.46 Adoption of Part 2 of the Local Government (Miscellaneous Provisions) Act 1976

Local authority publish notice of intention to
pass a resolution to adopt part 2 in a local
newspaper for 2 consecutive weeks
(section 45(3))

Local authority serves a copy of the notice
of intention on the Council of each parish
affected by the resolution (section 45(3))

Council passes a resolution to adopt part 2 of
the Local Government (Miscellaneous
Provisions) Act 1976 (section 45(2))

One month later the provisions of part 2
come into force (section 45(2))

Local Government (Miscellaneous Provisions) Act 1976 Part 2

Application of Part 2

The council can resolve that Part 2 of this Act applies to their area and if the council do resolve this, the provisions come into force in the relevant area on the day specified in the resolution at least 1 month after the date the resolution is passed (section 45(2)).

Before passing such a resolution the council must:

(a) publish in 2 consecutive weeks, in a local newspaper circulating in their area, notice of their intention to pass the resolution; and

(b) serve a copy of the notice before the notice above is first published, on the council of each parish or community which would be affected by the resolution or, in the case of such a parish which has no parish council, on the chairman of the parish meeting (section 45(3)).

Proprietor's responsibilities

The proprietor of any hackney carriage licensed by a district council must present the hackney carriage for inspection and testing by or on behalf of the council within a period and at a place within the area of the council as they may by notice reasonably require on no more than three times in any period of 12 months (section 50(1) LG(MP)A1976).

The proprietor of any hackney carriage:

(a) licensed by a district council; or

(b) where an application for a licence has been made to a district council;

must, within a period the district council may by notice reasonably require, state in writing the address of every place where such hackney carriage is kept when not in use, and must if the district council require allow facilities reasonably necessary to enable them to allow the hackney carriage to be inspected and tested there (section 50(2) LG(MP)A1976).

The proprietor of a hackney carriage licensed by a district council must report to them within 72 hours of the occurrence of any accident to the hackney carriage causing damage materially affecting the safety, performance or appearance of the hackney carriage or the comfort or convenience of persons carried (section 50(3) LG(MP)A1976).

The proprietor of any hackney carriage licensed by a district council must at the request of any authorised officer of the council produce a valid insurance policy for inspection (section 50(4) LG(MP)A1976).

If any person without reasonable excuse contravenes these provisions, he is guilty of an offence (section 50(5) LG(MP)A1976).

Stands for hackney carriages (FC 3.47)

A district council may appoint stands for hackney carriages for the whole or any part of a day in any highway in the district which is maintainable at the public

FC 3.47 Hackney carriage stands outside London – Section 63 Local Government (Miscellaneous Provisions) Act 1976

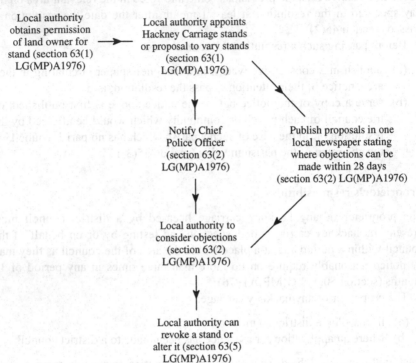

Local authority obtains permission of land owner for stand (section 63(1) LG(MP)A1976)

Local authority appoints Hackney Carriage stands or proposal to vary stands (section 63(1) LG(MP)A1976)

Notify Chief Police Officer (section 63(2) LG(MP)A1976)

Publish proposals in one local newspaper stating where objections can be made within 28 days (section 63(2) LG(MP)A1976)

Local authority to consider objections (section 63(2) LG(MP)A1976)

Local authority can revoke a stand or alter it (section 63(5) LG(MP)A1976)

expense and, with the consent of the owner, on any land in the district which does not form part of a highway and may vary the number of hackney carriages permitted to be at each stand (section 63(1) LG(MP)A1976).

Before appointing any stand for hackney carriages or varying the number of hackney carriages to be at each stand, a district council must give notice to the chief officer of police for the police area in which the stand is situated and also give public notice of the proposal by advertisement in at least one local newspaper circulating in the district and must take into consideration any objections or representations about the proposal which may be made to them in writing within 28 days of the first publication of such notice (section 63(2) LG(MP)A1976).

The district council must not appoint any stand which would:

(a) unreasonably prevent access to any premises;
(b) impede the use of any points authorised as points for the taking up or setting down of passengers, or in such a position as to interfere unreasonably with access to any station or depot of any passenger road transport operators, except with the consent of those operators;
(c) be on any highway except with the consent of the highway authority,

and in deciding the position of stands a district council must have regard to the position of any bus stops for the time being in use (section 63(3) LG(MP)A1976).

Previously designated stands for hackney carriages remain in operation (section 63(4) LG(MP)A1976). The council can revoke an appointment and alter any stand (section 63(5) LG(MP)A1976).

Prohibition of other vehicles on hackney carriage stands

No person can cause or permit any vehicle other than a hackney carriage to wait on any stand for hackney carriages during any period for which that stand has been appointed, or is deemed to have been appointed, by a district council (section 64(1) LG(MP)A1976).

Notice of the prohibition must be indicated by traffic signs prescribed or authorised by the Secretary of State under the Road Traffic Regulation Act 1984 (section 64(2) LG(MP)A1976).

If any person without reasonable excuse contravenes these provisions, he is guilty of an offence (section 64(3) LG(MP)A1976).

In any proceedings against the driver of a public service vehicle it shall be a defence to show that, by reason of obstruction to traffic or for other compelling reason, he allowed his vehicle to wait on a stand or part thereof and that he caused or permitted his vehicle to wait only for so long as was reasonably necessary for the taking up or setting down of passengers (section 64(4) LG(MP)A1976).

Fares (FC 3.48)

Fixing of fares for hackney carriages

A district council may fix the rates or fares within the district as well for time and distance, and all other charges in connection with the hire of a vehicle or

FC 3.48 Hackney carriage fares outside London – Section 65 Local Government (Miscellaneous Provisions) Act 1976 (LG(MP)A1976)

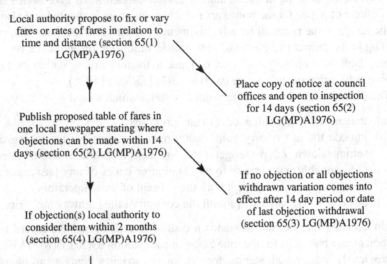

Local authority propose to fix or vary fares or rates of fares in relation to time and distance (section 65(1) LG(MP)A1976)

Place copy of notice at council offices and open to inspection for 14 days (section 65(2) LG(MP)A1976)

Publish proposed table of fares in one local newspaper stating where objections can be made within 14 days (section 65(2) LG(MP)A1976)

If no objection or all objections withdrawn variation comes into effect after 14 day period or date of last objection withdrawal (section 65(3) LG(MP)A1976)

If objection(s) local authority to consider them within 2 months (section 65(4) LG(MP)A1976)

Variation comes into effect within 2 months (section 65(4) LG(MP)A1976)

with the arrangements for the hire of a vehicle, to be paid in respect of the hire of hackney carriages by means of a table (table of fares) and can be varied (section 65(1) LG(MP)A1976).

When a district council make or vary a table of fares they must publish in at least one local newspaper circulating in the district a notice setting out the table of fares or variation and specifying a period more than 14 days from the date of the first publication of the notice, within which and the manner in which objections to the table of fares or variation can be made.

A copy of the notice must be deposited at the offices of the council which published the notice for more than 14 days, and be open to public inspection without payment at all reasonable hours (section 65(2) LG(MP)A1976).

If no objection to a table of fares or variation is made within the period specified in the notice, or if all objections so made are withdrawn, the table of fares or variation comes into operation on the date of the expiration of the period specified in the notice or the date of withdrawal of the objection or, if more than one, of the last objection, whichever date is the later (section 65(3) LG(MP)A1976).

If objection is made and is not withdrawn, the district council must set a further date, within 2 months after the first specified date, on which the table of fares comes into force with or without modifications as decided by them after consideration of the objections (section 65(4) LG(MP)A1976).

A table of fares made or varied has effect as if were in hackney carriage bye-laws made under the 1847 Act (section 65(5) LG(MP)A1976).

Fares for long journeys

No licensed driver of a hackney carriage undertaking for any hirer a journey ending outside the district and where no fare was agreed before the hiring, can require a fare greater than that indicated on the taximeter with which the hackney carriage is equipped or, if it is not equipped with a taximeter, greater than that in relevant bye-laws or the current table of fares (section 66(1) LG(MP)A1976). If any person knowingly contravenes this provision, he is guilty of an offence (section 66(2) LG(MP)A1976).

Prolongation of journeys

No driver of a licenced hackney carriage can, without reasonable cause, unnecessarily prolong, in distance or in time, the journey for which the hackney carriage has been hired (section 69(1) LG(MP)A1976). If any person contravenes this provision, he is guilty of an offence (section 69(2) LG(MP)A1976).

Hackney carriages used for private hire

No hackney carriage can be used in the district under a contract or purported contract for private hire except at a rate of fares or charges not greater than that fixed by the bye-laws or table of fares, and, when a hackney carriage is used, the fare or charge must be calculated from the point in the district at which the hirer starts the journey (section 67(2) LG(MP)A1976). Any person who knowingly contravenes this is guilty of an offence (section 67(1) LG(MP)A1976).

Fitness of hackney carriages

Any authorised officer of the council or any constable shall have power at all reasonable times to inspect and test, for the purpose of ascertaining its fitness, any hackney carriage licensed by a district council, or any taximeter fixed to a vehicle, and if he is not satisfied as to the fitness of the hackney carriage or the accuracy of its taximeter he may, by notice in writing, require the proprietor of the hackney carriage to make it or its taximeter available for further inspection and testing at a reasonable time and place specified in the notice and suspend the vehicle licence until the authorised officer or constable is satisfied, however, if the authorised officer or constable is not satisfied within 2 months, the licence will be deemed to have been revoked (section 68 LG(MP) A1976).

Touting for hire car services (Section 167 of the Criminal Justice and Public Order Act 1994, as amended)

It is an offence, in a public place, to solicit persons to hire vehicles to carry them as passengers (section 167(1)). The soliciting need not refer to any particular vehicle or the mere display of a sign on a vehicle that the vehicle is for hire soliciting (section 167(2)).

This does apply where soliciting persons to hire licensed taxis is permitted by a scheme under the Transport Act 1985 (schemes for shared taxis) (section 167(3)).

It is a defence for the accused to show that he was soliciting for passengers to be carried at separate fares by public service vehicles on behalf of the holder of a PSV operator's licence for those vehicles whose authority he had at the time of the alleged offence (section 167(4)).

A person guilty of an offence is liable on summary conviction to a fine not exceeding level 4 on the standard scale (section 167(5)).

Offences due to fault of other person, etc.

Where an offence is due to the act or default of another person, then, whether proceedings are taken against the charged person or not, the other person may be charged with and convicted of the offence, and liable on conviction to the same punishment as might have been imposed on the charged person if he had been convicted of the offence (section 72(1) LG(MP)A1976).

Obstruction of authorised officers

Any person who:

(a) wilfully obstructs an authorised officer or constable acting under the act or the Act of 1847; or

(b) without reasonable excuse fails to comply with any requirement properly made to him by such officer or constable; or

(c) without reasonable cause fails to give such an officer or constable any other assistance or information which he may reasonably require of such person for the purpose of the performance of his functions is guilty of an offence (section 73(1) LG(MP)A1976).

If any person, in giving any such information, makes any statement which he knows to be false, he is guilty of an offence (section 73(2) LG(MP)A1976).

Exempt vehicles

These provisions do not apply where a contract for the hire of the vehicle is made outside the district if the vehicle is not made available for hire within the district, or to funeral cars or wedding cars.

Penalties

Any person who commits an offence against this Act where no penalty is expressly stated is liable on summary conviction to a fine not exceeding level 3 on the standard scale (section 76 LG(MP)A1976).

Appeals

Sections 300 to 302 of the Public Health Act 1936 apply to appeals relating to these provisions (section 77(1) LG(MP)A1976).

If any requirement, refusal or other decision of a district council against which there is a right of appeal involves the execution of any work or the taking of any action; or makes it unlawful for any person to carry on a business which he was lawfully carrying on up to the time of the requirement, refusal or decision, then, until the time for appealing has expired, or, when an appeal is lodged, until the appeal is disposed of or withdrawn or fails for want of prosecution no proceedings can be taken in respect of any failure to execute the work, or take the action; and that person may carry on that business (section 77(2) LG(MP)A1976).

This does not apply to a decision under section 61(1) of this Act (section 77(3) LG(MP)A1976).

Authentication of licences

Any vehicle licence or driver's licence granted by a district council must be signed by an authorised officer of the council (section 79 LG(MP)A1976).

Local Services (Operation by Licensed Hire Cars) Regulations 2009

These Regulations apply where a licensed hire car is being used to provide a local service under a special licence. They prescribe the hire car code and documents, plates and marks for the holder of a special licence.

Taxi licensing: control of numbers

Section 37 of the Town Police Clauses Act 1847 enables the grant of a licence to be refused, for the purpose of limiting the number of hackney carriages in respect of which licences are granted, if, but only if, the person authorised to grant licences is satisfied that there is no significant demand for the services of hackney carriages (within the area to which the licence would apply) which is unmet (section 16 Transport Act 1985).

Taximeters

Local authorities usually require taximeters to be provided for hackney carriages. They are not required to be provided by law. Section 68 Local Government (Miscellaneous Provisions) Act 1976 allows inspection of taximeters and suspension powers if found to be irregular.

Definitions

controlled district means any area where a resolution has been passed by a district council under section 45 of this Act; or section 255(4) of the Greater London Authority Act 1999 (section 80 LG(MP)A1976).

commissioners means the commissioners, trustees, or other persons or body corporate intrusted by the special Act with powers for executing the purposes of that Act and the special Act means any Act which passed for the improvement or regulation of any defined town or district. This means the local authority given the powers to control taxis and mini cabs.

hackney carriage means a wheeled carriage, used in standing or plying for hire in any street within the prescribed distance. It must also have a numbered plate, resembling or intended to resemble a plate, fixed on it a hackney carriage. This does not apply to a licensed stage coach standing or plying for passengers to be carried for hire at separate fares (section 38 TPCA 1847).

London cab means hackney carriage which is any carriage for the conveyance of passengers which plies for hire and is neither a stage carriage nor a tramcar (Metropolitan Public Carriage Act 1869).

private hire vehicle means a motor vehicle constructed or adapted to seat fewer than nine passengers, other than a hackney carriage or public service vehicle or a London cab or tramcar, which is provided for hire with the services of a driver for the purpose of carrying passengers.

proprietor includes a part-proprietor and, in relation to a vehicle which is the subject of a hiring agreement or hire-purchase agreement, means the person in possession of the vehicle under that agreement.

taximeter means any device for calculating the fare to be charged in respect of any journey in a hackney carriage or private hire vehicle by reference to the distance travelled or time elapsed since the start of the journey, or a combination of both (section 80 LG(MP)A1976).

contract means:

(a) a contract made otherwise than while the relevant hackney carriage is plying for hire in the district or waiting at a place in the district which, when the contract is made, is a stand for hackney carriages appointed by the district council under section 63 of this Act; and

(b) a contract made, otherwise than with or through the driver of the relevant hackney carriage, while it is so plying or waiting (section 67(3)LG(MP) A 1976).

TAXI LICENCES – PRIVATE HIRE VEHICLES OUTSIDE LONDON

References

Local Government (Miscellaneous Provisions) Act 1976 (as amended) (LG(MP) A 1976).
Local Services (Operation by Licensed Hire Cars) Regulations 2009.

Extent

This procedure applies to England and Wales.

Introduction

Private hire vehicles (PHVs), more commonly referred to as minicabs, drivers and operators are subject to licensing if a district council has adopted Part II of the 1976 Act (most have) or has similar provisions contained in a local Act. The licensing conditions that are applied to taxi and PHV drivers and the local conditions of vehicle fitness are for each council to decide, so can vary considerably from area to area. Taxis, which are the only vehicles permitted to ply for hire, are normally subject to a stricter regime than the minicabs. The Department for Transport publishes best practice guidance for local licensing authorities.

Vehicle, driver's and operator's licences

The proprietor of any vehicle, not being a hackney carriage or London cab where a vehicle licence is in force, can use or permit a private hire vehicle to be used in a controlled district without having a current licence for the vehicle.

No person can act as driver of or operate any private hire vehicle in a controlled district without having a current licence.

The proprietor of a licensed private hire vehicle cannot employ a driver who does not have a current licence.

No licensed person in a controlled district can operate any vehicle as a private hire vehicle:

(i) if for the vehicle a current licence is not in force; or

(ii) if the driver does not have a current licence (section 46(1)).

If any person knowingly contravenes these provisions, he is guilty of an offence (section 46(2)).

Vehicles (FC 3.49)

Licensing of private hire vehicles

A district council receiving an application from the proprietor of any vehicle for use as a private hire vehicle, may grant a vehicle licence unless they are satisfied:

  (a) that the vehicle is:
      (i) suitable in type, size and design for use as a private hire vehicle;
      (ii) not of such design and appearance as to lead any person to believe that the vehicle is a hackney carriage;
      (iii) in a suitable mechanical condition;
      (iv) safe and comfortable;
  (b) it has an insurance policy or such security that complies with the requirements of Road Traffic Act 1988.

The Council cannot refuse a licence for the purpose of limiting the number of vehicles granted licences by the council (section 48(1)).

A district council may attach conditions, as they may consider reasonably necessary, requiring or prohibiting the display of signs on or from the vehicle (section 48(2)).

Every vehicle licence granted must specify:

  (a) the name and address of the applicant and every other person who is a proprietor of the private hire vehicle for which the licence is granted, or who is concerned, either solely or in partnership with any other person, in the keeping, employing or letting on hire of the private hire vehicle;
  (b) the number of the licence which must correspond with the number to be painted or marked on the plate or disc to be exhibited on the private hire vehicle;
  (c) the conditions attached to the grant of the licence; and
  (d) such other particulars as the district council consider reasonably necessary (section 48(3)).

Every licence granted must:

  (a) be signed by an authorised officer of the council which granted it;
  (b) relate to no more than one private hire vehicle; and
  (c) remain in force for no longer than 1 year as the district council may specify in the licence (section 48(4)).

Where a district council grant a vehicle licence for a private hire vehicle they must issue a plate or disc identifying that vehicle as a private hire vehicle in respect of which a vehicle licence has been granted (section 48(5)). The plate or disc issued must be exhibited on the vehicle in such manner as the district council prescribes by a condition attached to the licence. If any person without reasonable excuse contravenes these provisions he is guilty of an offence (section 48(6)).

Any person aggrieved by the refusal of a district council to grant a vehicle licence, or by any conditions specified in a licence, may appeal to a magistrates' court (section 48(7)).

FC 3.49 Licensing of private hire vehicles outside London – Local Government (Miscellaneous Provisions) Act 1976

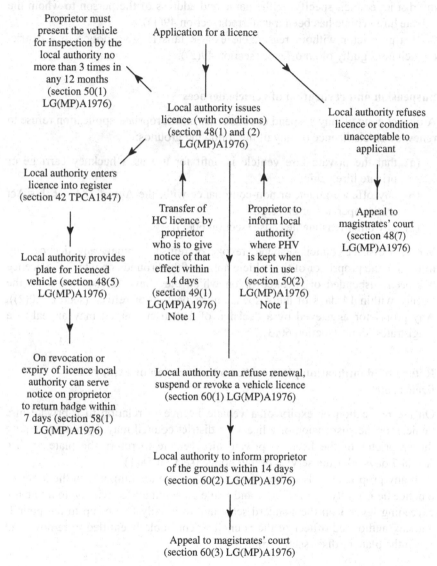

Proprietor must present the vehicle for inspection by the local authority no more than 3 times in any 12 months (section 50(1) LG(MP)A1976)

Application for a licence

Local authority issues licence (with conditions) (section 48(1) and (2) LG(MP)A1976)

Local authority refuses licence or condition unacceptable to applicant

Local authority enters licence into register (section 42 TPCA1847)

Transfer of HC licence by proprietor who is to give notice of that effect within 14 days (section 49(1) LG(MP)A1976) Note 1

Proprietor to inform local authority where PHV is kept when not in use (section 50(2) LG(MP)A1976) Note 1

Appeal to magistrates' court (section 48(7) LG(MP)A1976)

Local authority provides plate for licenced vehicle (section 48(5) LG(MP)A1976)

On revocation or expiry of licence local authority can serve notice on proprietor to return badge within 7 days (section 58(1) LG(MP)A1976)

Local authority can refuse renewal, suspend or revoke a vehicle licence (section 60(1) LG(MP)A1976)

Local authority to inform proprietor of the grounds within 14 days (section 60(2) LG(MP)A1976)

Appeal to magistrates' court (section 60(3) LG(MP)A1976)

Notes
1. Application must be accompanied with fee, vehicle insurance policy, etc., see page 365.
2. Fees – £25 per annum plus other sums determined by the local authority (Section 70 LG(MP)1976).
3. Appeals heard under sections 300 to 302 Public Health Act 1936.

Transfer of private hire vehicles

If the proprietor of a private hire vehicle with a licence transfers his interest in private hire vehicle to a person other than the proprietor whose name is specified in the licence, he must within 14 days after the transfer give notice in writing to the district council specifying the name and address of the person to whom the private hire vehicle has been transferred (section 49(1)).

If a proprietor without reasonable excuse fails to give notice to a district council he is guilty of an offence (section 49(2)).

Suspension and revocation of vehicle licences

A district council may suspend or revoke, or on appropriate application refuse to renew a vehicle licence on any of the following grounds:

(a) that the private hire vehicle is unfit for use as a hackney carriage or private hire vehicle;

(b) any offence under, or non-compliance with, the Act of 1847 or this Act by the operator or driver; or

(c) any other reasonable cause (section 60(1)).

Where a district council suspend, revoke or refuse to renew any licence they must give the proprietor of the vehicle notice of the grounds on which the licence has been suspended or revoked or on which they have refused to renew the licence within 14 days of the suspension, revocation or refusal (section 60(2)). Any proprietor aggrieved by a decision of a district council may appeal to a magistrates' court (section 60(3)).

Return of identification plate or disc on revocation or expiry of licence, etc.

On the revocation or expiry of a vehicle licence in relation to a private hire vehicle; or the suspension of a licence a district council may, by notice, require the proprietor of the licensed private hire vehicle to return the plate or disc within 7 days after the service of that notice (section 58(1)).

If any proprietor fails without reasonable excuse to comply with the terms of a notice he is guilty of an offence and liable on summary conviction to a fine not exceeding level 3 on the standard scale and to a daily fine of up to ten pounds and any authorised officer of the council or constable is entitled to remove and retain the plate or disc (section 58(2)).

Proprietors' responsibilities

The proprietor of any private hire vehicle licensed by a district council must present the private hire vehicle for inspection and testing by or on behalf of the council within a period and at a place within the area of the council as they may by notice reasonably require on no more than three times in any period of 12 months (section 50(1)).

The proprietor of any private hire vehicle:

(a) licensed by a district council, or
(b) where an application for a licence has been made to a district council,

must, within a period the district council may by notice reasonably require, state in writing the address of every place where such private hire vehicle is kept when not in use, and shall if the district council require allow facilities reasonably necessary to enable them to allow the private hire vehicle to be inspected and tested there (section 50(2)).

The proprietor of a private hire vehicle licensed by a district council must report to them within 72 hours of the occurrence of any accident to the hackney carriage or private hire vehicle causing damage materially affecting the safety, performance or appearance of the private hire vehicle or the comfort or convenience of persons carried (section 50(3)).

The proprietor of any private hire vehicle licensed by a district council must at the request of any authorised officer of the council produce a valid insurance policy for inspection (section 50(4)).

If any person without reasonable excuse contravenes these provisions, he is guilty of an offence (section 50(5)).

Drivers (FC 3.50)

Licensing of drivers of private hire vehicles

A district council must, on the receipt of an application from any person for the grant of a licence to drive private hire vehicles, grant to that person a driver's licence if they are satisfied that the applicant is a fit and proper person to hold a driver's licence, and if that person has for at least 12 months been authorised to drive a motor car (section 51).

A district council may require an applicant for a driver's licence in respect of a private hire vehicle:

(a) to produce a certificate signed by a registered medical practitioner that he is physically fit to be the driver of a private hire vehicle; and
(b) whether or not such a certificate has been produced, to submit to examination by a registered medical practitioner selected by the district council as to his physical fitness to be the driver of a private hire vehicle (section 57(2)).

If any person knowingly or recklessly makes a false statement or omits any material particular in giving information, he is guilty of an offence (section 57(3)).

Every licence granted by a district council to any person to drive a private hire vehicle remains in force for 3 years from the date of the licence or for a lesser period specified by the district council in the licence.

A district council may demand and recover for the grant of a licence to drive a private hire vehicle, a fee they consider reasonable with a view to recovering the costs of issue and administration (section 53(2)).

FC 3.50 Licensing of drivers of private hire vehicles outside London – Sections 51 to 54 and 61 Local Government (Miscellaneous Provisions) Act 1976 (LG(MP)A 1976)

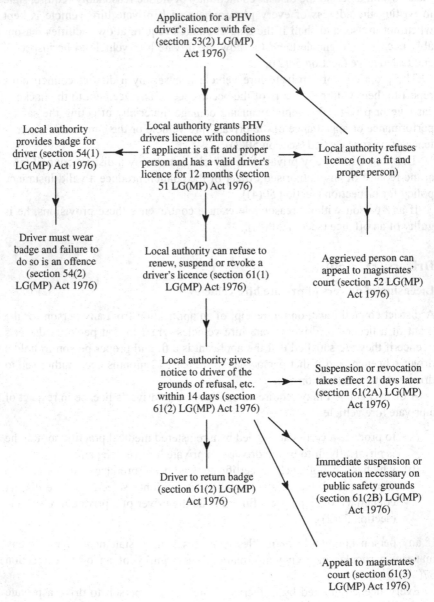

Application for a PHV driver's licence with fee (section 53(2) LG(MP) Act 1976)

Local authority grants PHV drivers licence with conditions if applicant is a fit and proper person and has a valid driver's licence for 12 months (section 51 LG(MP) Act 1976)

Local authority provides badge for driver (section 54(1) LG(MP) Act 1976)

Local authority refuses licence (not a fit and proper person)

Driver must wear badge and failure to do so is an offence (section 54(2) LG(MP) Act 1976)

Local authority can refuse to renew, suspend or revoke a driver's licence (section 61(1) LG(MP) Act 1976)

Aggrieved person can appeal to magistrates' court (section 52 LG(MP) Act 1976)

Local authority gives notice to driver of the grounds of refusal, etc. within 14 days (section 61(2) LG(MP) Act 1976)

Suspension or revocation takes effect 21 days later (section 61(2A) LG(MP) Act 1976)

Immediate suspension or revocation necessary on public safety grounds (section 61(2B) LG(MP) Act 1976)

Driver to return badge (section 61(2) LG(MP) Act 1976)

Appeal to magistrates' court (section 61(3) LG(MP) Act 1976)

The driver of any private hire vehicle licensed by a district council must on request of any authorised officer of the council or of any constable produce for inspection his driver's licence either immediately or:

(a) in the case of a request by an authorised officer, at the principal offices of the council within 5 days beginning with the day following that on which the request is made;

(b) in the case of a request by a constable, within 5 days at any police station in the area of the council and is nominated by the driver when the request is made (section 53(3)).

If any person without reasonable excuse contravenes these provisions, he is guilty of an offence (section 53(4)).

Issue of driver's badges

When granting a driver's licence a district council must issue a driver's badge in such a form prescribed by them (section 54(1)). A driver must at all times, acting within the driver's licence granted to him, wear the badge in such position and manner as to be plainly and distinctly visible.

If any person without reasonable excuse contravenes this, he is guilty of an offence (section 54(2)).

Suspension and revocation of driver's licences

A district council may suspend or revoke or on relevant application refuse to renew the licence of a driver of a private hire vehicle on any of the following grounds:

(a) that he has since the grant of the licence been convicted of an offence involving dishonesty, indecency or violence; or been convicted of an offence under or has failed to comply with the provisions of the 1847 Act or of this Act; or

(b) any other reasonable cause (section 61(1)).

Where a district council suspend, revoke or refuse to renew any licence they must give the driver notice of the grounds on which the licence has been suspended or revoked or on which they have refused to renew such licence within 14 days of such suspension, revocation or refusal and the driver must, on demand, return the driver's badge issued to him to the district council.

If any person without reasonable excuse contravenes this section he is guilty of an offence and liable on summary conviction to a fine not exceeding level 1 on the standard scale (section 61(2)).

A suspension or revocation of the licence of a driver takes effect 21 days after the day on which notice is given to the driver (section 61(2A)).

If it appears that the interests of public safety require the suspension or revocation of the licence to have immediate effect, and the notice given to the driver includes such a statement and an explanation why, the suspension or revocation takes effect when the notice is given to the driver (section 61(2B)).

Any driver aggrieved by a decision of a district council may appeal to a magistrates' court (section 61(3)).

Appeals

Any person aggrieved by the refusal of the district council to grant a private hire vehicle driver's licence or any conditions attached to the grant of a driver's licence may appeal to a magistrates' court (section 52).

Operators (FC 3.51)

Licensing of operators of private hire vehicles

A district council must, on receipt of an application from any person for the grant of a licence to operate private hire vehicles grant to that person an operator's licence if they are satisfied that the applicant is a fit and proper person to hold an operator's licence (section 55(1)).

These licences remain in force for up to 5 years, as a district council may specify in the licence (section 55(2)).

A district council may attach conditions to the licence as they may consider reasonably necessary (section 55(3)).

A district council may require an applicant for an operator's licence to submit to them information about:

(a) the name and address of the applicant;
(b) the address or addresses whether within the area of the council or not from which he intends to carry on business in connection with licensed private hire vehicles;
(c) any trade or business activities he has carried on before making the application;
(d) any previous application he has made for an operator's licence;
(e) the revocation or suspension of any operator's licence previously held by him;
(f) any convictions recorded against the applicant;

as they may reasonably consider necessary to enable them to determine whether to grant such licence and

(g) if the applicant is or has been a director or secretary of a company, information about any convictions recorded against that company at any relevant time; any trade or business activities carried on by that company; any previous application made by that company for an operator's licence; and any revocation or suspension of an operator's licence previously held by that company;
(h) if the applicant is a company, information as to any convictions recorded against a director or secretary of that company; any trade or business activities carried on by any such director or secretary; any previous application made by any such director or secretary for an operator's licence; and any revocation or suspension of an operator's licence previously held by such director or secretary;

FC 3.51 Licensing of operators of private hire vehicles outside London – Sections 55 and 62 Local Government (Miscellaneous Provisions) Act 1976

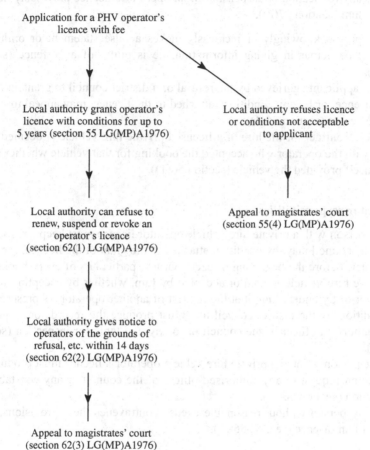

Application for a PHV operator's licence with fee

Local authority grants operator's licence with conditions for up to 5 years (section 55 LG(MP)A1976)

Local authority refuses licence or conditions not acceptable to applicant

Local authority can refuse to renew, suspend or revoke an operator's licence (section 62(1) LG(MP)A1976)

Appeal to magistrates' court (section 55(4) LG(MP)A1976)

Local authority gives notice to operators of the grounds of refusal, etc. within 14 days (section 62(2) LG(MP)A1976)

Appeal to magistrates' court (section 62(3) LG(MP)A1976)

Note

Fees – £25 per annum plus other sums determined by the Local Authority (section 70(1) LG(MP) Act 1976).

(i) if the applicant proposes to operate the vehicle in partnership with any other person, information as to any convictions recorded against that person; any trade or business activities carried on by that person; any previous application made by that person for an operator's licence; and any revocation or suspension of an operator's licence previously held by him (section 57(2)).

If any person knowingly or recklessly makes a false statement or omits any material particular in giving information, he is guilty of an offence (section 57(3)).

Any applicant aggrieved by the refusal of a district council to grant an operator's licence, or by any conditions attached to the licence, may appeal to a magistrates' court (section 55(4)).

Every contract for the hire of a licensed private hire vehicle is deemed to be made with the operator who accepted the booking for that vehicle whether or not he himself provided the vehicle (section 56(1)).

Operator responsibilities

Every person with a private hire vehicle operator licence must keep a record in a form the council may, by condition attached to the grant of the licence, prescribe and enter, before the beginning of each journey, particulars of every booking of a private hire vehicle invited or accepted by him, whether by accepting it from the hirer or by undertaking it at the request of another operator, as prescribed in a condition by the district council and must produce the records on request to any authorised officer of the council or to any constable for inspection (section 56(2)).

The person granted a private hire vehicle operator's licence must produce the licence on request to any authorised officer of the council or any constable for inspection (section 56(4)).

If any person without reasonable excuse contravenes these provisions, he is guilty of an offence (section 56(5)).

Suspension and revocation of operator's licences

A district council may suspend or revoke, or on relevant application refuse to renew an operator's licence on any of the following grounds:

(a) any offence under, or non-compliance with this Act;
(b) any conduct on the part of the operator which appears to the district council to render him unfit to hold an operator's licence;
(c) any material change since the licence was granted in any of the circumstances of the operator on the basis of which the licence was granted; or
(d) any other reasonable cause (section 62(1)).

Where a district council suspend, revoke or refuse to renew any licence they must give the operator notice of the grounds on which the licence has been suspended or revoked or on which they have refused to renew the licence within 14 days of such suspension, revocation or refusal (section 62(2)).

Any operator aggrieved by a decision of a district council may appeal to a magistrates' court (section 62(3)).

Fees for vehicle and operator's licences

A district council may charge fees for the grant of vehicle and operator's licences resolved by them from time to time, sufficient to cover:

(a) the reasonable cost of the carrying out by or on behalf of the district council of inspections of private hire vehicles to determine whether a licence should be granted or renewed;

(b) the reasonable cost of providing hackney carriage stands; and

(c) any reasonable administrative or other costs in connection with the control and supervision of private hire vehicles (section 70(1)).

The fees chargeable must not exceed for a vehicle licence for a private hire vehicle and an operator's licence, twenty-five pounds per annum or other sums as the council may, from time to time determine (section 70(1)).

If a district council determine that the maximum fees should be varied they must publish, in at least one local newspaper circulating in the district, a notice setting out the variation proposed, drawing attention to where the notice can be and specifying the period over 28 days from the date of the first publication of the notice, within which and how objections to the variation can be made. A copy of the notice must be deposited at the offices of the council which published the notice for 28 days from the date of the first publication and must be open to public inspection without payment at all reasonable hours (section 70(1)).

If no objection to a variation is made within the period specified in the notice or if all objections are withdrawn, the variation comes into operation on the date of the expiration of the period specified in the notice or the date of withdrawal of the objection or, if more than one, of the last objection, whichever date is the later (section 70(4)).

If objection is made and is not withdrawn, the district council must set a further date, within 2 months after the first specified date, on which the variation comes into force with or without modification as decided by the district council after consideration of the objections (section 70(5)).

A district council may remit (return) the whole or part of any fee chargeable for the grant of a licence in any case in which they think it appropriate to do so (section 70(6)).

Rules and powers

Power to require applicants to submit information

A district council may require any applicant for any relevant licence to submit information necessary to enable them to determine whether the licence should be granted and whether conditions should be attached to the licence (section 57(1)).

Prohibition of other vehicles on hackney carriage stands

No person can cause or permit any vehicle other than a hackney carriage to wait on any stand for hackney carriages during any period for which that stand has been appointed, or is deemed to have been appointed, by a district council (section 64(1)). Notice of the prohibition must be indicated by traffic signs prescribed or authorised by the Secretary of State under the Road Traffic Regulation Act 1984 (section 64(2)). If any person without reasonable excuse contravenes these provisions, he is guilty of an offence (section 64(3)).

In any proceedings against the driver of a public service vehicle it shall be a defence to show that, by reason of obstruction to traffic or for other compelling reason, he allowed his vehicle to wait on a stand or part thereof and that he caused or permitted his vehicle to wait only for so long as was reasonably necessary for the taking up or setting down of passengers (section 64(4)).

Hackney carriages used for private hire

No hackney carriage can be used in the district under a contract or purported contract for private hire except at a rate of fares or charges not greater than that fixed by the bye-laws or table of fares, and, when a hackney carriage is used, the fare or charge must be calculated from the point in the district at which the hirer starts the journey (section 67(2)). Any person who knowingly contravenes this is guilty of an offence (section 67(1)).

Fitness of private hire vehicles

Any authorised officer of the council or any constable has the power at all reasonable times to inspect and test, for the purpose of ascertaining its fitness, any private hire vehicle licensed by a district council or its taximeter, and if he is not satisfied as to the fitness of the private hire vehicle or the accuracy of its taximeter he may, by notice in writing, require the proprietor of the private hire vehicle to make it or its taximeter available for further inspection and testing at a reasonable time and place specified in the notice and suspend the vehicle licence until the authorised officer or constable is satisfied, however, if the authorised officer or constable is not satisfied within 2 months, the licence will be deemed to have been revoked (section 68).

Prolongation of journeys

No driver of a licenced private hire vehicle can, without reasonable cause, unnecessarily prolong, in distance or in time, the journey for which the private vehicle has been hired (section 69(1)). If any person contravenes this provisions, he is guilty of an offence (section 69(2)).

Local Services (Operation by Licensed Hire Cars) Regulations 2009

These Regulations apply where a licensed hire car is being used to provide a local service under a special licence. They prescribe the hire car code and documents, plates and marks for the holder of a special licence.

Taximeters

Private hire vehicles are not required equipped with any form of taximeter. But if a private hire vehicle is equipped with one and it is used for hire in a controlled district it must be tested and approved by or on behalf of the licensing district council (section 71(1)).

Any person who:

(a) tampers with any seal on any taximeter without lawful excuse; or
(b) alters any taximeter with intent to mislead; or
(c) knowingly causes or permits a vehicle of which he is the proprietor to be used in contravention above is guilty of an offence (section 71(2)).

Offences due to fault of other person, etc.

Where an offence is due to the act or default of another person, then, whether proceedings are taken against the charged person or not, the other person may be charged with and convicted of the offence, and liable on conviction to the same punishment as might have been imposed on the charged person if he had been convicted of the offence (section 72(1)).

Section 167 of the Criminal Justice and Public Order Act 1994, as amended

Touting for hire car services

It is an offence, in a public place, to solicit persons to hire vehicles to carry them as passengers (section 167(1)). The soliciting need not refer to any particular vehicle or the mere display of a sign on a vehicle that the vehicle is for hire soliciting (section 167(2)).

This does apply where soliciting persons to hire licensed taxis is permitted by a scheme under the Transport Act 1985 (schemes for shared taxis) (section 167(3)).

It is a defence for the accused to show that he was soliciting for passengers to be carried at separate fares by public service vehicles on behalf of the holder of a PSV operator's licence for those vehicles whose authority he had at the time of the alleged offence (section 167(4)).

A person guilty of an offence is liable on summary conviction to a fine not exceeding level 4 on the standard scale (section 167(5)).

Obstruction of authorised officers

Any person who:

(a) wilfully obstructs an authorised officer or constable acting under the act or the Act of 1847; or
(b) without reasonable excuse fails to comply with any requirement properly made to him by such officer or constable; or

(c) without reasonable cause fails to give such an officer or constable any other assistance or information which he may reasonably require of such person for the purpose of the performance of his functions is guilty of an offence (section 73(1)).

If any person, in giving any such information, makes any statement which he knows to be false, he is guilty of an offence (section 73(2)).

Exempt vehicles

These provisions do not apply to:

- where a contract for the hire of the vehicle is made outside the district if the vehicle is not made available for hire within the district;
- funeral cars;
- wedding cars;
- the display of any plate, disc or notice in or on any private hire is under a contract for the hire of the vehicle for a period more than 24 hours (section 75(1));
- where driver's and proprietor's licences for private hire vehicles are in force for the vehicle and driver of the vehicle (section 75(2)).

Where a private hire licence is in force for a vehicle, the council which issued the licence may, by a notice in writing given to the proprietor of the vehicle, provide that the display of a plate or disc is not required on any occasion specified in the notice or does not apply while the notice is carried in the vehicle and on any such occasion the driver's badge does not need to be worn (section 75(3)).

Penalties

Any person who commits an offence against this Act where no penalty is expressly stated is liable on summary conviction to a fine not exceeding level 3 on the standard scale (section 76).

Appeals

Sections 300 to 302 of the Public Health Act 1936, which applies to appeals relating to these provisions, shall have effect as if this part of this Act were part of that Act (section 77(1)).

If any requirement, refusal or other decision of a district council against which there is a right of appeal involves the execution of any work or the taking of any action; or makes it unlawful for any person to carry on a business which he was lawfully carrying on up to the time of the requirement, refusal or decision, then, until the time for appealing has expired, or, when an appeal is lodged, until the appeal is disposed of or withdrawn or fails for want of prosecution no proceedings can be taken in respect of any failure to execute the work, or take the action; and that person may carry on that business (section 77(2)).

This does not apply to a decision under section 61(1) of this Act (section 77(3)).

Authentication of licences

Any vehicle licence or driver's licence granted by a district council must be signed by an authorised officer of the council (section 79).

Definitions

controlled district means any area where a resolution has been passed by a district council under section 45 of this Act; or section 255(4) of the Greater London Authority Act 1999 (section 80).

private hire vehicle means a motor vehicle constructed or adapted to seat fewer than nine passengers, other than a hackney carriage or public service vehicle or a London cab or tramcar, which is provided for hire with the services of a driver for the purpose of carrying passengers.

proprietor includes a part-proprietor and, in relation to a vehicle which is the subject of a hiring agreement or hire-purchase agreement, means the person in possession of the vehicle under that agreement.

public service vehicle means a vehicle adapted to carry more than eight passengers used for carrying passengers for hire or reward; or a vehicle not so adapted, is used for carrying passengers for hire or reward at separate fares in the course of a business of carrying passengers (section 1 Public Passenger Vehicles Act 1981).

contract means:

(a) a contract made otherwise than while the relevant hackney carriage is plying for hire in the district or waiting at a place in the district which, when the contract is made, is a stand for hackney carriages appointed by the district council under section 63 of this Act; and

(b) a contract made, otherwise than with or through the driver of the relevant hackney carriage, while it is so plying or waiting (section 67(3)).

Chapter 4

HOUSING

SELECTIVE LICENSING OF RESIDENTIAL ACCOMMODATION

Reference

Housing Act 2004 sections 79 to 100.

Extent

These provisions apply to England and Wales.

Scope

The Housing Act 2004 provides for the mandatory licensing of certain Houses in Multiple Occupation (see page 388), and the discretionary licensing of other Houses in Multiple Occupation and other privately rented property. Here the provisions relate to selective licensing of privately rented dwellings.

Designation of selective licensing areas (FC 4.1)

Local authorities can designate all or part of their district for selective licensing of rented dwellings (section 80(1)).

Before making this designation it must consider:

(a) that the area is, or is likely to become, an area of low housing demand; and

(b) that making a designation will, when combined with other measures, contribute to the improvement of the social or economic conditions in the area (section 80(3)).

It must take into account (among other matters):

(a) the value of residential premises in the area, in comparison to the value of similar premises in other areas which the authority consider to be comparable (whether in terms of types of housing, local amenities, availability of transport or otherwise);

(b) the turnover of occupiers of residential premises;

(c) the number of residential premises which are available to buy or rent and the length of time for which they remain unoccupied (section 80(4));

(d) that the area is experiencing a significant and persistent problem caused by anti-social behaviour;

(e) that some or all of the private sector landlords who have let premises in the area are failing to take action to combat the problem that it would be appropriate for them to take; and

(f) that making a designation will, when combined with other measures taken in the area by the local authority, lead to a reduction in, or the elimination of, the problem (section 80(6)).

Before making a designation the local authority must:

(a) take reasonable steps to consult persons who are likely to be affected by the designation; and

(b) consider any representations made in accordance with the consultation and not withdrawn (section 80(9)).

The local authority must also ensure that any exercise of the power is consistent with the authority's overall housing strategy(section 81(2)) and must also seek to adopt a co-ordinated approach in dealing with homelessness, empty properties and anti-social behaviour.

It must consider whether there are any other courses of action available to them that might provide an effective method of achieving the objective or objectives that the designation would be intended to achieve, and they consider that making the designation will significantly assist them to achieve the objective or objectives (whether or not they take any other course of action as well).

The designation must be submitted to and be confirmed by the Secretary of State or comply with prior approved schemes (section 82).

If confirmed a designation comes into force on a date specified by the local authority, no earlier than 3 months after the date on which the designation is confirmed (section 80(3) and 80(4)).

Notification

As soon as the designation is confirmed or made, the local authority must publish a notice stating:

(a) that the designation has been made;

(b) whether the designation was required to be confirmed or that a general approval applied to it (giving details of the approval in question);

(c) the date on which the designation is to come into force; and

(d) any other information which may be prescribed (section 83(2)).

Copies of the designation, and prescribed information must be made available (section 83(3)).

Duration, review and revocation of designations

Unless previously revoked, a designation ceases to have effect at the time that is specified in the designation, no later than 5 years after the date on which the designation came into force. The local authority must from time to time review the operation of any designation made by them.

FC 4.1 Designation of areas for selective licensing of rented dwellings – Sections 80 to 84 Housing Act 2004

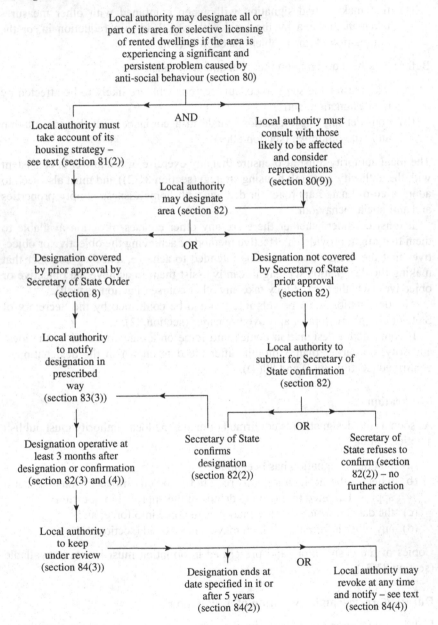

Local authority may designate all or part of its area for selective licensing of rented dwellings if the area is experiencing a significant and persistent problem caused by anti-social behaviour (section 80)

AND

Local authority must take account of its housing strategy – see text (section 81(2))

Local authority must consult with those likely to be affected and consider representations (section 80(9))

Local authority may designate area (section 82)

OR

Designation covered by prior approval by Secretary of State Order (section 8)

Designation not covered by Secretary of State prior approval (section 82)

Local authority to notify designation in prescribed way (section 83(3))

Local authority to submit for Secretary of State confirmation (section 82)

OR

Designation operative at least 3 months after designation or confirmation (section 82(3) and (4))

Secretary of State confirms designation (section 82(2))

Secretary of State refuses to confirm (section 82(2)) – no further action

Local authority to keep under review (section 84(3))

OR

Designation ends at date specified in it or after 5 years (section 84(2))

Local authority may revoke at any time and notify – see text (section 84(4))

On review if they consider it appropriate, the local authority may revoke the designation ceasing to have effect on the date specified.

The local authority must publish notice of the revocation (section 84).

The designation applies to all privately rented dwellings except houses in multiple occupation in the designated area. The local authority must take all reasonable steps to secure that applications for licences are made to them for houses in their area which are required to be licensed (section 85).

Temporary exemption

A person having control of or managing a house which is required to be licensed may notify the local authority of his intention to take steps to ensure the house is no longer required to be licensed. If the local authority thinks fit it may serve a temporary exemption notice in respect of the house.

A temporary exemption notice is in force for 3 months beginning with the date on which it is served. If the local authority receives a further notification of intention to take steps to ensure licensing is not required, a second temporary exemption notice may be served for a further 3 months after the date when the first notice ceases to be in force (section 86(5)).

If the authority decide not to serve a temporary exemption notice, they must without delay serve on the person concerned a notice informing him of:

(a) the decision;
(b) the reasons for it and the date on which it was made;
(c) the right to appeal against the decision; and
(d) the period within which an appeal may be made.

The person concerned may appeal to a residential property tribunal against the decision within the period of 28 days beginning with the date of the decision. The tribunal may confirm or reverse the decision of the authority, and if it reverses the decision, must direct the local authority to issue a temporary exemption notice with effect from such date as the tribunal directs (section 86).

Applications for licences (FC 4.3)

An application for a licence must be made to the local authority accompanied by a fee fixed by the authority. In fixing the fee for application the local authority should consider all costs incurred in carrying out their functions under the licensing scheme (section 87).

Grant or refusal of licence

On receiving an application for licensing the authority must either grant or refuse a licence.

If the local authority are satisfied:

(a) that the proposed licence holder is a fit and proper person and is the most appropriate person to be the licence holder;

FC 4.2 Selective licensing of rented dwellings – temporary exemption notices – Section 86 Housing Act 2004

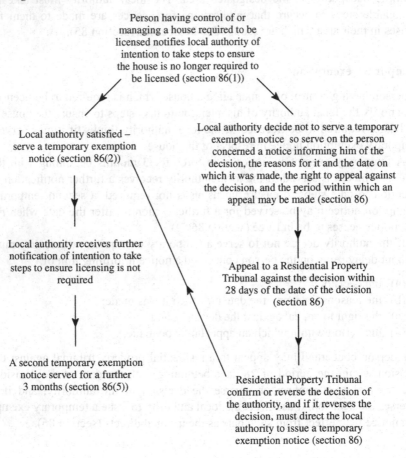

Person having control of or managing a house required to be licensed notifies local authority of intention to take steps to ensure the house is no longer required to be licensed (section 86(1))

Local authority satisfied – serve a temporary exemption notice (section 86(2))

Local authority decide not to serve a temporary exemption notice so serve on the person concerned a notice informing him of the decision, the reasons for it and the date on which it was made, the right to appeal against the decision, and the period within which an appeal may be made (section 86)

Local authority receives further notification of intention to take steps to ensure licensing is not required

Appeal to a Residential Property Tribunal against the decision within 28 days of the date of the decision (section 86)

A second temporary exemption notice served for a further 3 months (section 86(5))

Residential Property Tribunal confirm or reverse the decision of the authority, and if it reverses the decision, must direct the local authority to issue a temporary exemption notice (section 86)

(b) that the proposed manager of the house is either the person having control of the house, or a person who is an agent or employee of the person having control of the house;

(c) that the proposed manager of the house is a fit and proper person; and

(d) that the proposed management arrangements for the house are satisfactory,

they may grant a licence either to the applicant, or to some other person, if both he and the applicant agree (section 88).

Tests for fitness, etc. and satisfactory management arrangements

In deciding whether a person is a fit and proper person to be the licence holder or the manager of the house, the local authority must have regard to evidence that shows that the person has:

(a) committed offences involving fraud, other dishonesty, violence or drugs, or of a sexual nature;

(b) practised unlawful discrimination on grounds of sex, colour, race, ethnic or national origins or disability in, or in connection with, the carrying on of any business; or

(c) contravened any provision of the law relating to housing or of landlord and tenant law.

They must also consider evidence that shows that any person associated or formerly associated with the person (whether on a personal, work or other basis) has done any of the things above and it appears to the authority that the evidence is relevant to the question whether the person is a fit and proper person to be the licence holder or the manager of the house.

In deciding whether the proposed management arrangements for the house are satisfactory, the local authority must have regard to:

(a) whether any person proposed to be involved in the management of the house has a sufficient level of competence to be involved;

(b) whether any person proposed to be involved in the management of the house (other than the manager) is a fit and proper person to be involved; and

(c) whether any proposed management structures and funding arrangements are suitable (section 89).

Licences

The local authority may attach conditions to a licence and they may include:

(a) restrictions or prohibitions on the use or occupation of particular parts of the house by persons occupying it;

(b) requiring the taking of reasonable and practicable steps to prevent or reduce anti-social behaviour by persons occupying or visiting the house;

(c) requiring facilities and equipment to be made available in the house;

(d) requiring such facilities and equipment to be kept in repair and proper working order; or

(e) requiring works needed for any such facilities or equipment to be made available (section 90(1) to (3)).

Local authorities must not use licensing to remove or reduce category 1 or category 2 hazards in the house but this does not prevent the authority from imposing licence conditions relating to the installation or maintenance of facilities or equipment (section 90).

Licences apply to only one house and comes into force at the specified time, and unless previously terminated or revoked continues in force for the specified period for at least 5 years. A licence cannot be transferred to another person. If the holder of the licence dies while the licence is in force, the licence ceases to be in force on his death. But for the 3 months after the licence holder's death, the house is to be treated as if on that date a temporary exemption notice had been served in respect of the house. If, at any time during that period, the personal representatives of the licence holder request the local authority to do so, the authority may serve on them a notice which has the same effect as a temporary exemption notice (section 91).

Variation of licences

The local authority may vary a licence with the agreement of the licence holder, or if they consider that there has been a change of circumstances since the time when the licence was granted including any discovery of new information. A variation made with the agreement of the licence holder takes effect at the time when it is made. Otherwise, it comes into force 28 days later unless there is an appeal against the variation.

The local authority can vary the licence on its own initiative or on application from the licence holder or a 'relevant person' (who has an estate or interest in the house concerned), or a person managing or having control of the house or a person who is subject to any restriction or obligation imposed by the licence (section 92).

Revocation of licences

The local authority may revoke a licence:

(a) with the agreement of the licence holder;
(b) where the licence holder or any other person has committed a serious breach of a condition of the licence;
(c) where the licence holder is no longer considered a fit and proper person to be the licence holder; or
(d) where the management of the house is no longer being carried on by persons who are fit and proper persons to be involved in its management.

A revocation made with the agreement of the licence holder takes effect at the time when it is made. Otherwise, 28 days from the notification of revocation unless there is an appeal against the revocation to the Residential Property Tribunal.

FC 4.3 Selective licensing of rented dwellings – Sections 87 to 93 Housing Act 2004

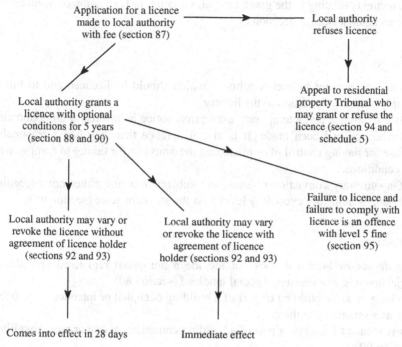

Application for a licence made to local authority with fee (section 87) ──────▶ Local authority refuses licence

Local authority grants a licence with optional conditions for 5 years (section 88 and 90)

Appeal to residential property Tribunal who may grant or refuse the licence (section 94 and schedule 5)

Local authority may vary or revoke the licence without agreement of licence holder (sections 92 and 93)

Local authority may vary or revoke the licence with agreement of licence holder (sections 92 and 93)

Failure to licence and failure to comply with licence is an offence with level 5 fine (section 95)

Comes into effect in 28 days   Immediate effect

Notes

1. Local authorities must not use licensing to remove or reduce category 1 or category 2 hazards in the house (section 90).
2. The local authority can vary the licence on its own initiative or on application from the licence holder or a 'relevant person', or a person managing or having control of the house or a person who is subject to any restriction or obligation imposed by the licence (section 92).

The local authority can revoke a licence on its own initiative or on application made by the licence holder or a 'relevant person' (see above) (section 93).

The procedural requirements and appeals against licence decisions and contained in Schedule 5 of the Housing Act 2004 which deals with procedural requirements relating to the grant, refusal, variation or revocation of licences and appeals against licence decisions.

Offences

It is an offence not to licence a house which should be licenced and to fail to comply with any condition of the licence.

It is a defence that a temporary exemption notice is in force or an application for a licence had been made. It is also a defence that they had a reasonable excuse for having control of or managing the house or for failing to comply with the condition.

On summary conviction offences are subject to a fine either not exceeding £20,000, or a fine not exceeding level 5 on the standard scale (section 95).

Definitions

private sector landlord does not include a non-profit registered provider of social housing or a registered social landlord (section 80).

dwelling means a building or part of a building occupied or intended to be occupied as a separate dwelling.

house means a building or part of a building consisting of one or more dwellings (section 99).

DESIGNATION OF AREAS BY LOCAL AUTHORITIES FOR ADDITIONAL LICENSING OF HOUSES IN MULTIPLE OCCUPATION

References

Housing Act 2004 sections 55 to 60.
The Licensing and Management of Houses in Multiple Occupation and other Houses (Miscellaneous Provisions) (England) Regulations 2006.

Extent

These provisions apply to England and Wales.

Scope

This procedure allows a local authority to designate the whole or part of its district as an area in which all or specified types of houses in multiple occupation will require to be registered in addition to those already designated by the Licensing of Houses in Multiple Occupation (Prescribed Descriptions) (England)

Order 2006 (section 56(1)). Following designation, the procedure for their licensing is the same as in FC 4.5.

Before making a designation the local authority must be satisfied that a significant proportion of the houses in multiple occupation to be included are being managed sufficiently ineffectively as to give rise, or be likely to give rise, to one or more particular problems either for their occupants or for members of the public, e.g. anti-social behaviour. In forming this opinion the local authority must have regard to any Codes of Practice issued under section 233 (section 56(2)–(6)).

Designation (FC 4.4)

Before making a designation the local authority must consult with those most likely to be affected and take account of any representations made (section 56(3)).

There are further matters for the local authority to consider:

(a) that designation is consistent with the local authority's overall housing strategy;

(b) there must be a co-ordinated approach to dealing with homelessness, empty properties and anti-social behaviour in the private rented sector which considers both combining additional licensing with other possible courses of action and measures to be taken by others;

(c) the local authority must also consider alternative actions; and

(d) the local authority must be satisfied that designation will significantly assist in dealing with the problems of the area (section 57).

Confirmation by Secretary of State and operation

Unless covered by a general approval (see The Housing Act 2004: Licensing of Houses in Multiple Occupation and Selective Licensing of Other Residential Accommodation (England) General Approval 2010) the designation requires confirmation by the Secretary of State (section 58(1)). There is no specified procedure for this part of the process. The Secretary of State may confirm or refuse to confirm a designation. The designation comes into effect no earlier than 3 months after it has been confirmed or, in the case of general approvals, 3 months after the designation was made (section 58).

Notification of designation

As soon as designation is confirmed or made the local authority must publicise it by:

1. Within 7 days:

(a) place notices on municipal buildings in or closest to the area;

(b) publish the notice on the internet site;

(c) publish it in the next edition of two local newspapers and 5 times more at intervals of between 2 and 3 weeks.

2. Within 2 weeks send a copy of the notice to:

   (a) those who responded to the consultation;
   (b) organisations representing landlords, tenants, agents;
   (c) organisations providing advice on landlord and tenant matters, e.g. housing advice centres.

The notice must contain:

   (a) the date of designation;
   (b) if the designation requires confirmation;
   (c) a brief description of the area;
   (d) contact details for the local authority, the place where the designation may be inspected, and where applications for licences may be obtained;
   (e) a statement advising persons affected to seek advice from the local authority;
   (f) a warning of the consequences of not applying for a licence.

(section 59(2) and the Licensing and Management of Houses in Multiple Occupation and other Houses (Miscellaneous Provisions) (England) Regulations 2006 regulation 9).

Duration

The designation must be reviewed from time to time and may be revoked at any time at the discretion of the local authority but ends no later than 5 years after it has been made (section 60(1) – (3)). Revocation does not require a confirmation by the Secretary of State.

Revocation

When a scheme is revoked the local authority must within 7 days:

   (a) place notices on municipal buildings in or closest to the area;
   (b) publish the notice on the internet site;
   (c) publish it in the next edition of two local newspapers.

The notice must give:

   (a) a brief description of the area;
   (b) a summary of the reasons for revocation;
   (c) the date revocation takes effect;
   (d) contact details for the local authority.

(section 60(6) and The Licensing and Management of Houses in Multiple Occupation and other Houses (Miscellaneous Provisions) (England) Regulations 2006 regulation 10).

**FC 4.4 Designation of areas for additional licensing of houses in multiple occupation (HMOs)
– Sections 55 to 60 Housing Act 2004**

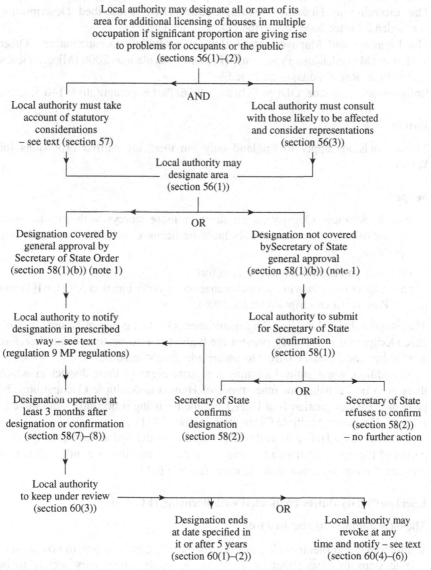

Local authority may designate all or part of its
area for additional licensing of houses in multiple
occupation if significant proportion are giving rise
to problems for occupants or the public
(sections 56(1)–(2))

AND

Local authority must take
account of statutory
considerations
– see text (section 57)

Local authority must consult
with those likely to be affected
and consider representations
(section 56(3))

Local authority may
designate area
(section 56(1))

OR

Designation covered by
general approval by
Secretary of State Order
(section 58(1)(b)) (note 1)

Designation not covered
bySecretary of State
general approval
(section 58(1)(b)) (note 1)

Local authority to notify
designation in prescribed
way – see text
(regulation 9 MP regulations)

Local authority to submit
for Secretary of State
confirmation
(section 58(1))

OR

Designation operative at
least 3 months after
designation or confirmation
(section 58(7)–(8))

Secretary of State
confirms
designation
(section 58(2))

Secretary of State
refuses to confirm
(section 58(2))
– no further action

Local authority
to keep under review
(section 60(3))

OR

Designation ends
at date specified in
it or after 5 years
(section 60(1)–(2))

Local authority may
revoke at any
time and notify – see text
(section 60(4)–(6))

Notes
1. The Housing Act 2004: Licensing of Houses in Multiple Occupation and Selective Licensing of
   Other Residential Accommodation (England) General Approval 2010.
2. MP regulations in this flowchart – The Licensing and Management of Houses in Multiple Occupation and other Houses (Miscellaneous Provisions) (England) Regulations 2006.

LICENSING OF HOUSES IN MULTIPLE OCCUPATION

References

Housing Act 2004 part 2.
Management of Houses in Multiple Occupation (England) Regulations 2006.
The Licensing of Houses in Multiple Occupation (Prescribed Descriptions) (England) Order 2006 (Prescribed Description Order).
The Licensing and Management of Houses in Multiple Occupation and Other Houses (Miscellaneous Provisions) (England) Regulations 2006 (Miscellaneous Provisions Regulations) as amended.
Enforcement Guidance, ODPM February 2006. Part 6. paragraphs 6.1–6.5.

Extent

These provisions apply to England only but there are similar provisions for Wales.

Scope

Houses in Multiple Occupation on three or more storeys with five or more tenants in two or more households must be licensed with the local authority except those:

(a) where an exemption notice is in force;
(b) subject to an interim or final management order (section 55(2), 61(1) and Prescribed Description Order 2006).

The intention behind the licensing requirements for this group of Houses in Multiple Occupation is that they present the highest risk to occupiers and therefore need to have additional controls to ensure adequate standards.

In addition, local authorities may designate areas of their district in which there is to be licensing for other types of Houses in Multiple Occupation. The procedure for designation is at FC 4.1 but the licensing procedure is the same as for other Houses in Multiple Occupation (section 56(1)).

A licence for a House in Multiple Occupation is defined as 'authorising occupation of the house concerned by not more than a maximum number of households or persons specified in the licence' (section 61(2)).

Local authority duties connected with licensing (FC 4.5)

The local authority is required to:

1. Promote the implementation of licensing. This includes a duty to take reasonable steps to ensure that applications are made where they appear to be required;
2. Ensure that it deals with applications promptly;
3. Satisfy itself that there are no part 1 enforcement functions remaining to be discharged in respect of a licensed House in Multiple Occupation as soon as possible but in any event not longer than 5 years from the first application (sections 55(5) and 61(4)).

FC 4.5 Licensing of houses in multiple occupation (HMOs) – Sections 55 to 71 Housing Act 2004

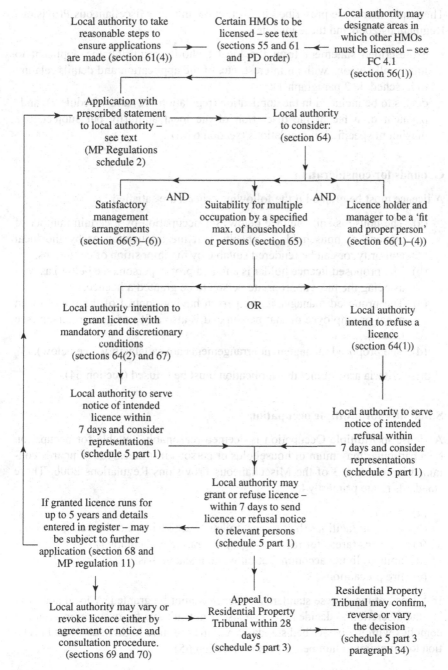

Notes

1. PD order – The Licensing of Houses in Multiple Occupation (Prescribed Descriptions) (England) Order 2006.
2. MP regulations – The Licensing and Management of Houses in Multiple Occupation and Other Houses (Miscellaneous Provisions) (England) Regulations 2006.

Applications for licences

The requirements are prescribed in Section 63 and the Miscellaneous Provisions Regulations 2006 and these include:

- a prescribed statement from the local authority requiring the applicant to inform all persons with an interest, etc. of the application and details relating to it (schedule 2 paragraph 1);
- details to be included in the application (regulation 7(2) and schedule 2); and
- payment of a fee at the discretion of the local authority but subject to a maximum specified in regulations (section 63(3)).

Grounds for consideration

A licence must be granted if the following criteria are satisfied:

(a) The house is suitable (see below) for occupation by a certain number of persons or households as specified in the application or by the local authority, or can be rendered suitable by the imposition of conditions.

(b) The proposed licence holder is a fit and proper person (see below) as well as being the most appropriate person to be granted a licence.

(c) The proposed manager is the person having control of the house or an agent or employee of that person and is also a fit and proper person (see below).

(d) The proposed management arrangements are satisfactory (see below).

If these criteria are not met the application must be refused (section 64).

Suitability for multiple occupation

A House in Multiple Occupation is deemed reasonably suitable for occupation by a particular maximum of households or persons if it meets the standards contained in schedule 3 of the Miscellaneous Provisions Regulations 2006. These standards relate primarily to:

(a) heating;

(b) washing facilities (bathrooms, toilets, washbasins and showers);

(c) kitchens (areas for food storage, preparation and cooking);

(d) units of living accommodation without shared basic facilities; and

(e) fire precautions.

If it does not meet these standards a licence cannot be granted. A local authority has the discretion to decide that the house is not suitable for occupation even if it complies with the prescribed standards and may require higher standards in relation to a maximum number of persons (section 65).

Fit and proper person

In determining whether a person is a fit and proper person to be a licence holder or manager, the local authority must have regard to whether that person, or a relevant associate, e.g. a spouse or business partner, has committed offences involving fraud, dishonesty, violence, drugs or sexual offences, unlawful discrimination in business, contravention of housing law or breach of an approved code of management, e.g. Housing (Codes of Management Practice) (Student Accommodation) (England) Order 2010 (section 66(1)–(4)).

Satisfactory management arrangements

In deciding if the proposed management arrangements are satisfactory the local authority must have regard, amongst other matters, to:

(a) the competence of the manager;
(b) fitness of other persons involved in the management structure;
(c) management structure; and
(d) funding (section 66(5) and (6)).

Grant and refusal of licences

Before granting a licence the local authority must:

(a) serve notice on the applicant and other relevant persons of its intention to grant the licence together with a copy of it;
(b) consider any representations made within the period specified.

If, following this consultation, the local authority intend to modify the proposed licence they must again serve notice and consider representations.
Before refusing to grant a licence the local authority must:

(a) serve notice on the applicant and other relevant persons of its intention; and
(b) state the reasons for refusal and indicate the length of the consultation period;
(c) consider representations made.

Following a grant or refusal the local authority must serve notice within 7 days on the applicant and other relevant persons setting out:

(a) the local authority's decision;
(b) the reasons for the decision and the date on which it was made; and
(c) rights of appeal and the appeal period (section 65 and schedule 5 part 1).

Licence conditions

Mandatory conditions

These are set out in schedule 4 and deal mainly with the safety of occupiers including furniture and electrical, gas and fire safety appliances. These must be included in each licence (section 67(3) and schedule 4).

Discretionary conditions

The local authority may include any conditions relating to the management, use and occupation of the house and its contents and condition. These might include:

(a) restrictions on the use of particular parts;
(b) the reduction of anti-social behaviour;
(c) the installation and maintenance of facilities and equipment to meet the standards required under section 65 above; and
(d) carrying out works to facilities and equipment within specified periods.

Local authorities are not to use licence conditions to deal with matters of health and safety but should use the enforcement procedures of part 1 (section 67).

General requirements and duration

A licence can only relate to a single House in Multiple Occupation and is valid for the period specified in it, which cannot exceed 5 years. They are not transferable but on death of a licence holder a 3 months' temporary exemption from the licensing requirement is given (section 68) (see page 379). If the house is to continue in a licensable status after the expiration of the licence, there must be a further application for a new licence which will be subject to the same procedure.

Variation

A local authority may vary a licence either by agreement with the owner or if there is a change of circumstances, e.g. the discovery of new information about issues present before the licence was granted or there is a need to vary the appropriate number of households or persons. Variation may be instigated by the local authority or on application from the owner or a relevant person. The procedure to be used in varying a licence is contained in part 2 of schedule 5 and provisions for appeal against refusals to vary in part 3.

Variations with the agreement of the owner come into effect immediately otherwise when the period for appeal has elapsed or any appeal has disposed of (section 69).

Revocation

A local authority may revoke a licence:

- with the agreement of the licence holder;
- where the licence holder has committed a serious breach of a licence condition or there are repeated breaches of conditions;
- where the local authority consider that the holder is no longer a fit and proper person; and
- where the local authority believes that the property no longer meets the standards required for a licence.

The procedure to be adopted is set out in part 2 of schedule 5 with appeals dealt with in part 3 of schedule 5. Revocations with the agreement of the owner come into effect immediately otherwise when the period for appeal has elapsed or any appeal has disposed of (section 70).

Appeals to the Residential Property Tribunal

Appeals against licensing decisions are dealt with in part 3 of schedule 5 and deal with appeals:

   (a) against the refusal or grant of a licence including licence conditions; and
   (b) decisions to vary or revoke or refusals to do so.

Generally appeals are to be made within 28 days of the date specified in the local authority notice.

Register of licences

A local authority must keep a register of the licences granted containing the details laid out in regulation 11 of the Miscellaneous Provisions Regulations. The register must be available to the public at the head office of the local authority and copies of it, or extracts from it, must be provided by the local authority on payment of a fee (section 232).

Offences

It is an offence, punishable by a fine of up to £20,000, to own or manage a House in Multiple Occupation without a licence or knowingly allow occupation by more persons or households than permitted. Breaching the licence conditions carries a fine of up to level 5. There is a defence of 'reasonable excuse' in each case (section 72).

There is also a provision for the making of rent repayment orders by the Residential Property Tribunal in relation to unlicensed Houses in Multiple Occupation whereby there is a penalty equivalent to the rent received during unlicensed operation of up to 12 months (sections 73 to 74).

Management regulations

The Management of Houses in Multiple Occupation (England) Regulations 2006, as amended, apply to all Houses in Multiple Occupation except those converted blocks of flats under section 257. They impose duties on managers concerning:

* provision of information to occupiers;
* safety measures including fire safety;
* maintenance of water, drainage, gas and electricity;
* maintaining common parts and living accommodation; and
* providing waste disposal facilities. They also place a duty on occupiers to ensure that the manager can effectively carry out these duties.

There are no provisions for the service of notices by a local authority to secure compliance but a person in breach of these requirements is liable to a fine of up to level 5.

It should be noted that there are duties placed on the manager of a House in Multiple Occupation through the Licensing and Management of Houses in Multiple Occupation (Additional Provisions) (England) Regulations 2007.

LICENSING OF CAMPING SITES

Reference

Public Health Act 1936 section 269 (as amended by Caravan Sites and Control of Development Act 1960).

Extent

These provisions apply to England and Wales.

Scope

The procedure applies to all moveable dwellings (other than caravans) and licences are required for:

(a) the use of land for camping purposes on more than 42 consecutive days or more than 60 days in any 12 consecutive months; or

(b) the keeping of a moveable dwelling on any one site, or two or more sites in succession if any of those sites is within 100 yards of another of them, for more than the same periods as in (a).

In respect of (a), land which is in the occupation of the same person, and within 100 yards of a site on which a moveable dwelling is stationed during any part of a day, is regarded as being used for camping on that day (section 269(1)–(3)). The removal of a moveable dwelling for not more than 48 hours does not constitute an interruption of the 42 days period (section 269(8)).

Exemptions

Apart from the use of land or moveable dwellings for periods less than those in (a) and (b) above, the following situations do not require licensing by the local authority:

(a) moveable dwellings kept by the owner on land occupied by him in connection with his dwelling if used only by him or members of his household;

(b) moveable dwellings kept by the owner on agricultural land occupied by him and used for habitation only at certain times of the year and only by persons employed in farming operations on that land;

(c) moveable dwellings not in use for human habitation kept on premises by an occupier who does not permit moveable dwellings to be kept there for habitation;

FC 4.6 Licensing of camping sites – Section 269 Public Health Act 1936

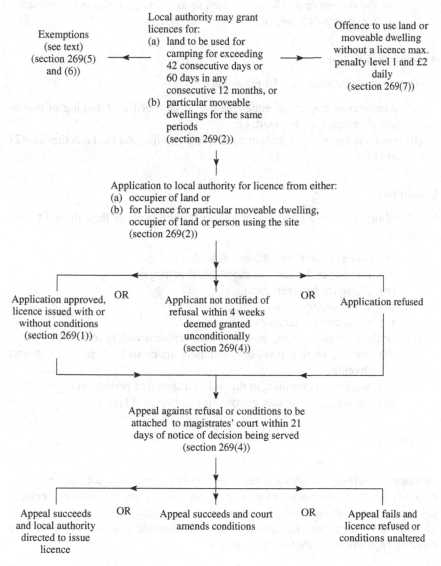

Exemptions
(see text)
(section 269(5)
and (6))

Local authority may grant
licences for:
(a) land to be used for
     camping for exceeding
     42 consecutive days or
     60 days in any
     consecutive 12 months, or
(b) particular moveable
     dwellings for the same
     periods
     (section 269(2))

Offence to use land or
moveable dwelling
without a licence max.
penalty level 1 and £2
daily
(section 269(7))

Application to local authority for licence from either:
(a) occupier of land or
(b) for licence for particular moveable dwelling,
     occupier of land or person using the site
     (section 269(2))

Application approved,
licence issued with or
without conditions
(section 269(1))

OR

Applicant not notified of
refusal within 4 weeks
deemed granted
unconditionally
(section 269(4))

OR

Application refused

Appeal against refusal or conditions to be
attached to magistrates' court within 21
days of notice of decision being served
(section 269(4))

Appeal succeeds
and local authority
directed to issue
licence

OR

Appeal succeeds and court
amends conditions

OR

Appeal fails and
licence refused or
conditions unaltered

Note
This procedure does not apply to caravans. For licensing of caravan sites, see FC 4.7.

(d) organisations granted a certificate of exemption by the Minister, being satisfied that:
- (i) the camping sites belonging to, provided by or used by the organisation are properly managed and kept in good sanitary condition; and
- (ii) the moveable dwellings are used so as not to give rise to any nuisance (section 269(5) and (6)).

Applicants (FC 4.6)

Applications for licences may be made by:

(a) occupiers in respect of either the use of land or the stationing of moveable dwellings on that land; or
(b) a person intending to station a moveable dwelling on land (section 269(2) and (3)).

Conditions

Local authorities may attach to licences such conditions as they think fit with respect to:

(a) for licences authorising the use of land:
- (i) number and classes of moveable dwellings;
- (ii) the space between them;
- (iii) water supply; and
- (iv) for securing sanitary conditions.
(b) for licences authorising the use of a moveable dwelling for:
- (i) the use of that dwelling, including space to be kept free between dwellings;
- (ii) securing its removal at the end of a specified period; and
- (iii) for securing sanitary conditions (section 269(1)).

Definitions

caravan, see page 405.
moveable dwelling includes any tent, any van or other conveyance (other than a caravan) whether on wheels or not, and ... any shed or similar structure, being a tent, conveyance or structure which is used either regularly or at seasons only, or intermittently, for human habitation but does not include a structure to which the building regulations apply (section 269(8)).

LICENSING OF CARAVAN SITES

References

Caravan Sites and Control of Development Act 1960 Part I (as amended).
Model Standards 2008 for Caravan Sites in England: Caravan Sites and Control of Development Act 1960 – Section 5 April 2008.

FC 4.7 Licensing of caravan sites – Caravan Sites and Control of Development Act 1960

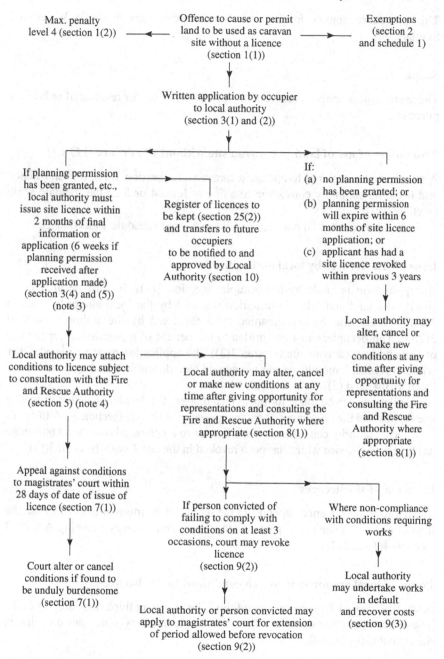

Max. penalty
level 4 (section 1(2))

Offence to cause or permit
land to be used as caravan
site without a licence
(section 1(1))

Exemptions
(section 2
and schedule 1)

Written application by occupier
to local authority
(section 3(1) and (2))

If planning permission
has been granted, etc.,
local authority must
issue site licence within
2 months of final
information or
application (6 weeks if
planning permission
received after
application made)
(section 3(4) and (5))
(note 3)

Register of licences to
be kept (section 25(2))
and transfers to future
occupiers
to be notified to and
approved by Local
Authority (section 10)

If:
(a) no planning permission
has been granted; or
(b) planning permission
will expire within 6
months of site licence
application; or
(c) applicant has had a
site licence revoked
within previous 3 years

Local authority may
alter, cancel or
make new
conditions at any
time after giving
opportunity for
representations and
consulting the Fire
and Rescue
Authority where
appropriate
(section 8(1))

Local authority may attach
conditions to licence subject
to consultation with the Fire
and Rescue Authority
(section 5) (note 4)

Local authority may alter, cancel
or make new conditions at any
time after giving opportunity for
representations and consulting the
Fire and Rescue Authority where
appropriate (section 8(1))

Appeal against conditions
to magistrates' court within
28 days of date of issue of
licence (section 7(1))

If person convicted of
failing to comply with
conditions on at least 3
occasions, court may revoke
licence
(section 9(2))

Where non-compliance
with conditions requiring
works

Court alter or cancel
conditions if found to
be unduly burdensome
(section 7(1))

Local authority
may undertake works
in default
and recover costs
(section 9(3))

Local authority or person convicted may
apply to magistrates' court for extension
of period allowed before revocation
(section 9(2))

Notes
1. For licensing of camping sites (other than for caravans), see FC 4.6.
2. There is no power for the local authority to refuse the licence for reasons other than these.
3. The licence is not to be issued for a limited period unless this is the case with the planning consent, in which case the two must be brought together (section 4).
4. Maximum penalty for failure to comply with conditions level 4 (section 9(1)).

Extent

This procedure applies in England. Similar provisions apply in Wales and Scotland.

Scope

These provisions apply to the licensing of caravan sites for residential or holiday purposes.

Prohibition of use of land as caravan site without site licence (FC 4.7)

A site licence is required to operate a caravan site (section 1(1)). It is an offence and liable on summary conviction to a fine of level 4 or 5 on the standard scale (section 1(2)).

Some caravan sites do not require a licence – see schedule 1 (section 2).

Issue of site licences by local authorities

An application is made by the occupier (section 3(1)). It must be in writing, specify the land and other information required by the local authority (section 3(2)). It must also be accompanied by a fee fixed by the authority (section 3(2A)). The occupier must be entitled to the benefit of a permission for the use of the land as a caravan site (section 3(3)).The application must be issued either within 6 weeks, 2 months or another agreed date depending on the above benefit (section 3(4) and (5)).

Regulations can be made for prescribed matters for deciding whether to issue a site licence and the procedures to follow in issuing a licence (section 3(5A to 5F)).

A local authority cannot issue a site licence to a person who to their knowledge has held a site licence which has been revoked in the last 3 years (section 3(6)).

Duration of site licences

The duration of a licence depends on the length of permission granted to use the land as a caravan site under part 3 of the Town and Country Planning Act 1947 (section 4(1) and (2)).

Power of local authority to attach conditions to site licences

A site licence may be subject to conditions the authority think necessary or desirable to impose on the occupier of the land in the interests of persons dwelling in the caravans (section 5).

Relevant protected sites: annual fee

A fee is payable for a site licence for a relevant protected site (section 5A(1)).

A local authority must inform the licence holder of the matters to which they have had regard in fixing that fee for the year in question (section 5A(2)).

Where an annual fee has become overdue, the local authority may apply to the tribunal for an order requiring the licence holder to pay the amount due by the date specified in the order (section 5A(3)).

Where a licence holder fails to comply with an order within 3 months, the local authority may apply to the tribunal for an order revoking the site licence (section 5A(4)).

Failure by local authority to issue site licence

Where a local authority fail to issue a site licence within the relevant period it is not an offence to operate the site without a licence (section 6).

Appeal to magistrates' court against conditions attached to site licence

Any person aggrieved by any condition in a site licence may, within 28 days of the date on which the licence was issued, appeal to a magistrates' court or, in a case relating to land in England, to the tribunal. The court or tribunal, if satisfied that the condition is unduly burdensome, may vary or cancel the condition (section 7(1)).

Where the tribunal varies or cancels a condition, it may attach a new condition to the licence (section 7(1A)).

If the effect of a condition requires the carrying out of works on the land, the condition does not have effect during the period of appeal (section 7(2)).

Power of local authority to alter conditions attached to site licences

The local authority may alter conditions but before doing so they must allow the holder of the licence an opportunity of making representations (section 8). A fee is payable if the holder of a site licence applies for alteration of conditions (section 8(1B)).

Where the holder of a site licence is aggrieved by any alteration of the conditions or by the refusal of the local authority of an application by him for the alteration of conditions, he may, within 28 days appeal to a magistrates' court or, in a case relating to land in England, to the tribunal. The court or tribunal may, if they allow the appeal, give to the local authority directions necessary to give effect to their decision (section 8(2)).

The alteration of the conditions does not have effect until written notification has been received by the holder of the licence. If any such alteration imposes a requirement on the holder of the licence to carry out on the land any works which he would not otherwise be required to carry out, the alteration does not have effect during the appeal period within (section 8(3)).

The local authority must consult the fire and rescue authority before exercising these powers on related matters (section 8(4)).

Breach of condition: land other than relevant protected sites in England

Breach of conditions is an offence and liable on summary conviction, to a fine not exceeding level 4 of the standard scale. Further offences may be subject to revocation of the site licence which is subject to appeal. Where works are required the local authority may carry them out themselves and recover the costs (section 9).

Compliance notices (FC 4.8)

Where there is failure to comply with a condition on a site licence for a relevant protected site the local authority may serve a compliance notice on the occupier (section 9A(1)).

A compliance notice sets out the condition and details of the failure to comply with it; requires the occupier of the land to take such steps the local authority consider appropriate specified in the notice in order to ensure that the condition is complied with; specifies the period within which those steps must be taken, and explains the right of appeal (section 9A(2) and (3)).

A local authority may revoke a compliance notice or vary it by extending the period specified (section 9A(4)).

The power to revoke or vary a compliance notice is exercisable by the local authority on an application made by the occupier of land on whom the notice was served, or on the authority's own initiative (section 9A(5)).

Where a local authority revoke or vary a compliance notice, they must notify the occupier of the land of the decision as soon as is reasonably practicable (section 9A(6)).

The offences and powers of the court related to a compliance notice are detailed in section 9B. Expenses can be claimed by the local authority for work in preparing and administering compliance notices (section 9C).

Power to take action following conviction of occupier

Where an occupier of land is convicted of failure to take steps required by a compliance notice, the local authority who issued the compliance notice may take any steps required, but not taken, by the compliance notice to be taken by the occupier, and take such further action the authority consider appropriate for ensuring that the condition is complied with (section 9D(1)). They must serve a notice on the occupier of the land describing the action the authority intend to take (section 9D(2) to (5)).

Power to take emergency action (FC 4.9)

Emergency action may be taken by a local authority on a relevant protected site where the occupier of the land is failing or has failed to comply with a condition, and as a result of that failure there is an imminent risk of serious harm to the health or safety of any person who is or may be on the land.

The authority must serve a notice on the occupier of the land which describes the emergency action the authority intend to take on the land and within 7 days of the authority starting to take the emergency action, the authority must serve on the occupier of the land a notice which:

FC 4.8 Caravan sites licensing – compliance notices

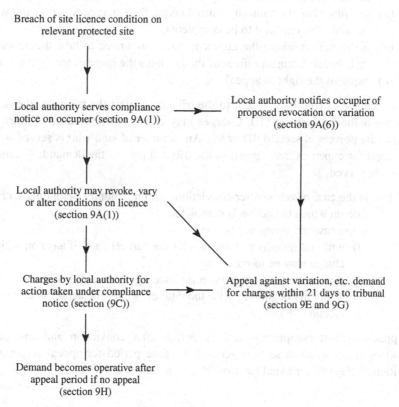

Breach of site licence condition on
relevant protected site

↓

Local authority serves compliance
notice on occupier (section 9A(1)) ⟶ Local authority notifies occupier of
proposed revocation or variation
(section 9A(6))

↓

Local authority may revoke, vary
or alter conditions on licence
(section 9A(1))

↓

Charges by local authority for
action taken under compliance
notice (section (9C)) ⟶ Appeal against variation, etc. demand
for charges within 21 days to tribunal
(section 9E and 9G)

↓

Demand becomes operative after
appeal period if no appeal
(section 9H)

Note
Power to revoke or vary a compliance notice is exercisable by the local authority on an application
made by the occupier on whom the notice was served, or on the authority's own initiative (section
9A(5)).

(a) describes the imminent risk of serious harm to the health or safety of persons who are or may be on the land;

(b) describes the emergency action which has been, and any emergency action which is to be, taken by the authority on the land;

(c) sets out when the authority started taking the emergency action and when the authority expect it to be completed;

(d) if the person whom the authority have authorised to take the action on their behalf is not an officer of theirs, states the name of that person; and

(e) explains the right of appeal.

An occupier of land may appeal to the tribunal against the taking of the action by the authority (section 9E). Charges may be made by the local authority in using the powers in section 9D or 9E. An occupier of land who is served with a demand for expenses may appeal to the tribunal against the demand. A demand must be served:

(a) in the case of action after conviction, within 2 months beginning with the date on which the action is completed;

(b) in the case of emergency action:
   (i) within 2 months beginning with the earliest date (if any) on which a charge may be imposed, or
   (ii) if the action has not been completed by the end of that period, within 2 months beginning with the date on which the action is completed (section 9F).

Appeals against compliance notices, action after conviction and emergency notices are detailed in section 9G and the time period for appeal is generally within 21 days to a tribunal (section 9G).

When compliance notice or expenses demand becomes operative

Where no appeal against a compliance notice is brought within the appeal period, the notice and any demand which was served with it become operative at the end of that period.

   Where no appeal action after a conviction is brought within the appeal period, the demand becomes operative at the end of that period.

   Where an appeal against a compliance notice is brought, and a decision on the appeal confirms the compliance notice, the notice and any demand which was served with it become operative:

(a) where the period within which an appeal to the Upper Tribunal may be brought expires without such an appeal having been brought, at the end of that period;

(b) where an appeal to the Upper Tribunal is brought and a decision on the appeal is given which confirms the notice, at the time of the decision.

Where an appeal against an emergency notice is brought, and a decision on the appeal confirms the demand, the demand becomes operative:

FC 4.9 Caravan sites licensing – emergency notices

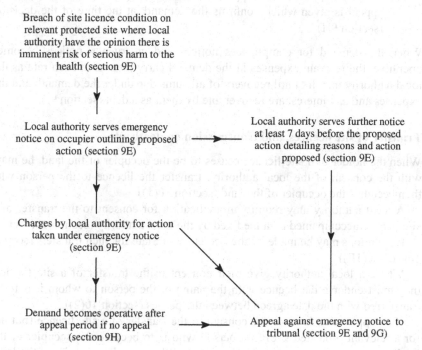

Breach of site licence condition on relevant protected site where local authority have the opinion there is imminent risk of serious harm to the health (section 9E)

↓

Local authority serves emergency notice on occupier outlining proposed action (section 9E) → Local authority serves further notice at least 7 days before the proposed action detailing reasons and action proposed (section 9E)

↓

Charges by local authority for action taken under emergency notice (section 9E)

↓

Demand becomes operative after appeal period if no appeal (section 9H) → Appeal against emergency notice to tribunal (section 9E and 9G)

Note

Power to revoke or vary a compliance notice is exercisable by the local authority on an application made by the occupier on whom the notice was served, or on the authority's own initiative (section 9A(5)).

(a) where the period within which an appeal to the Upper Tribunal may be brought expires without such an appeal having been brought, at the end of that period;

(b) where an appeal to the Upper Tribunal is brought and a decision on the appeal is given which confirms the demand, at the time of the decision (section 9H).

When the demand for compliances notices and emergency notices becomes operative, the relevant expenses in the demand carry interest at such rate as the local authority may fix until recovery of all sums due under the demand; and the expenses and any interest are recoverable by them as a debt (section 9I).

Transfer of site licences, and transmission on death, etc.

When the holder of a site licence ceases to be the occupier of the land, he may, with the consent of the local authority, transfer the licence to the person who then becomes the occupier of the land (section 10(1)).

A local authority may require an application for consent to the transfer of a site licence accompanied by a fee fixed by the local authority (section 10(1A)).

Regulations may be made by the Secretary of State for such transfers (section 10(1B) to (1F)).

Where a local authority give their consent to the transfer of a site licence, they must endorse the licence with the name of the person to whom it is to be transferred with the date agreed between the parties (section 10(2)).

If an application is made for consent to the transfer of a site licence (but not for a relevant protected site) to a person who is to become the occupier of the land, that person may apply for a site licence if he were the occupier of the land, and if the local authority at any time before issuing a site licence in compliance with that application give their consent to the transfer they need not proceed with the application for the site licence (section 10(3)).

Where any person becomes entitled to an estate or interest in land in respect of which a site licence is in force and becomes the occupier of the land he is treated as having become the holder of the licence on the day on which he became the occupier of the land, and the local authority must, if an application is made to them, endorse his name and date on the licence (section 10(4)).

Powers to charge fees

Before charging the fee, the local authority must prepare and publish a fees policy.

When fixing a fee the local authority:

(a) must act in accordance with their fees policy;

(b) may fix different fees for different cases or descriptions of case;

(c) may determine that no fee is required to be paid in certain cases or descriptions of case.

The local authority may revise their fees policy and, where they do so, must publish the revised policy (section 10A).

Duty of licence holder to surrender licence for alteration

A local authority which have issued a site licence may require the holder to deliver it up to enable them to enter in it any alteration of the conditions or other terms of the licence. If the holder of a site licence fails without reasonable excuse to comply with a requirement he is liable on summary conviction to a fine not exceeding level 1 on the standard scale (section 11).

Power of local authorities to provide sites for caravans

A local authority has power to provide caravan sites for use as permanent residences, and to manage the sites or lease them to some other person. Before exercising this power they must consult the fire and rescue authority about the measures to be taken for preventing and detecting the outbreak of fire on the site and the provision and maintenance of means of fighting fire on it (section 24).

Registers of site licences

Every local authority must keep a register of caravan site licences and open for inspection by the public at all reasonable times. Endorsed licences must be recorded with the name, and the date entered in the licence (section 25).

Power of entry

Any authorised officer of a local authority, on producing, if required, duly authenticated document showing his authority, has a right at all reasonable hours to enter any land which is used as a caravan site or in respect of which an application for a site licence has been made for determining conditions to be attached to a site licence or whether conditions should be altered; for ascertaining whether there is, or has been, on or in connection with the land any contravention or for whether circumstances exist which would authorise the local authority to take any action, or execute any work, but 24 hours' notice of the intended entry must be given to the occupier.

A warrant for enter may be obtained from a justice of the peace where admission to any land has been refused, or that refusal is apprehended, or that the occupier of the land is temporarily absent and the case is one of urgency, or that an application for admission would defeat the object of the entry and that there is reasonable ground for entering on the land.

A person who wilfully obstructs an authorised person is liable on summary conviction where the wilful obstruction occurs in relation to land, to a fine not exceeding level 4 on the standard scale (section 26).

Definitions

caravan means any structure designed or adapted for human habitation which is capable of being moved from one place to another (whether by being towed, or by being transported on a motor vehicle or trailer) and any motor vehicle so designed or adapted, but does not include any railway rolling stock which is for the time being on rails forming part of a railway system, or any tent (section 29(1)).

caravan site means land on which a caravan is stationed for the purposes of human habitation and land which is used in conjunction with land on which a caravan is so stationed (section 1(4)).

tribunal means the First-tier Tribunal or where determined by or under Tribunal Procedure Rules, the Upper Tribunal (section 29(1)).

relevant protected site means land in respect of which a site licence is required under this Part, other than land in respect of which the relevant planning permission under Part 3 of the Town and Country Planning Act 1990 or the site licence is, subject to subsection (6):

    (a) expressed to be granted for holiday use only, or

    (b) otherwise so expressed or subject to such conditions that there are times of the year when no caravan may be stationed on the land for human habitation (section 5A(5)).

Chapter 5

ENVIRONMENT

ENVIRONMENTAL PERMITTING

Introduction

The Environmental Permitting system implements the requirements of the Integrated Pollution Prevention and Control (IPPC) Directive which applies to the more complex and potentially polluting industries and deals with any likely significant pollutant emission to air, water or land. It also addresses noise, energy efficiency, waste minimisation, prevention of accidental emissions and site protection and restoration.

The Environmental Permitting system is designed to allow regulators to focus on medium- and high-risk operations whilst continuing to protect the environment and human health.

The Environmental Permitting (England and Wales) Regulations 2010 provide industry, regulators and others with a permitting and compliance system including systems for discharge, consenting, groundwater authorisations and radioactive substances regulation.

Larger industrial installations are subject to integrated controls on all likely significant emissions. Most are regulated by the Environment Agency and are often referred to as 'Part A(1)' activities. These are regulated for emissions to air, land, water and other environmental considerations.

Local authorities regulate premises with 'Part A(2)' (comparatively less polluting) installations for air, land, water and other environmental considerations and 'Part B' activities (smaller installations) which are generally limited to air pollution control.

The Part B regime is known as Local Air Pollution Prevention and Control (LAPPC), and the A2 regime as Local Authority Industrial Pollution Prevention and Control (LA-IPPC). The port health authority may be the regulator in port areas (regulation 32).

Part A(1) and A(2) installations

The IPPC system as enforced by both the Environment Agency and local authorities for the installations allocated to them applies an integrated approach to ensure a high level of protection to the environment as a whole and to prevent and, where that is not practicable, reduce emissions to acceptable levels. The conditions set in granting permits are to be based on the use of 'Best Available Techniques' (BAT) – see below. The Regulations also include provisions relating to energy efficiency, site restoration, noise, odour, waste minimisation, accident prevention and heat and vibrations.

Part B installations

Regulated by local authorities under LAPPC, these are also subject to conditions based on BAT but only so far as these are related to emissions to air.

Definitions

activities 'are industrial activities forming part of an installation' and are defined as 'activities of any nature, whether:

  (a) industrial or commercial or other activities, or
  (b) carried on on particular premises or otherwise,

and includes (with or without other activities) the depositing, keeping or disposal of any substance' (Pollution Prevention and Control Act 1999 section 1(2)).

emission limit value means the mass, expressed in terms of specific parameters, concentration or level of an emission, which may not be exceeded during one or more periods of time.

environmental pollution means pollution of the air, water or land which may give rise to any harm; and for the purposes of this definition . . .:

  (a) pollution includes pollution caused by noise, heat or vibrations or any other kind of release of energy, and
  (b) air includes air within buildings and air within other natural or man-made structures above or below ground.

In this definition harm means:

  (a) harm to the health of human beings or other living organisms;
  (b) harm to the quality of the environment, including:
      (i) harm to the quality of the environment taken as a whole;
      (ii) harm to the quality of the air, water or land; and
      (iii) other impairment of, or interference with, the ecological systems of which any living organisms form part;
  (c) offence to the senses of human beings;
  (d) damage to property; or
  (e) impairment of, or interference with, amenities or other legitimate uses of the environment (expressions used in this paragraph having the same meaning as in Council Directive 96/61/EC).

(Pollution Prevention and Control Act 1999 section 1(2) and (3)).

emission means:

  (a) in relation to a Part A installation, the direct or indirect release of substances, vibrations, heat or noise from individual or diffuse sources in the installation into the air, water or land;

(b) in relation to a Part B installation, the direct release of substances or heat from individual or diffuse sources in the installation into the air;

(c) in relation to Part A mobile plant, the direct or indirect release of substances, vibrations, heat or noise from the mobile plant into the air, water or land;

(d) in relation to Part B mobile plant, the direct release of substances or heat from the mobile plant into the air;

(e) in relation to a waste operation, the direct or indirect release of substances, vibrations, heat or noise from individual or diffuse sources related to the operation into the air, water or land;

(f) in relation to a mining waste operation, the direct or indirect release of substances, vibrations, heat or noise from individual or diffuse sources related to the operation into the air, water or land; and

(g) in relation to a radioactive substances activity, the direct or indirect release of radioactive material or radioactive waste (Regulation 2(1)).

installation means (except where used in the definition of 'excluded plant'):

(a) a stationary technical unit where one or more activities are carried on, and

(b) any other location on the same site where any other directly associated activities are carried on, and references to an installation include references to part of an installation (paragraph 1 of Part 1 of Schedule 1).

mobile plant means any of the following:

(a) Part A mobile plant
(b) Part B mobile plant
(c) waste mobile plant (Regulation 2).

operator, in relation to a regulated facility, means:

(a) the person who has control over the operation of the regulated facility;

(b) if the regulated facility has not yet been put into operation, the person who will have control over the regulated facility when it is put into operation; or

(c) if a regulated facility authorised by an environmental permit ceases to be in operation, the person who holds the environmental permit (Regulation 7).

pollution, in relation to a water discharge activity or groundwater activity, means the direct or indirect introduction, as a result of human activity, of substances or heat into the air, water or land which may:

(a) be harmful to human health or the quality of aquatic ecosystems or terrestrial ecosystems directly depending on aquatic ecosystems;

(b) result in damage to material property; or

(c) impair or interfere with amenities or other legitimate uses of the environment (Regulation 2(1)).

pollution, other than in relation to a water discharge activity or groundwater activity, means any emission as a result of human activity which may:

(a) be harmful to human health or the quality of the environment;

(b) cause offence to a human sense;

(c) result in damage to material property; or

(d) impair or interfere with amenities or other legitimate uses of the environment (Regulation 2(1)).

pollutant means any substance liable to cause pollution (Regulation 2(1)).

regulated facility means any of the following:

(a) an installation

(b) mobile plant

(c) a waste operation

(d) a mining waste operation

(e) a radioactive substances activity

(f) a water discharge activity

(g) a groundwater activity (Regulation 8).

substantial change means a change in operation of an installation which in the regulator's opinion may have significant negative effects on human beings or the environment and includes:

(a) in relation to a Part A installation, a change in operation which in itself meets the thresholds, if any, set out in Part 2 of Schedule 1, and

(b) in relation to an incineration plant or co-incineration plant for non-hazardous waste, a change in operation which would involve the incineration or co-incineration of hazardous waste (paragraph 5 schedule 5).

change in operation – a change in the nature or functioning, or an extension, of the installation which may have consequences for the environment. A change in operation could entail either technical alterations or modifications in operational or management practices, including changes to raw materials or fuels used and to the installation throughput (schedule 5 paragraph 5).

Best available technique (BAT)

Operators of Part A(1) installations are required to use BAT to achieve a high level of protection of the environment taken as a whole and, for Part B, in relation to emissions into the air. BAT is defined in the IPPC directive as

> the most effective and advanced stage in the development of activities and their methods of operation which indicates the practical suitability of particular techniques for providing in principle the basis for emission limit values designed to prevent and, where that is not practicable, generally to reduce emissions and the impact on the environment as a whole.

In essence BAT is a technique that balances the costs to the operator with the benefits to the environment, with a consideration of local circumstances and provides the main basis for the setting of emission limit values – see definition on page 408.

Guidance

1. *Integrated Pollution Prevention and Control – a practical guide*, fourth edn. 2005 – DEFRA.
   This a general guide to the whole of the IPPC system and covers all aspects. It is primarily intended for Environment Agency regulators in relation to A(1) installations and persons affected by it.
2. Environmental permitting guidance relevant to the Local Authority Pollution Control regime, in particular the *LAPC General Guidance Manual* and *Controlling Pollution from Industry: Regulation by Local Authorities – a short guide*.
   These manuals are the principal guidance to local authorities in exercising their functions under the Environmental Permitting Regulations, i.e. LA-IPPC and LAPPC and is applicable to all procedures in this chapter.
3. Process Guidance Notes (PGNs) and Sector Guidance Notes (SGNs).
   These are issued by the Secretary of State under Regulation 64 and form statutory guidance on what constitutes Best Available Technique for LA-IPPC installations for each of the main sectors regulated.
4. AQ Notes.
   These are issued by DEFRA and form additional guidance to LAs on a wide range of both technical and administrative issues.
5. Explanatory Memorandum to Environmental Permitting (England and Wales) Regulations 2010.
6. Environmental Permitting Guidance: The IPPC Directive Part A(1) Installations and Part A(1) Mobile Plant For the Environmental Permitting (England and Wales) Regulations 2010.
7. Environmental Permitting Guidance: Core Guidance for the Environmental Permitting (England and Wales) Regulations 2010.
8. Environmental Permitting General Guidance Manual on Policy and Procedures for A2 and B Installations Local Authority Integrated Pollution Prevention and Control (LA-IPPC) and Local Authority Pollution Prevention and Control (LAPPC).
9. General Guidance Manual on Policy and Procedures for A2 and B Installations Part B of Manual Annexes and Parts C and D.

All of the publications above are available on the following website at https://www.gov.uk/government/publications/environmental-permitting-guidance-core-guidance–2.

Instruments

An instrument means a notice, notification, certificate, direction or form under these Regulations. An instrument must be in writing.

Instruments may be served on or given to a person by personal delivery, leaving it at the person's proper address or sending it by post or electronic means to the person's proper address.

In the case of a body corporate, an instrument may be served on or given to the secretary, clerk or director of that body. In the case of a partnership, an instrument may be served on or given to a partner or a person having control or management of the partnership business.

The proper address is the last known address except:

(a) in the case of a body corporate or their secretary or clerk the registered or principal office of that body, or the e-mail address of the secretary or clerk;

(b) in the case of a partnership or a partner or person having control or management of the partnership business the principal office of the partnership, or the e-mail address of a partner or a person having that control or management (regulation 10).

Application to the Crown

Whilst the Crown is bound by the Environmental Permitting Regulations it is not criminally responsible and cannot be prosecuted for non-compliance with notices, etc. However, local authorities may apply to the High Court to have the Crown's actions, or lack of them, declared unlawful (regulation 11 and schedule 4).

Obtaining of information

By notice served on any person, local authorities may require to be given such information as is specified in the notice and in such form and time/period as is specified (regulation 60(1)).

PERMITTING OF ACTIVITIES BY LOCAL AUTHORITIES

References

Environment Act 1995.
Pollution Prevention and Control Act 1999.
Environmental Permitting Regulations (England and Wales) Regulations 2010 as amended.
Environmental Permitting Guidance Core Guidance for the Environmental Permitting (England and Wales) Regulations 2010. Last revised: March 2013.
Environmental Permitting: General Guidance Manual on Policy and Procedures for A2 and B Installations (Revised April 2012).

Extent

These provisions apply to England and Wales.

Scope

These procedures apply to activities, installations and mobile plant, prescribed in parts A(2) and B of schedule 1 to the Environmental Permitting Regulations 2010. These activities, etc. are controlled and permitted by the local authority. In each case the operator (definitions, page 408) is required to obtain a permit from the local authority before operating the installation or plant. All installations, etc. existing before March 2007 should now have permits so that these procedures relate to the supervision of those plants and to the permitting of new ones.

Applications (FC 5.1)

Applications are to be made in writing to the local authority, with the prescribed fee (see page 404), and must contain all of the information specified in regulation 14 and paragraph 2 of part 1 of schedule 5. Application forms are in part C of the Secretary of State Guidance Manual.

For novel or complex installations staged applications are possible by agreement between the operator and the local authority.

Local authorities are empowered to request by notice additional information sufficient for them to determine the application. If the information requested has not been provided within the time specified in the notice the application is deemed to have been withdrawn (schedule 5 part 1 paragraph 4).

Consultations

Subject to the exceptions in schedule 4, the local authority must send copies of the application within the communication consultation period to the public consultees (a person whom the regulator considers is affected by, is likely to be affected by, or has an interest in, an application). The consultation period is 30 working days starting on the day the regulator receives an application; unless national security or confidentiality is an issue the 30 days starts on the determination date.

The local authority must have regard to make representations from the public consultees before determining the application (schedule 5 paragraph 11).

The Environment Agency should always be a public consultee as should the local authority when the Environment Agency is dealing with applications for part A(1). There is a specimen consultation letter in the part C guidance manual.

Determination of applications

The general principles against which all applications, parts A and B, are to be considered are:

(a) that all appropriate preventative measures must be taken against pollution and in particular the application of BAT (see page 410);
(b) that no significant pollution will be caused.

The additional general principles to be considered for part A activities are:

(a) that satisfactory methods will be in place to avoid waste production and, where it is produced, for its recovery and disposal;
(b) there is efficient use of energy;

FC 5.1 Permitting of scheduled activities by local authorities – Regulation 14 and Schedule 5 Environmental Permitting Regulations 2010

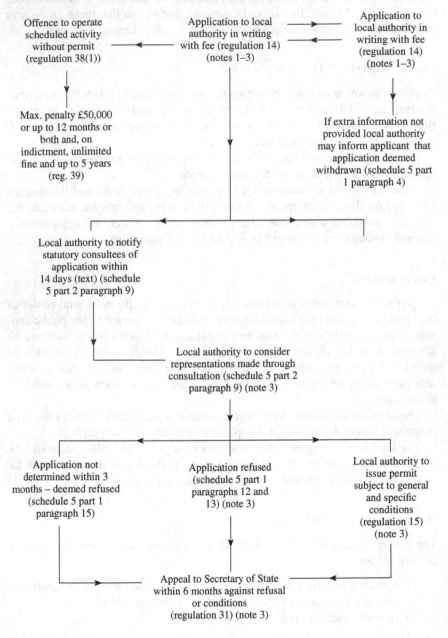

Notes

1. For details to be included see regulation 14 and part 1 schedule 5.
2. For consolidation of a number of regulated facilities with a single permit (regulation 18).
3. Details of these events to be included in public register (regulation 46 and paragraph 1 schedule 24).
4. Request for Information notice form in part D of General Manual.

(c) that measures are taken to prevent accidents and to limit their consequences;

(d) that, on the cessation of operation of the installation, satisfactory measures will be taken to avoid pollution and return the site to a satisfactory state.

The local authority must determine applications generally within 3 months of submission unless a longer period is agreed. In the case of transfer it is 2 months and where national security and confidentiality is an issue it is 4 months. Failure to do so is deemed to be a refusal (schedule 5 part 1 paragraph 15).

Content and form of the permit

An environmental permit must specify the operator and the regulated facility whose operation it authorises. It can be in electronic form and must include a map, plan or other description of the site showing the geographical extent of the site of the regulated facility, but the map, etc. does not apply to mobile plant or radioactive substances activities (regulation 14).

Conditions may be attached to the permit requiring the operator to carry out works or do other things on land which the operator is not entitled to do without obtaining the consent of another person (regulation 15).

A regulator cannot authorise the operation of more than one regulated facility under a single environmental permit but can authorise a single environmental permit for the operation by the same operator more than one mobile plant, standard facility, regulated facility on the same site or more than one radioactive substances activity (regulation 17).

If the local authority decides to issue a permit it may attach conditions (paragraph 12 part 1 schedule 5) appropriate to the following issues and in accordance with sectoral technical guidance. For all activities these are:

(a) Emission limit values (EMLs) or equivalent parameters for pollutants, in and likely to be emitted in significant quantities. These will normally be based on BAT (see page 410) taking account of the particular characteristics and the local environment of the plant.

(b) Best Available Technique for preventing or reducing emissions.

It is suggested that content and conditions should cover:

(a) long distance and transboundary pollution;

(b) the protection of soil and groundwater and the management of waste;

(c) precautions to protect the environment when the installation is not operating normally, e.g. during start up;

(d) site monitoring and remediation;

(e) the ongoing monitoring of emissions and the submission of reports to the local authority;

(f) notification procedures to deal with incidents or accidents;

(g) must take account of conditions for emissions to water specified by the Environment Agency; and

(h) must avoid conflict with other legislation prescribing release levels.

Standard rules

The Secretary of State is empowered to make Standard Rules (generally known as process and sector guidance notes) for certain types of installation which can be used by a local authority instead of site-specific conditions. Standard Rules will, by their nature, be suitable for industry sectors where installations share similar characteristics (regulations 26 to 30).

Refusals

A permit must be granted or refused following an application. The permit cannot be granted by the local authority unless it considers that the applicant will be the person who will have control after the permit is issued and will ensure that the installation or mobile plant is operated so as to comply with the environmental permit (schedule 5 part 1 paragraph 12 and 13). This may be, for example, where there is reason to believe that the operator lacks the management systems or competence to run the installation according to the application or any permit conditions.

There is an appeal procedure against refusal – see below.

Transfers (FC 5.2)

Transfers of permits between operators must be the subject of a joint application to the local authority who may agree to the transfer unless it considers that the conditions attached will not be complied with (regulation 21). There is an appeal against refusal (regulation 31) (see below).

Permit reviews

The local authority must periodically review the environmental permits and make periodic inspections of regulated facilities (regulation 34).

It is suggested that the reviews be carried out where:

(a) the installation causes such significant pollution that the local authority must change the Emission Limit Values;

(b) substantial changes in Best Available Technique make it possible to reduce emissions significantly without excessive costs; and

(c) operators must change techniques for reasons of safety.

Variation of conditions (FC 5.3)

Operators of permitted installations and mobile plant must make an application for variation to the local authority of their intention to make a 'substantial change' and the rules for consultation and public participation apply (form in part C of the general guidance manual) (schedule 5).

The local authority may vary the conditions at any time on its own initiative. The prescribed fee must accompany applications (see 'Charges' below). The

FC 5.2 Transfer of permits – Regulation 21 and Schedule 5 Environmental Permitting Regulations 2010

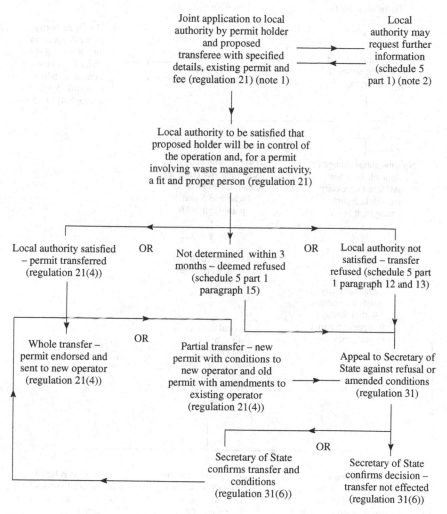

Notes
1. Application form in part C of General Manual.
2. Specimen Request for Information Notice in part D of the General Manual.
3. Information to be included in public register (regulation 46 and schedule 24 paragraph 1).

FC 5.3 51 Variations to permit conditions Regulation 20 and Schedule 5 Environmental Permitting Regulations 2010

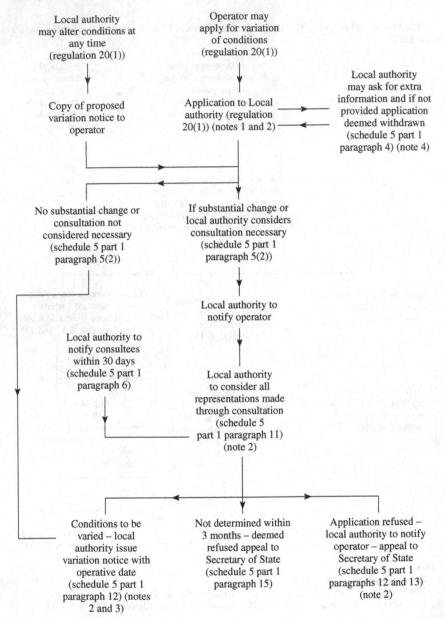

Notes
1. The details to be included in the application are set out in regulation 14 and paragraph 2 of part 1 of schedule 5.
2. Information to be included in public register (regulation 46 and schedule 24 paragraph 1).
3. Model form in part C of General Manual.
4. Specimen Request for Information Notice in part D of General Manual.

application or local authority led variation is subject to the same procedures as an initial application including a period of public participation and consultation before determination.

There is an appeal against the change of conditions and against a refusal to grant an application for change (see below) (regulation 20).

Surrender of permits (FC 5.4 and 5.5 notification and applications for part A and part B)

An environmental permit may be surrendered by the operator of an activity where he intends to cease operation. The operator must notify the local authority about a Part B installation, except a waste operation, mobile plant or stand-alone water discharge activity or stand-alone groundwater activity. In all other cases an application must be made to the local authority. The local authority may accept the surrender of the permit if it is satisfied that the steps to be taken to implement the closure are appropriate to avoid risk of pollution and will return the site to a satisfactory state.

In the case of partial surrenders the regulator must serve a notice on the operator specifying that it is necessary to vary the conditions, the variation, and the date the variation takes effect.

Applications must be determined in the same way as initial applications, variations and transfers and failure to do so is deemed to be a refusal. There is an appeal against refusal to accept surrender (regulations 24 and 25).

Enforcement notices (FC 5.6)

If the local authority is of the opinion that an environmental permit has been is or is likely to be contravened they may serve an enforcement notice on the operator. The notice must specify:

(a) the contravention
(b) the steps necessary to remedy it
(c) the time period allowed.

Non-compliance is an offence – see below (regulation 36).

Suspension notices (FC 5.6)

Where the local authority believes that the continued operation of the activities will involve a risk of serious pollution it may serve a suspension notice on the operator. This applies whether or not the activities are permitted and may deal with a failure to comply with conditions of a permit. Alternatively the situation may be dealt with by use of its powers under regulation 57 – see below.

The suspension notice must:

(a) specify what the risk of serious pollution is;
(b) specify the steps to be taken;
(c) specify the period within which the steps must be taken;

FC 5.4 Application for surrender of an environmental permit – Regulation 25 Environmental Permitting Regulations 2010

Operator ceases or intends to cease operating whole or part of an installation other than part B installation, mobile plant, stand alone water discharge or groundwater activity (regulation 25(1))

Operator makes application to local authority with specified details as if an initial application is being made (text) and a date of surrender not less than 20 days from notification (regulation 24(3)) (note 1)

Permit ceases to be effective on date specified in notification (reg. 24(4))

For partial surrenders local authority may vary conditions for remaining permit by serving a variation notice (see FC 5.3) (regulation 24(6))

Notes
1. Application form in part C of General Manual.
2. Schedule 5 part 1 of the Environmental Permitting Regulations 2010 apply to surrender of an environmental permit.
3. The local authority can:
   – request more information (schedule 5 part 1 paragraph 4).
   – refuse an application (schedule 5 part 1 paragraph 12).
4. If application refuse or deemed to be refused the operator can appeal to the secretary of state (regulation 31).

FC 5.5 Notification of surrender of an environmental permit – Regulation 24 Environmental Permitting Regulations 2010

Operator ceases or intends to cease operating whole or
part of a part B installation, mobile plant, stand alone
water discharge or groundwater activity
(regulation 24(1))

↓

Operator to notify local authority with specified details as if an
initial application is being made (text)
and a date of surrender not less than 20 days from notification
(regulation 24(3)) (note 1)

↓

Permit ceases to be effective on date specified in
notification (reg. 24(4))

↓

For partial surrenders local authority may vary
conditions for remaining permit
by serving a variation notice (see FC 5.3)
(regulation 24(6))

Notes
1. Application form in part C of General Manual.
2. Schedule 5 part 1 of the Environmental Permitting Regulations 2010 apply to surrender of an environmental permit.

FC 5.6 Enforcement and suspension notices – Regulations 36 and 37 Environmental Permitting Regulations 2010

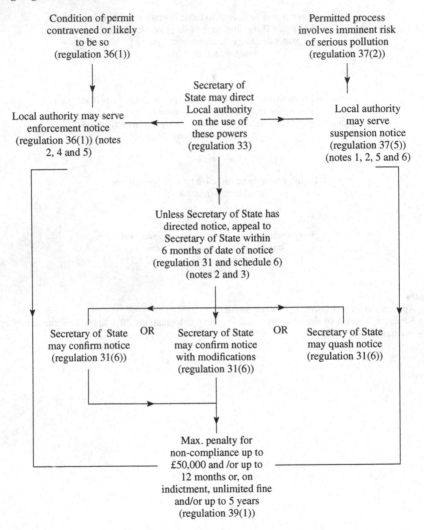

Notes
1. Where these circumstances exist the local authority may serve a suspension notice unless taking action itself under regulation 37.
2. Details to be entered in public register (regulation 46 and schedule 24 paragraph 1).
3. The notices are not suspended during the appeal process (regulation 31(9)).
4. Instead of serving an enforcement notice the local authority may prosecute for the contravention (regulation 38).
5. An enforcement or suspension notice may be withdrawn at any time (regulations 36(4) and 37(8)).
6. Specimen Suspension Notice in part D of the General Manual.

(d) state that any permit shall cease to have effect to the extent specified in the notice until the notice is withdrawn;

(e) where the activities will continue in part, state any additional measures to (b) above that must be taken.

The local authority may withdraw the notice at any time and must do so when the risk has been removed (regulation 37).

Prevention or remedying of pollution (FC 5.7)

Where the local authority believes that the operation of an activity where an environmental permit is in place involves a risk of serious pollution or there has been the commission of an offence by the operator that causes pollution, it may arrange itself for steps to be taken to remove that risk or pollution. Before taking any steps the local authority must give the operator 5 days' notice.

The costs of the necessary works may be recovered from the operator unless the operator can show that there was no risk or the costs, or part of them, were incurred unnecessarily (regulation 57).

Revocation notices (FC 5.8)

The local authority may revoke a permit at any time, in whole or in part, by the service of a revocation notice. This is a wide power that can be used whenever the local authority considers it to be appropriate.

The notice must be served on the operator and specify:

(a) the reasons for the revocation;

(b) in the case of a partial revocation:

   (i) the extent to which the permit is being revoked, and

   (ii) any variation to the conditions of the permit; and

(c) the date on which the revocation will take place, at least 20 working days after the notice is served (regulations 22 and 23).

Unless the local authority withdraws a revocation notice, an environmental permit ceases to have effect on the date specified in the notice:

(a) in the case of a revocation in whole, entirely; or

(b) in the case of a partial revocation, to the extent of the part revoked.

Public registers

Local authorities must maintain public registers containing information on all Part A(2) and Part B permits issued by them for installations in their areas and also details of installations in their area regulated by the Environment Agency (regulation 46). The content of the register is prescribed by schedule 24, paragraph 1. The register may be in any form and the necessary content of registers is set out in Annex 13 of the general guidance manual.

FC 5.7 Local authority powers to prevent or remedy pollution – Regulation 57 Environmental Permitting Regulations 2010

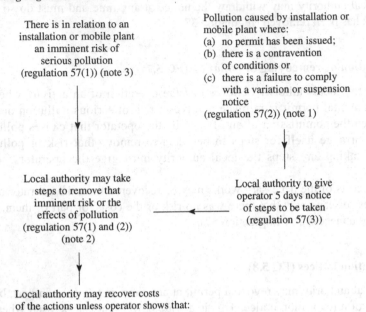

There is in relation to an installation or mobile plant an imminent risk of serious pollution (regulation 57(1)) (note 3)

Pollution caused by installation or mobile plant where:
(a) no permit has been issued;
(b) there is a contravention of conditions or
(c) there is a failure to comply with a variation or suspension notice
(regulation 57(2)) (note 1)

Local authority may take steps to remove that imminent risk or the effects of pollution (regulation 57(1) and (2)) (note 2)

Local authority to give operator 5 days notice of steps to be taken (regulation 57(3))

Local authority may recover costs of the actions unless operator shows that:
(a) there was no risk of serious pollution or
(b) costs were unnecessarily incurred
(regulation 57(4) and (5))

Notes
1. This power may be implemented in addition to the prosecution of committed offences.
2. There is no provision for appeal against these actions, only a challenge to the recovery of costs.
3. Where these circumstances exist the local authority may implement this procedure or serve a suspension notice – see FC 5.6.

FC 5.8 Revocation of permits by local authority – Regulations 22 and 23 Environmental Permitting Regulations 2010

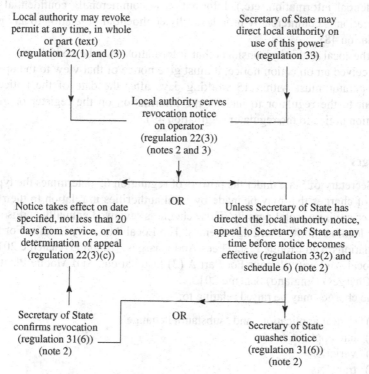

Notes
1. The revocation is suspended pending determination of the appeal or withdrawal of the notice (regulation 31(9)(b)).
2. Details to be entered in public register (regulation 46 and schedule 24 paragraph 1).
3. Specimen Revocation Notice in part D of the General Manual.

Commercial confidentiality

The local authority must exclude what it considers commercially confidential information or it receives a notice from the operator that the information is confidential (FC 5.9 and FC 5.10, application for exclusion of info and handling confidential information, etc.). Information is commercially confidential if the information is commercially or industrially confidential in relation to any person (regulation 48).

If the local authority considers that information may be confidential but has not received an objection notice, it must give notice of that view to the operator. The operator must within 15 working days after the date of the notice give consent to the regulator to include the information on the register or give an objection notice to the regulator (regulation 49).

Charges

The Secretary of State under the powers of regulation 65 determines the type and level of charges that may be made by local authorities in relation to their activities for environmental permits. This scheme is reviewed annually and is set out in the local authority charging schemes: The Local Authority Permits For Part B Installations And Mobile Plant (Fees And Charges) (England) Scheme 2012 and the Local Authority Permits For Part A (2) Installations And Mobile Plant (Fees And Charges) (England) Scheme 2012.

The charges may be raised relating to:

(a) permit applications and 'substantial change'
(b) late applications
(c) variations
(d) transfers
(e) surrender
(f) subsistence charges on an annual basis.

The fees are based on a risk-based approach in line with the 'polluter pays' principle. All processes, except waste oil burners and petrol stations, are required to be risk-rated to establish the inspection frequency and hours that the local authority needs to spend on them.

Appeals

Unusually for environmental health enforcement work, rights of appeal are to the Secretary of State and not to a court of law. The operator may appeal against local authority decisions relating to:

(a) refusal of an application for a permit (regulation 13);
(b) refusal of an application for the variation of a permit (regulation 20);
(c) the service of a revocation, enforcement or suspension notice (regulations 22, 36, 37);
(d) determination that information is not commercially confidential (regulation 53);

FC 5.9 57 Applications to LA to exclude commercially confidential information from public register

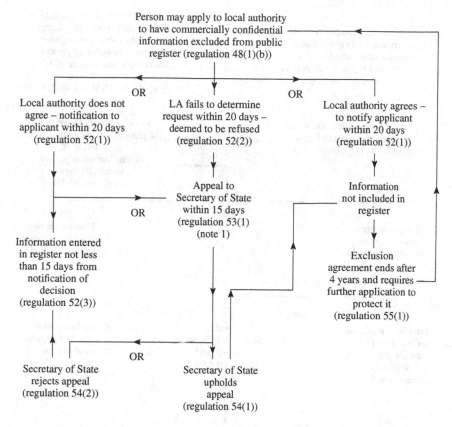

Person may apply to local authority to have commercially confidential information excluded from public register (regulation 48(1)(b))

OR — OR

Local authority does not agree – notification to applicant within 20 days (regulation 52(1))

LA fails to determine request within 20 days – deemed to be refused (regulation 52(2))

Local authority agrees – to notify applicant within 20 days (regulation 52(1))

Appeal to Secretary of State within 15 days (regulation 53(1)) (note 1)

OR

Information not included in register

Information entered in register not less than 15 days from notification of decision (regulation 52(3))

Exclusion agreement ends after 4 years and requires further application to protect it (regulation 55(1))

OR

Secretary of State rejects appeal (regulation 54(2))

Secretary of State upholds appeal (regulation 54(1))

FC 5.10 Handling of commercially confidential and national security information by local authorities – Regulations 47 to 55 Environmental Permitting Regulations 2010

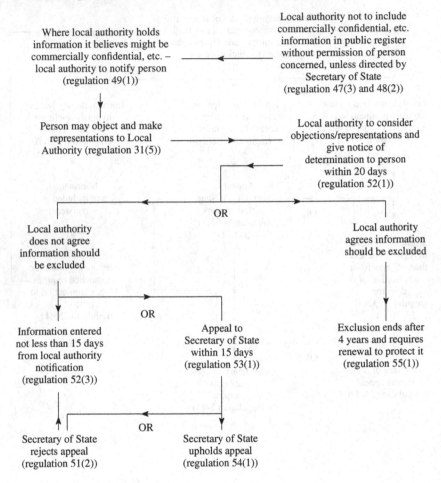

Where local authority holds information it believes might be commercially confidential, etc. – local authority to notify person (regulation 49(1))

Local authority not to include commercially confidential, etc. information in public register without permission of person concerned, unless directed by Secretary of State (regulation 47(3) and 48(2))

Person may object and make representations to Local Authority (regulation 31(5))

Local authority to consider objections/representations and give notice of determination to person within 20 days (regulation 52(1))

OR

Local authority does not agree information should be excluded

Local authority agrees information should be excluded

OR

Information entered not less than 15 days from local authority notification (regulation 52(3))

Appeal to Secretary of State within 15 days (regulation 53(1))

Exclusion ends after 4 years and requires renewal to protect it (regulation 55(1))

OR

Secretary of State rejects appeal (regulation 51(2))

Secretary of State upholds appeal (regulation 54(1))

(e) refusal of an application to transfer or surrender a permit (regulations 21, 24 and 25);

(f) the service of a variation notice on the local authority's initiative (regulation 20).

The full procedure with timescales is set out in schedule 6 to the Environmental Permitting Regulations. The Secretary of State has the power to affirm or quash the local authority decisions and to alter the terms of any conditions (regulation 31(6) and schedule 6).

Offences

Offences committed against any of the requirements of the Environmental Permitting Regulations are punishable, on summary conviction, of a fine up to £50,000 and/or up to 12 months' imprisonment. Conviction in the Crown Court may lead to a fine and/or to imprisonment for up to 5 years (regulations 38 and 39).

LICENSING OF SCRAP METAL DEALERS

References

Scrap Metal Dealers Act 2013.

Scrap Metal Dealers Act 2013 (Prescribed Documents and Information for Verification of Name and Address) Regulations 2013.

Scrap Metal Dealers Act 2013 (Prescribed Relevant Offences and Relevant Enforcement Action) Regulations 2013.

Determining suitability to hold a scrap metal dealer's licence Home Office October 2013.

Scrap Metal Dealer Act 2013: licence fee charges Home Office August 2013.

Scrap Metal Dealers Act 2013: supplementary guidance Home Office October 2013.

Requirement for licence to carry on business as scrap metal dealer

No person may carry on business as a scrap metal dealer unless they have an authorised scrap metal licence (section 1(1)). A person who carries on business as a scrap metal dealer without a licence is guilty of an offence and liable on summary conviction to a fine not exceeding level 5 on the standard scale (section 1(3)).

Form and effect of licence (FC 5.11a and FC 5.11b)

A scrap metal licence issued by a local authority must be either a site licence, or a collector's licence (section 2(1) and (2)).

A site licence authorises the licensee to carry on business at any site in the authority's area which is identified in the licence (section 2(3)). It must:

(a) name the licensee;

(b) name the authority;

(c) identify all the sites in the authority's area at which the licensee is authorised to carry on business;

(d) name the site manager of each site; and

(e) state the date on which the licence is due to expire (section 2(4)).

A collector's licence authorises the licensee to carry on business as a mobile collector in the authority's area (section 2(5)). A collector's licence must name the licensee and the authority, and state the date on which the licence is due to expire (section 2(6)).

The Secretary of State has made regulations prescribing further requirements about the form and content of licences (section 2(8)).

A person may hold more than one licence issued by different local authorities, but may not hold more than one licence issued by any one authority (section 2(9)).

Issue of licence

A local authority must not issue or renew a scrap metal licence unless it is satisfied that the applicant is a suitable person to carry on business as a scrap metal dealer (section 3(1)).

In determining whether the applicant is a suitable person, the authority may have regard to any information which it considers to be relevant, including in particular:

(a) whether the applicant or any site manager has been convicted of any relevant offence or has been the subject of any relevant enforcement action;

(b) any previous refusal of an application for the issue or renewal of a scrap metal licence or for a relevant environmental permit or registration (and the reasons for the refusal);

(c) any previous revocation of a scrap metal licence (and the reasons for the revocation); and

(d) whether the applicant has demonstrated that there will be in place adequate procedures to ensure that the provisions of this Act are complied with (section 3(2)).

In determining whether a company is a suitable person to carry on business as a scrap metal dealer, a local authority is to have regard, in particular, to whether any of the following is a suitable person:

(a) any director of the company;

(b) any secretary of the company; or

(c) any shadow director of the company (any person in accordance with whose directions or instructions the directors of the company are accustomed to act) (section 3(5)).

In determining whether a partnership is a suitable person to carry on business as a scrap metal dealer, a local authority is to have regard, in particular, to whether each of the partners is a suitable person (section 3(5)).

The authority must also have regard to any guidance on determining suitability issued by the Secretary of State (section 3(6)).

The authority may consult other persons regarding the suitability of an applicant, including in particular:

(a) any other local authority
(b) the Environment Agency
(c) the Natural Resources Body for Wales
(d) an officer of a police force (section 3(7)).

If the applicant or any site manager has been convicted of a relevant offence, the authority may include in the licence one or both of the following conditions:

(a) that the dealer must not receive scrap metal except between 9 a.m. and 5 p.m. on any day;
(b) that all scrap metal received must be kept in the form in which it is received for a specified period, not exceeding 72 hours, beginning with the time when it is received (section 3(8)).

Term of licence

A licence expires 3 years after the day on which it is issued (schedule 1 paragraph 1(1)).

But if an application to renew a licence is received before the licence expires, the licence continues in effect.

If the application is withdrawn, the licence expires at the end of the day on which the application is withdrawn. If the application is refused, the licence expires when no appeal is possible in relation to the refusal or any such appeal is finally determined or withdrawn. If the licence is renewed, it expires 3 years after the day on which it is renewed or (if renewed more than once) the day on which it is last renewed (schedule 1 paragraph 1(2)).

Applications

A licence is to be issued or renewed on an application, which must be accompanied by:

(a) if the applicant is an individual, the full name, date of birth and usual place of residence of the applicant;
(b) if the applicant is a company, the name and registered number of the applicant and the address of the applicant's registered office;
(c) if the applicant is a partnership, the full name, date of birth and usual place of residence of each partner;
(d) any proposed trading name;
(e) the telephone number and e-mail address (if any) of the applicant;
(f) the address of any site in the area of any other local authority at which the applicant carries on business as a scrap metal dealer or proposes to do so;
(g) details of any relevant environmental permit or registration in relation to the applicant;

FC 5.11a Scrap metal dealers – Scrap Metal Dealers Act 2013 – Site and collectors licences (1)

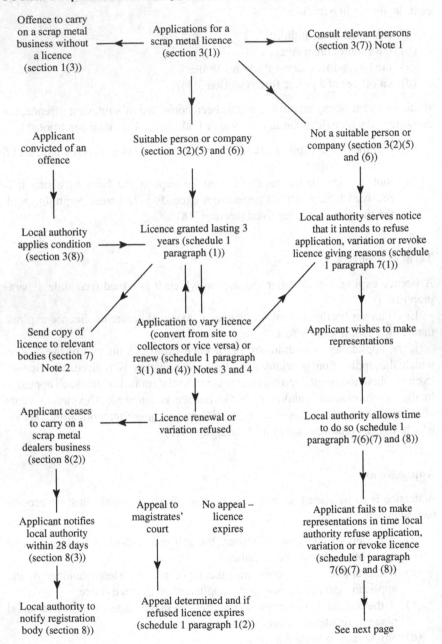

Notes
1. Relevant persons – any other local authority; the Environment Agency; the Natural Resources Body for Wales; an officer of a police force (section 3(7)).
2. Relevant bodies – the Environment Agency; the Natural Resources Body for Wales (section 7).
3. Offence to fail to apply for variation (schedule 1, paragraph 3(5)).
4. If renewal or variation granted licence for 3 years.

FC 5.11b Scrap metal dealers – Scrap Metal Dealers Act 2013 – Site and collectors licences (2)

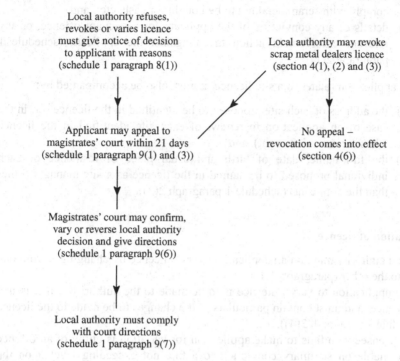

Local authority refuses,
revokes or varies licence
must give notice of decision
to applicant with reasons
(schedule 1 paragraph 8(1))

Local authority may revoke
scrap metal dealers licence
(section 4(1), (2) and (3))

Applicant may appeal to
magistrates' court within 21 days
(schedule 1 paragraph 9(1) and (3))

No appeal –
revocation comes into effect
(section 4(6))

Magistrates' court may confirm,
vary or reverse local authority
decision and give directions
(schedule 1 paragraph 9(6))

Local authority must comply
with court directions
(schedule 1 paragraph 9(7))

Notes
1. Local authority can vary licence if it thinks necessary (section 4(7)).
2. Local authority can vary if licensee or site manager is convicted of an offence (section 3(8)).
3. Applicant must notify authority of changes to the licence (section 8(1)).

(h) details of any other scrap metal licence issued (whether or not by the local authority) to the applicant within the period of 3 years ending with the date of the application;

(i) details of the bank account which is proposed to be used in order to comply with scrap metal not to be bought for cash, etc.; and

(j) details of any conviction of the applicant for a relevant offence, or any relevant enforcement action taken against the applicant (schedule 1 paragraph 2(1)).

If the application relates to a site licence, it must also be accompanied by:

(a) the address of each site proposed to be identified in the licence (or, in the case of an application to renew, of each site identified in the licence whose renewal is sought), and

(b) the full name, date of birth and usual place of residence of each individual proposed to be named in the licence as a site manager (other than the applicant) (schedule 1 paragraph 2(2)).

Variation of licence

A local authority may, on an application, vary a licence by changing it from one type to the other (paragraph 3(1)).

An application to vary a licence is to be made to the authority which issued the licence, and must contain particulars of the changes to be made to the licence (schedule 1 paragraph 3(4)).

A licensee who fails to make application for variation is guilty of an offence and is liable on summary conviction to a fine not exceeding level 3 on the standard scale (schedule 1 paragraph 3(5)).

It is a defence for a person charged with an offence to prove that the person took all reasonable steps to avoid committing the offence (schedule 1 paragraph 3(6)).

Further information

The local authority may request that the applicant provide further information the authority considers relevant for the purpose of considering the application (paragraph 4(1)). If an applicant fails to provide information requested, the authority may decline to proceed with the application (schedule 1 paragraph 4(2)).

Offence of making false statement

An applicant who in an application or in response to a request for further information makes a statement knowing it be false in a material particular, or recklessly makes a statement which is false in a material particular, is guilty of an offence and liable on summary conviction to a fine not exceeding level 3 on the standard scale (schedule 1 paragraph 5).

Fee

An application must be accompanied by a fee set by the authority. In setting a fee, the authority must have regard to any guidance issued by the Secretary of State (schedule 1 paragraph 6).

Right to make representations

If a local authority proposes to refuse an application, or revoke or vary a licence the authority must give the applicant or licensee a notice which sets out what the authority proposes to do and the reasons for it (schedule 1 paragraph 7(1)).

A notice must also state that, within the period specified in the notice, applicant or licensee (A) may either:

(a) make representations about the proposal, or
(b) inform the authority that A wishes to do so (schedule 1 paragraph 7(3)).

The period specified in the notice must be not less than 14 days beginning with the date on which the notice is given to A (schedule 1 paragraph 7(4)).

The authority may refuse the application, or revoke or vary the licence, if:

(a) within the period specified in the notice, A informs the authority that A does not wish to make representations, or
(b) the period specified in the notice expires and A has neither made representations nor informed the authority that A wishes to do so (schedule 1 paragraph 7(5)).

If, within the period specified in the notice, A informs the authority that A wishes to make representations, the authority:

(a) must allow A a further reasonable period to make representations, and
(b) may refuse the application, or revoke or vary the licence, if A fails to make representations within that period (schedule 1 paragraph 7(6)).

If A makes representations, the authority must consider the representations (schedule 1 paragraph 7(7)).

If A informs the authority that A wishes to make oral representations, the authority must give A the opportunity of appearing before, and being heard by, a person appointed by the authority (schedule 1 paragraph 7(8)).

Notice of decision

If the authority refuses the application, or revokes or varies the licence, it must give A a notice setting out the decision and the reasons for it (schedule 1 paragraph 8(1)).

A notice must also state:

(a) that A may appeal against the decision;
(b) the time within which such an appeal may be brought; and
(c) in the case of a revocation or variation, the date on which the revocation or variation is to take effect (schedule 1 paragraph 8(2)).

Appeals

An applicant may appeal to a magistrates' court against the refusal of an application (schedule 1 paragraph 9(1)).

A licensee may appeal to a magistrates' court against:

(a) the inclusion in a licence of a condition, or
(b) the revocation or variation of a licence (schedule 1 paragraph 9(2)).

An appeal is to be made within the period of 21 days beginning with the day on which notice of the decision to refuse the application, to include the condition, or to revoke or vary the licence was given (schedule 1 paragraph 9(3)).

An appeal is to be by way of complaint for an order (schedule 1 paragraph 9(4)). The magistrates' court may:

(a) confirm, vary or reverse the authority's decision, and
(b) give such directions as it considers appropriate (schedule 1 paragraph 9(6)).

The authority must comply with any directions given by the magistrates' court (schedule 1 paragraph 9(7)).

The authority need not comply with any such directions:

(a) until the time for making an application by way of case stated has passed, or
(b) if such an application is made, until the application is finally determined or withdrawn (schedule 1 paragraph 9(8)).

Revocation of licence and imposition of conditions

The authority may revoke a scrap metal licence if it is satisfied that the licensee does not carry on business at any of the sites identified in the licence (section 4(1)) or if it is satisfied that a site manager named in the licence does not act as site manager at any of the sites identified in the licence (section 4(2)). It may also revoke a licence if it is no longer satisfied that the licensee is a suitable person to carry on business as a scrap metal dealer (section 4(3)).

If the licensee or any site manager named in a licence is convicted of a relevant offence, the authority may vary the licence by adding one or both of the conditions (section 4(5)).

A revocation or variation comes into effect when no appeal is possible in relation to the revocation or variation, or when any such appeal is finally determined or withdrawn (section 4(6)).

But if the authority considers that the licence should not continue in force without conditions, it may by notice provide:

(a) that, until a revocation under this section comes into effect, the licence is subject to one or both of the conditions set out in section 3(8), or
(b) that a variation under this section comes into effect immediately (section 4(7)).

Schedule 1 makes further provision about licences.

Supply of information by authority

The local authority must supply any information to any other local authority; the Environment Agency; the Natural Resources Body for Wales or an officer of a police force (section 6).

Register of licences

The Environment Agency must maintain a register of scrap metal licences issued by authorities in England. The Natural Resources Body for Wales must maintain a register of scrap metal licences issued by authorities in Wales. Each entry in the registers must record the name of the authority which issued the licence, the name of the licensee, any trading name of the licensee, the address of any site identified in the licence, the type of licence, and the date on which the licence is due to expire.

The registers are to be open for inspection to the public (section 7).

Notification requirements

An applicant for a scrap metal licence, or for the renewal or variation of a licence, must notify the authority to which the application was made of any changes which materially affect the accuracy of the information which the applicant has provided in connection with the application (section 8(1)).

A licensee who is not carrying on business as a scrap metal dealer in the area of the authority which issued the licence must notify the authority (section 8(2)).

This notification must be given within 28 days of the beginning of the period in which the licensee is not carrying on business in that area while licensed (section 8(3)).

If a licensee carries on business under a trading name, the licensee must notify the authority which issued the licence of any change to that name (section 8(4)). Notification must be given within 28 days of the change occurring (section 8(5)).

An authority must notify the relevant environment body of any notification given to the authority, any variation made by the authority, and any revocation by the authority of a licence (section 8(6)). This notification must be given within 28 days of the notification, variation or revocation (section 8(7)).

Where an authority notifies the relevant environment body, the body must amend the register accordingly (section 8(8)).

An applicant or licensee who fails to comply with this is guilty of an offence and is liable on summary conviction to a fine not exceeding level 3 on the standard scale (section 8(9)).

It is a defence for a person charged with this offence to prove that the person took all reasonable steps to avoid committing the offence (section 8(10)).

Closure of unlicensed sites

Schedule 2 makes provision for the closure of sites at which a scrap metal business is being carried on without a licence (section 9).

Closure notice

Where a constable or the local authority is satisfied that premises are being used by a scrap metal dealer in the course of business, and the premises are not a licensed site (not a residential premises) (paragraph 2(1) and (2)) they may issue a closure notice which:

(a) states that the constable or authority is satisfied above;
(b) gives the reasons for that;
(c) states that the constable or authority may apply to the court for a closure order; and
(d) specifies the steps which may be taken to ensure that the alleged use of the premises ceases (paragraph 2(3)).

The constable or authority must give the closure notice to:

(a) the person who appears to the constable or authority to be the site manager of the premises, and
(b) any person who appears to the constable or authority to be a director, manager or other officer of the business and to any person who has an interest in the premises (paragraph 2(4) to (7)).

Cancellation of closure notice

A closure notice may be cancelled by a cancellation notice issued by a constable or the local authority (paragraph 3(1)).

A cancellation notice takes effect when it is given to any one of the persons to whom the closure notice was given (paragraph 3(2)).

The cancellation notice must also be given to any other person to whom the closure notice was given (paragraph 3(3)).

Application for closure order

Where a closure notice has been given, a constable or the local authority may make a complaint to a justice of the peace for a closure order (paragraph 4(1)).

A complaint must be made between 7 days and 6 months after the date on which the closure notice was given (paragraph 4(2)).

A complaint cannot be made if the constable or authority is satisfied that the premises are not (or are no longer) being used by a scrap metal dealer in the course of business, and there is no reasonable likelihood that the premises will be so used in the future (paragraph 4(3)).

Where a complaint has been made, the justice may issue a summons to answer to the complaint (paragraph 4(4)). The summons must be directed to any person to whom the closure notice was given (paragraph 4(5)).

If a summons is issued, notice of the date, time and place at which the complaint will be heard must be given to all the persons to whom the closure notice was given (paragraph 4(6)).

Closure order

If the court is satisfied that the closure notice was given and that the premises continue to be used by a scrap metal dealer in the course of business, or there is a reasonable likelihood that the premises will be so used in the future they may make a closure order (paragraph 5(1) and (2)).

A closure order may, in particular, require:

(a) that the premises be closed immediately to the public and remain closed until a constable or the local authority makes a certificate under paragraph 6;

(b) that the use of the premises by a scrap metal dealer in the course of business be discontinued immediately;

(c) that any defendant pay into court (to a designated officer) a sum the court determines and the sum will not be released by the court to that person until the other requirements of the order are met (paragraph 5(3) and (7)).

4. A closure order including a requirement on immediate closure may include conditions the court considers appropriate relating to:

(a) the admission of persons onto the premises;

(b) the access by persons to another part of any building (paragraph 5(4)).

A closure order may include provisions the court considers appropriate for dealing with the consequences if the order should cease to have effect under paragraph 6 (paragraph 5(5)).

As soon as practicable after a closure order is made, the complainant must fix a copy of it in a conspicuous position on the premises in respect of which it was made (paragraph 5(6)).

Termination of closure order by certificate of constable or authority

Where a closure order has been made, but a constable or the local authority is satisfied that the need for the order has ceased the constable or authority may make a certificate to that effect (paragraph 6(1) and (2)). The closure order ceases to have effect when the certificate is made (paragraph 6(3)).

Any sum paid into court under the order is to be released by the court to the defendant (paragraph 6(4)).

As soon as practicable after making a certificate, the constable or authority must:

(a) give a copy of it to any person against whom the closure order was made;

(b) give a copy of it to the designated officer for the court which made the order;

(c) fix a copy of it in a conspicuous position on the premises in respect of which the order was made (paragraph 6(5)); and

(d) give a copy of the certificate to any person who requests one (paragraph 6(6)).

Discharge of closure order by court

Any of the following persons may make a complaint to a justice of the peace for a discharge order:

(a) any person to whom the relevant closure notice was given, or
(b) any person who has an interest in the premises but to whom the closure
     notice was not given (paragraph 7(1)).

The court may not make a discharge order unless it is satisfied that there is no
longer a need for the closure order (paragraph 7(2)).

The justice may issue a summons directed to:

(a) such constable as the justice considers appropriate, or
(b) the local authority,

requiring that person to appear before the magistrates' court to answer to the
complaint (paragraph 7(3)).

If a summons is issued notice of the date, time and place at which the complaint will be heard must be given to all the persons to whom the closure notice
was given (paragraph 7(4)).

Appeals

An appeal may be made to the Crown Court against:

(a) a closure order;
(b) a decision not to make a closure order;
(c) a discharge order;
(d) a decision not to make a discharge order (paragraph 8(1)).

Any appeal must be made within 21 days of the day on which the order or the
decision in question was made (paragraph 8(2)).

An appeal against a closure order or a decision not to make a discharge order
may be made by:

(a) any person to whom the relevant closure notice was given;
(b) any person who has an interest in the premises but to whom the closure
     notice was not given (paragraph 8(3)).

An appeal against a decision not to make a closure order or against a discharge
order may be made by a constable or (as the case may be) the local authority
(paragraph 8(4)).

On an appeal the Crown Court may make an order it considers appropriate
(paragraph 8(5)).

Enforcement of closure order

A person is guilty of an offence if the person, without reasonable excuse:

(a) permits premises to be open in contravention of a closure order, or
(b) otherwise fails to comply with, or does an act in contravention of, a
     closure order (paragraph 9(1)).

If a closure order has been made, a constable or an authorised person may (if
necessary using reasonable force):

(a) enter the premises at any reasonable time, and

(b) having entered the premises, do anything reasonably necessary for the purpose of securing compliance with the order (paragraph 9(2)).

Where a constable or an authorised person seeks to exercise his powers and if the owner, occupier or other person in charge of the premises requires the officer to produce evidence of the officer's identity, or evidence of the officer's authority to exercise those powers, the officer must produce that evidence (paragraph 9(3) and (4)).

A person who intentionally obstructs a constable or an authorised person in the exercise of powers is guilty of an offence (paragraph 9(5)).

A person guilty of an offence is liable on summary conviction to a fine not exceeding level 5 on the standard scale (paragraph 9(6)).

Definitions

carrying on business as a scrap metal dealer and scrap metal

A person carries on business as a scrap metal dealer if the person:

(a) carries on a business which consists wholly or partly in buying or selling scrap metal, whether or not the metal is sold in the form in which it was bought, or

(b) carries on business as a motor salvage operator (so far as that does not fall within paragraph (a)) (section 21(2)).

A person carries on business as a motor salvage operator if the person carries on a business which consists:

(a) wholly or partly in recovering salvageable parts from motor vehicles for re-use or sale and subsequently selling or otherwise disposing of the rest of the vehicle for scrap;

(b) wholly or mainly in buying written-off vehicles and subsequently repairing and reselling them;

(c) wholly or mainly in buying or selling motor vehicles which are to be the subject (whether immediately or on a subsequent re-sale) of any of the activities mentioned in paragraphs (a) and (b); or

(d) wholly or mainly in activities falling within paragraphs (b) and (c) (section 21(5)).

scrap metal includes:

(a) any old, waste or discarded metal or metallic material, and

(b) any product, article or assembly which is made from or contains metal and is broken, worn out or regarded by its last holder as having reached the end of its useful life (section 21(6)).

Chapter 6

MISCELLANEOUS

PREMISES APPROVAL FOR CIVIL MARRIAGE OR CIVIL PARTNERSHIP

References

Marriages and Civil Partnerships (Approved Premises) Regulations 2005.
Marriage Act 1949.
Civil Partnerships Act 2004.

Extent

These regulations apply to England and Wales.

Scope

These Regulations provide a system for the approval of premises in which civil marriage ceremonies and the formation of civil partnerships can take place.
The Regulations relate to:

(a) the procedure for the making of an application for approval;

(b) the consultation procedure following the making of an application for approval;

(c) the procedure for granting or refusing an application for approval including the requirements which have to be satisfied for the grant of an approval;

(d) the conditions which may be attached by the authority to the grant of approval;

(e) the procedure for renewing and revoking an approval;

(f) the procedure for reviewing a decision to refuse an approval or to attach conditions to the grant of an approval;

(g) the procedure for the keeping of registers of approved premises by the authority; and

(h) the setting of fees for an application or renewal of an approval and for the attendance of a civil partnership registrar when two people sign a civil partnership schedule on approved premises.

FC 6.1 Premises approval for civil marriage or civil partnership – Marriages and Civil Partnerships (Approved Premises) Regulations 2005

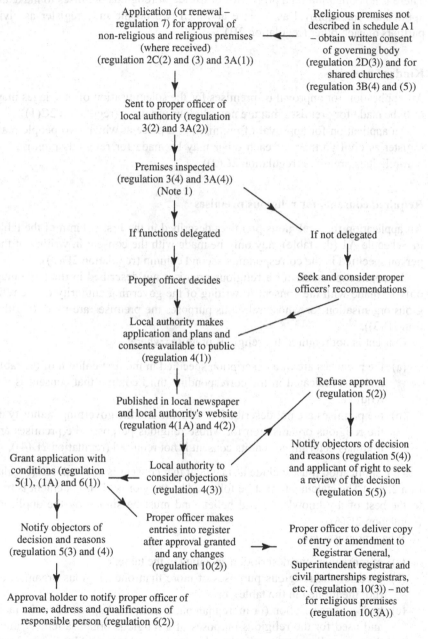

Application (or renewal –
regulation 7) for approval of
non-religious and religious premises
(where received)
(regulation 2C(2) and (3) and 3A(1))

Religious premises not
described in schedule A1
– obtain written consent
of governing body
(regulation 2D(3)) and for
shared churches
(regulation 3B(4) and (5))

Sent to proper officer of
local authority (regulation
3(2) and 3A(2))

Premises inspected
(regulation 3(4) and 3A(4))
(Note 1)

If functions delegated

If not delegated

Proper officer decides

Seek and consider proper
officers' recommendations

Local authority makes
application and plans and
consents available to public
(regulation 4(1))

Refuse approval
(regulation 5(2))

Published in local newspaper
and local authority's website
(regulation 4(1A) and 4(2))

Notify objectors of decision
and reasons (regulation 5(4))
and applicant of right to seek
a review of the decision
(regulation 5(5))

Grant application with
conditions (regulation
5(1), (1A) and 6(1))

Local authority to
consider objections
(regulation 4(3))

Notify objectors of
decision and reasons
(regulation 5(3) and (4))

Proper officer makes
entries into register
after approval granted
and any changes
(regulation 10(2))

Proper officer to deliver copy
of entry or amendment to
Registrar General,
Superintendent registrar and
civil partnerships registrars,
etc. (regulation 10(3)) – not
for religious premises
(regulation 10(3A))

Approval holder to notify proper officer of
name, address and qualifications of
responsible person (regulation 6(2))

Notes
1. Inspection of religious premises may be dispensed with (regulation 3A(5)).
2. If original approval expires during the time for determination of renewal application – original application remains in force (regulation 7(5)).
3. If no application for renewal and original expires and application made within one month of expiry the original continues until the application is determined (regulation 7(6)).

Religious premises (FC 6.1)

There is no obligation on a proprietor or trustee of religious premises to make an application for approval as a place at which two people may register as civil partners of each other (regulation 2B).

Kinds of premises

An application for approval of premises for the solemnisation of marriages may only be made for premises that are not religious premises (regulation 2C(1)).

An application for approval of premises as a place at which two people may register as civil partners of each other may be made for religious premises or non-religious premises (regulation 2C(2)).

Required consents for religious premises

An application for a religious premises described in the first column of the table in Schedule A1 (the table) may only be made with the consent in writing of the person specified in the corresponding second column (regulation 2D(2)).

An application made for a religious premises not described in the table may only be made with the consent in writing of the governing authority of the religious organisation for whose religious purposes the premises are used. (regulation 2D(3)).

Consent is not required for religious premises where:

(a) the premises are of a description specified in the first column of the table and it is indicated in the corresponding third column that consent is not required,

(b) the premises are not described in the table and the governing authority of the religious organisation for whose religious purposes the premises are used has determined that its consent is not required (regulation 2D(4)),

and the application must include a statement that consent is not required (regulation 2D(5)).The statement must be followed by the words 'This statement is true to the best of my knowledge and belief' and must be signed by the applicant (regulation 2D(6)).

Where premises are:

(a) of more than one description specified in the table, or

(b) used for the religious purposes of more than one religious organisation (not mentioned in the table), or

(c) both of a description (or more than one description) specified in the table and used for the religious purposes of a religious organisation (or more than one religious organisation) (not mentioned in the table),

the reference to consent is for each applicable person specified in the table or the governing authority of each applicable religious organisation, or both (regulation 2D(8)). This does not apply to religious premises where there is a sharing agreement or to shared buildings (regulation 2D(9)).

Application procedure for non-religious premises

An application for approval may be made by a proprietor or a trustee of premises (regulation 3(1)). The application must be sent to the proper officer of the authority in writing, including the name and address of the applicant, a plan of the premises which clearly identifies the room or rooms in which the proceedings will take place if approval is granted and a fee if the authority requires it, or an amount on account of that fee (regulation 3(2)).

The premises must comply with the following conditions:

1. Having regard to their primary use, situation, construction and state of repair, the premises must, in the opinion of the authority, be a seemly and dignified venue for the proceedings.
2. The premises must be regularly available to the public for use for:
   (a) the solemnisation of marriages, or
   (b) the formation of civil partnerships.
3. The premises must have the benefit of fire precautions reasonably required by the authority, having consulted with the fire and rescue authority, and other reasonable provision for the health and safety of persons employed in or visiting the premises as the authority considers appropriate.
4. The premises must not be:
   (a) religious premises, or
   (b) a register office.
5. The room or rooms in which the proceedings are to take place if approval is granted must be identifiable by description as a distinct part of the premises (schedule 1).

The applicant must provide the authority with any additional information it may reasonably require in order to determine the application (regulation 3(3)).

After receiving an application, the authority must arrange for the premises to be inspected; and if the functions of the authority have not been delegated to the proper officer, seek and consider his recommendation on the application (regulation 3(4)).

Application procedure for religious premises

An application for approval may be made by a proprietor or a trustee of religious premises (regulation 3A(1)). The application must be sent to the proper officer of the authority in writing, including the name and address of the applicant, a plan of the premises which clearly identifies the room or rooms in which the proceedings will take place if approval is granted, the required consent(s) and a fee if the authority requires it, or an amount on account of that fee (regulation 3A(2)).

The premises must comply with the following conditions:

1. Having regard to their primary use, situation, construction and state of repair, the premises must, in the opinion of the authority, be a seemly and dignified venue for the proceedings.

2. Except so far as section 196 of, and paragraph 2 of Schedule 23 to, the Equality Act 2010 applies, the premises must be regularly available to the public for the formation of civil partnerships.
3. The premises must have the benefit of fire precautions reasonably required by the authority, having consulted with the fire and rescue authority, and other reasonable provision for the health and safety of the persons employed in or visiting the premises as the authority considers appropriate.
4. The room or rooms in which the proceedings are to take place if approval is granted must be identifiable by description as a distinct part of the premises (Schedule 1A).

The applicant must provide the authority with any additional information it may reasonably require in order to determine the application (regulation 3A(3)).

After receiving an application, the authority must arrange for the premises to be inspected; and if the functions of the authority have not been delegated to the proper officer, seek and consider his recommendation on the application (regulation 3A(4)). If the authority considers that it is not necessary for the premises to be inspected (because, for example, they are premises where marriages may be solemnised), it may dispense with the requirement to do so (regulation 3A(5)).

Application procedure for shared church or other buildings

The procedure is the same for non-religious premises, however the application for premises where there is a sharing agreement must include the required consent in respect of each of the sharing Churches (regulation 3B(4)).

An application for a shared building must include the required consent in respect of each of the Churches that uses the shared building (regulation 3B(5)).

Public consultation

After receiving an application the authority must make the application, the plan accompanying it and, if applicable, the required consent, available to members of the public for inspection at all reasonable hours during the working day until the application has been finally determined or withdrawn and give public notice of the application (regulation 4(1)).

The notice must be published:

(a) in a newspaper (which may be a newspaper distributed free of charge) which is in general circulation at intervals of not more than 1 week in the area in which the premises are situated, or

(b) on the authority's website (in which case it must include the date of publication),

and may also be published in such other ways as the authority considers necessary (regulation 4(1A)).

The notice on the authority's website must:

(a) identify the premises and the applicant;

(b) indicate the address at which the application, the plan accompanying it and, if applicable, the required consent can be inspected;

(c) state that any person may give notice in writing of an objection to the grant of approval, with reasons for the objection, within 21 days from the date on which the notice is published in the newspaper or on the authority's website; and

(d) state the address of the offices of the authority to which such notice of objection should be given (regulation 4(2)).

Before reaching a decision on the application, the authority must consider any notice of objection (regulation 4(3)).

Grant or refusal of approval

The authority may grant approval of premises that are not religious premises only if it is satisfied:

(a) that the application has been made in accordance with the Regulations;

(b) that the premises fulfil the requirements set out in Schedule 1; and

(c) that the premises fulfil any other reasonable requirements which the authority considers appropriate so the facilities provided at the premises are suitable (regulation 5(1)).

The authority may grant approval of religious premises only if it is satisfied that:

(a) the application has been made in accordance with the Regulations;

(b) the premises are religious premises (see definitions);

(c) the premises fulfil the requirements set out in Schedule 1A; and

(d) the premises fulfil any other reasonable requirements which the authority considers appropriate so the facilities provided at the premises are suitable (regulation 5(1A)).

The authority may refuse to grant approval if, it considers, having regard to the number of other approved premises in its area, that the superintendent registrar and a registrar or a civil partnership registrar, are unlikely to be available regularly to attend proceedings on the premises (regulation 5(2)).

The authority must notify the applicant and any person who has given notice of objection in writing of its decision, including any conditions (regulation 5(3)).

If approval is refused, or conditions other than those specified in Schedule 2 or Schedule 2A are attached to the approval, or approval is granted after a person has given notice of objection, the authority must set out in any notification its reasons for reaching that decision (regulation 5(4)).

If approval is refused or conditions other than those specified in Schedule 2 or Schedule 2A are attached to the approval, the authority shall notify the applicant of the right to seek a review of its decision (regulation 5(5)).

Conditions

On grant of an approval the authority:

1. must attach to the approval of premises that are not religious premises the standard conditions in Schedule 2;

2. must attach to the approval of religious premises the standard conditions contained in Schedule 2A; and
3. may attach to the approval further conditions it considers reasonable in order to ensure that the facilities provided at the premises are suitable and that proceedings on the premises do not give rise to a nuisance of any kind (regulation 6(1)).

Immediately after the grant of an approval the holder of that approval must notify the proper officer of the name, address and qualification of the responsible person (regulation 6(2)).

Expiry and renewal of approval

An approval is valid for 3 years or more as the authority may determine (regulation 7(1)).

An approval remains in force even if the holder ceases to have a proprietary interest in the premises and the person to whom his interest is transferred will be deemed to be the holder in his place (regulation 7(2)).

Where religious premises to which a sharing agreement relates have been approved and one of the sharing Churches withdraws from the sharing agreement, but the religious premises continue to be used by the other sharing Church (or Churches); or

Where a shared building has been approved and one of the Churches that uses the shared building ceases to do so; but the building continues to be used by the other Church (or Churches), the approval remains in force (regulation 7(2A)).

An application for renewal of an approval may be made by the holder within 12 months before it is due to expire (regulation 7(3)).

The same requirements (name and address, plan, fee, etc.) for an application apply to an application to renew an approval (regulation 7(4)).

If an application for renewal has been made and that application has not been finally determined or withdrawn before the date on which the approval would expire, the approval will continue in effect until the application is finally determined or withdrawn (regulation 7(5)).

Where the holder fails to apply for the renewal of approval and the approval expires due to his failure, an application for renewal is made within 1 month of the expiry; this will reinstate the approval and it will continue in effect until the application is finally determined or withdrawn (regulation 7(6)).

Revocation of approval (FC 6.2)

An authority which has granted an approval may revoke it if it is satisfied that:

(a) the holder has failed to comply with one or more of the conditions attached to the approval, or
(b) the use or structure of the premises has changed so that having regard to the requirements set out in Schedule 1 or Schedule 1A and any requirements set by the authority, the premises are no longer suitable for any proceedings (regulation 8(1)).

Before revoking an approval the authority must deliver to the holder of the approval a notice in writing specifying the ground or grounds which it proposes to revoke the approval and inviting the holder to make written representations as to the proposed revocation within the period, not less than 14 days, specified in the notice (regulation 8(2)).

The authority must deliver a copy of the notice to the superintendent registrar for the district in which the premises are situated, and the civil partnership registrars and employees or officers or other persons provided by a registration authority who are authorised to attest notices of proposed civil partnership for the area in which the premises are situated (regulation 8(3)).

Where the authority proposes to revoke an approval of religious premises, it need not deliver a copy of the notice to the superintendent registrar for the district in which the premises are situated (regulation 8(3A)).

Before reaching a final decision on the proposed revocation, the authority must take into account any representations made to it within the period specified in the notice by or on behalf of the holder of the approval (regulation 8(4)).

If the authority decides to revoke the approval, it must deliver a further notice in writing to the holder, stating the date on which the approval will cease and the procedure the decision may be subject to review (regulation 8(5)).

The Registrar General may direct the authority to revoke any approval if, in her opinion, there have been breaches of the law relating to the proceedings on the approved premises (regulation 8(6)).

Before directing revocation the Registrar General must notify the holder of the grounds on which she proposes to direct that the approval be revoked and deliver a notice in writing to the holder inviting him to make representations in writing as to the proposed revocation within 14 days or other period she may determine (regulation 8(7)).

Before reaching a final decision on the proposed direction, the Registrar General must take account of any representations made to her within the period above by or on behalf of the holder of the approval (regulation 8(8)).

The authority must immediately revoke any approval, with immediate effect, if directed to in writing by the Registrar General and deliver a notice of revocation in writing to the holder (regulation 8(9)).

The authority must revoke any approval with immediate effect as soon as possible after being requested to do so by the holder and deliver a notice of revocation in writing to the holder (regulation 8(10)).

On receipt of notice of revocation the holder of an approval must immediately give notice of the revocation to all parties who have made arrangements for any proceedings to take place in the premises but have not yet taken place (regulation 8(11)).

Revocation of approval: withdrawal of required consent (FC 6.3)

An authority that has granted an approval of religious premises must revoke that approval if it is notified by the holder that:

(a) a required consent in respect of the approved premises has been or will be withdrawn, or

FC 6.2 Revocation of licence for civil marriage or civil partnership premises – Marriages and Civil Partnerships (Approved Premises) Regulations 2005

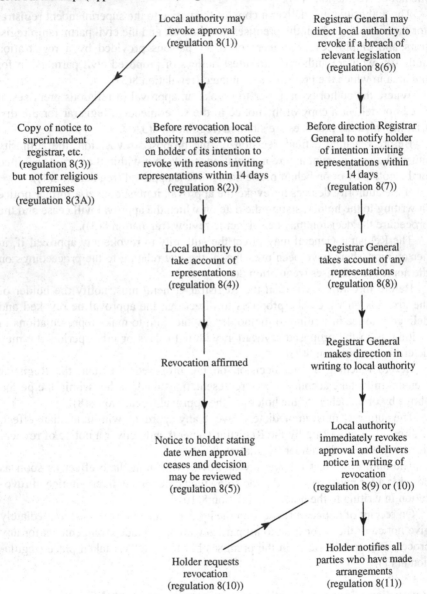

Local authority may revoke approval (regulation 8(1))

Registrar General may direct local authority to revoke if a breach of relevant legislation (regulation 8(6))

Copy of notice to superintendent registrar, etc. (regulation 8(3)) but not for religious premises (regulation 8(3A))

Before revocation local authority must serve notice on holder of its intention to revoke with reasons inviting representations within 14 days (regulation 8(2))

Before direction Registrar General to notify holder of intention inviting representations within 14 days (regulation 8(7))

Local authority to take account of representations (regulation 8(4))

Registrar General takes account of any representations (regulation 8(8))

Registrar General makes direction in writing to local authority

Revocation affirmed

Notice to holder stating date when approval ceases and decision may be reviewed (regulation 8(5))

Local authority immediately revokes approval and delivers notice in writing of revocation (regulation 8(9) or (10))

Holder requests revocation (regulation 8(10))

Holder notifies all parties who have made arrangements (regulation 8(11))

Note
Approval for three years and remains in force if proprietor change or if in shared buildings and churches change circumstances (regulation 7).

(b) in respect of premises that did not previously require it, consent is now or will be required (regulation 8A(1)).

Where an authority receives such notification on a day that is not a working day it is to be taken to have been received on the next day that is a working day (regulation 8A(10)).

The authority must then immediately revoke the approval to take effect on the day following that on which it received the notification or on the day on which consent will be withdrawn, whichever is later (regulation 8A(2)and (3)).

The authority need not revoke the approval mentioned if the holder includes the required consent with his or her notification (regulation 8A(4)).

The authority may cancel the revocation if, prior to it taking effect, the authority is provided with the required consent (regulation 8A(5)).

Immediately after revoking an approval, the authority must deliver a notice of revocation to the holder of the approval (regulation 8A(6)).

This notice must be in writing and specify the date on which the revocation takes effect (regulation 8A(7)).

If the authority cancels a revocation it must immediately deliver a notice of cancellation in writing to the holder of the approval (regulation 8A(8)).

On receipt of the notice, the holder of the approval must give notice of the revocation to all the parties who have made arrangements for the formation of their civil partnerships to take place on those premises on or after the day on which the revocation takes effect (regulation 8A(9)).

Reviews (FC 6.4)

An applicant who is aggrieved about a refusal, refusal of a renewal or to attach conditions other than those specified in Schedule 2 or Schedule 2A may request a review of that decision, or to revoke that approval (regulation 9(1) and (2)).

A person requesting a review must pay a fee if required by the authority (regulation 9(3)).

The proper officer must immediately arrange for a review of the decision and neither an officer nor any member of a committee or sub-committee of the authority which made the decision on behalf of the authority can take part in the decision on the review (regulation 9(4)).

On a review of a decision the authority may:

(a) confirm the original decision;
(b) vary an original decision to grant or renew approval, in particular by removing conditions or by attaching new or different conditions; or
(c) substitute a different decision, which may be subject to new or different conditions than those which were previously attached to it (regulation 9(5)).

The authority must give notice in writing to the applicant or holder of its decision on review, stating its reasons for that decision and (except where the original decision is confirmed) the date from which it takes effect (regulation 9(6)).

FC 6.3 Withdrawal of consent in religious premises – Marriages and Civil Partnerships (Approved Premises) Regulations 2005

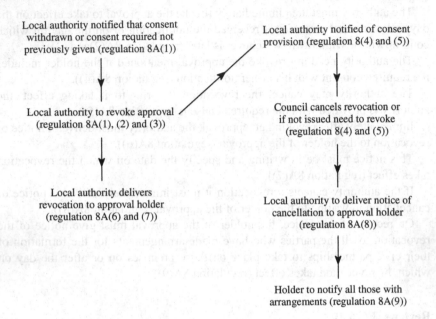

Local authority notified that consent withdrawn or consent required not previously given (regulation 8A(1))

Local authority notified of consent provision (regulation 8(4) and (5))

Local authority to revoke approval (regulation 8A(1), (2) and (3))

Council cancels revocation or if not issued need to revoke (regulation 8(4) and (5))

Local authority delivers revocation to approval holder (regulation 8A(6) and (7))

Local authority to deliver notice of cancellation to approval holder (regulation 8A(8))

Holder to notify all those with arrangements (regulation 8A(9))

FC 6.4 Review of premises licence – Marriages and Civil Partnerships (Approved Premises) Regulations 2005

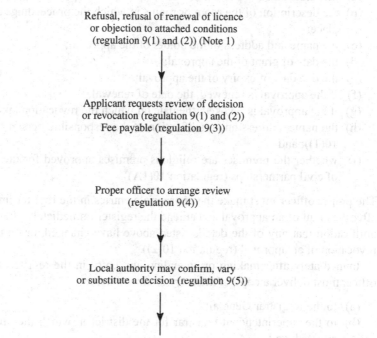

Refusal, refusal of renewal of licence
or objection to attached conditions
(regulation 9(1) and (2)) (Note 1)

Applicant requests review of decision
or revocation (regulation 9(1) and (2))
Fee payable (regulation 9(3))

Proper officer to arrange review
(regulation 9(4))

Local authority may confirm, vary
or substitute a decision (regulation 9(5))

Local authority to give notice in writing to
applicant or holder of decision and reasons
(regulation 9(6))

Note
1. Other than those specified in Schedule 2 or Schedule 2A.

Registers of approved premises

Each authority must keep a register of all premises which are approved by the authority, containing:

(a) the name and full postal address of the approved premises;
(b) the description of the room or rooms in which the proceedings are to take place;
(c) the name and address of the holder of the approval;
(d) the date of grant of the approval;
(e) the due date of expiry of the approval;
(f) if the approval is renewed, the date of renewal;
(g) if the approval is revoked, the date on which the revocation takes effect;
(h) the name, address and qualification of the responsible person (regulation 10(1)); and
(i) whether the premises are religious premises approved for the formation of civil partnerships (regulation 10(1A)).

The proper officer must make the appropriate entries in the register immediately after the grant of an approval and amend the register immediately after receiving notification that any of the details listed above have changed, or on renewal or revocation of an approval (regulation 10(2)).

Immediately after making or amending any entry in the register, the proper officer must deliver a copy of the entry or amendment:

(a) to the Registrar General;
(b) to the superintendent registrar for the district in which the premises are situated; and
(c) to the civil partnership registrars and employees or officers or other persons provided by a registration authority who are authorised to attest notices of proposed civil partnership for the area in which the premises are situated (regulation 10(3)).

For religious premises the proper officer need not deliver a copy of the entry or any amendment to the entry to the superintendent registrar for the district in which the premises are situated (regulation 10(3A)).

The register must be open to public inspection during normal working hours (regulation 10(4)) and it must be kept in permanent form which may include its maintenance on a computer (regulation 10(5)).

Fees

An authority may determine a fee for an application for or the renewal of an approval (regulation 12(1)).

A fee determined for a particular application or renewal cannot exceed the amount which reasonably represents the costs incurred or to be incurred by the authority for that application or renewal or which reasonably represents the average costs incurred in respect of an application or renewal of a particular class (regulation 12(2) and (3)).

Fees cannot include review costs unless and until such a review is requested, but where such a review is requested an authority may determine an additional fee taking into account only the additional costs of the review (regulation 12(4)).

Any authority may charge a fee even though it may not yet have incurred any cost in respect of that application or renewal (regulation 12(5)).

The superintendent registrar in whose presence persons are married on approved premises can charge a fee determined by the authority reasonably representing all the costs to it of providing a registrar and superintendent registrar to attend at a solemnisation (regulation 12(6)).

Where a civil partnership registrar for any area attends when two people sign the civil partnership schedule on approved premises, the authority is entitled to receive a fee reasonably representing all the costs to it of providing the civil partnership registrar to attend at the formation (regulation 12(7)).

An authority may set different fees for different cases or circumstances (regulation 12(8)).

Definitions

proper officer means the proper officer referred to in section 13(2)(h) of the Registration Service Act 1953.

religious premises means:

(a) a church or chapel of the Church of England;

(b) a church or chapel of the Church in Wales;

(c) a place of meeting for religious worship included in the list of certified places maintained by the Registrar General under section 7 of the Places of Worship Registration Act 1855 including religious premises with a sharing agreement or shared building and which are registered as a place of meeting for religious worship of any Church sharing the premises (other than the Church of England or the Church in Wales) (regulation 5(1C));

(d) a place of meeting for members of the Society of Friends; or

(e) a Jewish synagogue (regulation 5(1B)).

WEIGHBRIDGE OPERATOR REGISTRATION

Reference

Weights and Measures Act 1985.

Extent

This act applies to England, Scotland, Wales and Northern Ireland.

Scope

This procedure applies to anyone who wants to be a certified public weighman.

Keepers of public equipment to hold certificate

No person can attend to any weighing or measuring equipment available for use by the public, demanded by a member of the public and for which a charge is made, other than a weighing or measuring of a person, unless he holds a certificate from a chief inspector that he has sufficient knowledge for the proper performance of his duties (section 18(1)). Any person who contravenes, or who causes or permits any other person to contravene this requirement, is guilty of an offence (section 18(3)).

Any person refused a certificate by a chief inspector can appeal against the refusal to the Secretary of State, who may if he thinks fit direct the chief inspector to grant the certificate (section 18(2)).

Provision of public equipment by local authorities

A local authority which is a local weights and measures authority may provide and maintain for use by the public weighing or measuring equipment (section 19(1)).

A local authority may employ persons to attend to any weighing or measuring equipment provided for use by the public (section 19(2)) and they can charge for its use as they may think fit (section 19(3)).

Offences

Any person appointed to attend to weighing or measuring equipment is guilty of an offence if they:

(a) without reasonable cause fail to carry out the weighing or measuring on demand;

(b) carry out the weighing or measuring unfairly;

(c) fail to deliver to the person demanding the weighing or measuring or to his agent a statement in writing of the weight or other measurement found; or

(d) fail to make a record of the weighing or measuring, including the time and date of it and, in the case of the weighing of a vehicle, the particulars of the vehicle and any load on the vehicle that identifies the vehicle and load (section 20(2)).

This applies where any article, vehicle (whether loaded or unloaded) or animal has been brought for weighing or measuring equipment is provided for the purpose of weighing or measuring articles, vehicles or animals of the description in question (section 20(1)).

If, in connection with any equipment above:

(a) any person appointed to attend to weighing or measuring equipment gives a false statement of any weight or other measurement found or makes a false record of any weighing or measuring, or

(b) any person commits any fraud in connection with any, or any purported, weighing or measuring equipment, he is guilty of an offence (section 20(3)).

If, in the case of a weighing or measuring of any article, vehicle or animal, the person bringing the article, vehicle or animal for weighing or measuring, fails to give his name and address, or gives a name or address which is incorrect, he is guilty of an offence (section 20(4)).

The person making any weighing or measuring equipment available for use by the public ('the responsible person') must retain any records for 2 years (section 20(5)).

An inspector, subject to the production of his credentials if so requested, may require the responsible person to produce any record for inspection at any time while it is retained by him (section 20(6)).

If the responsible person fails to retain any record or fails to produce it he is guilty of an offence (section 20(7)).

If any person wilfully destroys or defaces any record within the 2 year period from the date when it was made, he is guilty of an offence (section 20(8)).

Definitions

chief inspector means a chief inspector of weights and measures.
inspector means an inspector of weights and measures.

Printed in the United States
by Baker & Taylor Publisher Services

Printed in the United States
by Baker & Taylor Publisher Services